T0182030

Problems in Classical and Quantum Mechanics

J. Daniel Kelley • Jacob J. Leventhal

Problems in Classical and Quantum Mechanics

Extracting the Underlying Concepts

 Springer

J. Daniel Kelley
The Boeing Company (retired)
University of Missouri-St. Louis
St. Louis, MO, USA

Jacob J. Leventhal
Department of Physics and Astronomy
University of Missouri-St. Louis
St. Louis, MO, USA

ISBN 978-3-319-83557-0   ISBN 978-3-319-46664-4   (eBook)
DOI 10.1007/978-3-319-46664-4

Printed on acid-free paper

This Springer imprint is published by Springer Nature
The registered company is Springer International Publishing AG
The registered company address is: Gewerbestrasse 11, 6330 Cham, Switzerland

This book is dedicated to our families.

Preface

This book is not a textbook. It is a collection of problems intended to aid students in their undergraduate and graduate level courses in physics. The book was, however, formulated with students who are preparing for the PhD qualifying exam in mind. Thus, the problems that are included are of the type that could be on this exam or are problems that are meant to elucidate an important principle.

There are many compilations of physics problems available to students, so it is reasonable to ask why this one is different. It is different because the aim is to place the problems in the broader context of the subject. The book is meant to facilitate the development of problem-solving skills to aid in the understanding of physics. We state the problem and then present the solution in detail. Further, we note and discuss the significance of the problem in the context of the subject under study. We analyze the broader implications of the solution including limiting cases and the relation to other problems. Many of the solutions are accompanied by a tutorial on their meaning and the route to solution. We stress that the solution of the problem is just the beginning of the learning process. As the subtitle infers, manipulation of the solution and changing the associated parameters can provide a great deal of insight.

Our approach is to make the discussions of each problem seem as though the student has come to one of our offices and asked for help solving it. It is our belief that when students come for help, it is not enough to simply show them how to arrive at the solution. We discuss with them the physics that they should be learning from the problem. That is, after all, why problems are assigned. We want the students to ask themselves "What physics can I learn from the problem?" not "How can I work this problem and go on to the next one as quickly as possible?"

Another feature of this book is the inclusion of more mathematical detail in the solutions than is usually provided. We have done this because the book is meant to be an aid in learning *physics*. Thus, at the risk of lengthening the book, we have attempted, when possible, to relieve the reader of the burden of spending a lot of time on mathematical detail after the plan of attack has been formulated.

We could not include every aspect of the subjects that we treat. Our choice of material was predicated on the notion that introductory concepts are the most vital. Some will argue with our choices, but we have made them. Another consideration we used when choosing material was, as noted above, our perception of those subjects most often included in PhD qualifying examinations. On occasion, we have provided introductions to chapters, sometimes merely for the sake of notation. None of these introductions should be considered substitutes for the appropriate textbooks.

Advice for Students

Solving problems is an integral part of learning physics. Over the years, we have heard many times "I understand the physics, but I can't work the problems." There is no polite response to this beyond "Nonsense—you do not yet understand the physics." This does not mean that you should have been able to work every problem the first time you encountered it. On the contrary, it is our experience that the concepts retained best are those that you comprehend only after having struggled with them. As in the weight room, "no pain, no gain." A related excuse is "I understand the physics, but I can't do the math." This is tantamount to saying that you are a very good auto mechanic, but you really have trouble using wrenches. Mathematics is the essential tool of physics.

We offer several hints for learning and retaining physics concepts. First, neatness does count! When you arrive at a satisfactory solution to a problem, we urge you to rewrite the solution in a clear comprehensible form. This permits you to review the solution and to decipher it weeks later when you are studying for an exam. Additionally, writing the solution in a coherent fashion is good practice for the exam. After all, you want to transmit to the grader what you know. It does you no good to know the material if the exam grader does not know that you know it.

Physics is not like some other disciplines. Rule: You cannot cram for a physics test. At least the vast majority of people cannot cram for a physics test. You can probably memorize the presidents and vice presidents of the United States or some list of dignitaries or geographical locations the night before an examination (and likely forget it after the exam), but learning a number of new *concepts* in a short time is very difficult. As students we have tried it (unsuccessfully) and wish to pass along our sad experiences. In short, it is worthwhile to stay as up to date as possible in physics courses.

Finally, we offer some advice on the best way to use instructor-provided solutions to assigned problems, a common aid in contemporary education. In our view, these solutions are a two-edged sword because they can lull a student into a false sense of security, thinking that they understand the material when, in fact, they do not. Often students will simply read the problem, think it over for a short time, and then peruse the solution. After a short digestion period, they think that they understand the solution and go on to the next problem. They have ignored the "no pain, no gain" rule. Glancing at the solution is like watching someone do push-ups. It does not benefit the spectator. Proper use of provided solutions takes a great deal of discipline. We recommend that the problem first be attempted without consulting the solution. If the student reaches an impasse, after perhaps a half hour, the solution should be consulted. *Do not, however, look beyond the portion of the solution that resolved the impasse.* Discipline! The student should then continue this procedure until the solution has been attained and (in all likelihood) understood. We contend that this method will lead to a better understanding of physics and permanent retention of the concepts that the problem was designed to illustrate. Yes, it takes more time this way, but the rewards are worth it.

St. Louis, MO, USA   J. Daniel Kelley
                                                                                         Jacob J. Leventhal

Contents

Part I Classical Mechanics

1 Newtonian Physics 3
    1. Simple pendulum 3
    2. Free fall with drag 5
    3. Collisions—2 balls and a brick wall 8
    4. Collisions—two blocks and a spring 10
    5. Two unequal masses attached to a spring 11
    6. Hole dug through a diameter of the earth 12
    7. Hole dug through a chord of the earth 14
    8. Sphere with a spherical (off center) hole 15
    9. Moving inclined plane... 18
    10. Particle moving in a cosine potential 19
    11. Particle moving in inverted Gaussian potential 21
    12. Hard sphere scattering—classical 22

2 Lagrangian and Hamiltonian Dynamics 25
    1. Brachistochrone 27
    2. Lagrange/Newton equivalence 30
    3. SHO - Lagrangian and Hamiltonian 31
    4. Simple pendulum - Lagrangian 33
    5. Simple pendulum with vertically moving pivot point 35
    6. Sliding pendulum 37
    7. Atwood's machine I 40
    8. Mass on table, pulley, hanging mass - Lagrangian.............. 41
    9. Mass on table, pulley, hanging mass - Hamiltonian 43
    10. Projectile motion 44
    11. Hanging disk - Hamiltonian 47
    12. Hanging disk - Lagrangian 49
    13. Rod pivoted at the end 49
    14. Moving inclined plane 52
    15. Rotating massless rod - Lagrangian 54

16. Rotating massless rod - Hamiltonian 55
17. Atwood's machine II 58
18. Sphere rolling on inclined plane 59
19. Bead on a wire 62
20. Two unequal masses attached to a spring 64

3   Central Forces and Orbits 67
1. Conservation of angular momentum I 68
2. Conservation of angular momentum II 68
3. Kepler's laws 69
4. Newton's gravitational law deduced from Kepler's laws 71
5. Total energy for a central potential 72
6. Equation of the orbit 73
7. Total energy for a circular orbit 74
8. Spiral orbit I 76
9. Spiral orbit II 77
10. Inverse cube force law 79
11. Isotropic oscillator I 81
12. Isotropic oscillator II 82
13. Kepler effective potential 83
14. Kepler orbits 85
15. Runge–Lenz vector I 89
16. Runge–Lenz vector II 92

4   Normal Modes and Coordinates 95
1. Two pendulums coupled by a spring 95
2. Three springs, two masses 102
3. Double pendulum 103
4. Pendulum with oscillating pivot 108
5. Two springs with thin rod attached 111
6. Three particles, two springs (CO_2) 114
References 117

Part II   Quantum Mechanics

5   Introductory Concepts 121
1. Bohr atom 121
2. Gravitational bohr radius 124
3. deBroglie wavelength of an electron in Bohr orbit 125
4. Uncertainty principle and the H-atom 126
5. Classical radius of the electron 128
6. Magnetic dipole moment of spinning charged sphere 129
7. Uncertainty principle and the SHO 130
8. Measurement, expectation values and probabilities 131
9. Expectation values: SHO 132
10. Expectation values: L-box 135
11. Rectangular barrier 137

12. Transmission and reflection over a square well 141
13. Delta function barrier I 142
14. Delta function barrier II 144
15. Semi-infinite barrier 145

6   Bound States in One Dimension 149
6.1   Degeneracy 149
6.2   Parity 150
6.3   Characteristics of the Eigenfunctions 151
6.4   Superposition Principle 151
       1. Stationary states 152
       2. Characteristics of eigenfunctions 153
       3. Delta function potential 156
       4. Half-well potential 158
       5. Sudden approximation I 159
       6. Sudden approximation II 162
       7. Wave function, probabilities I 164
       8. Wave function, probabilities II 166
       9. Sudden approximation III 169

7   Ladder Operators for the Harmonic Oscillator 173
       1. Commutation of x and p 174
       2. General relations 175
       3. Expectation values I 179
       4. Expectation values II 181
       5. Expectation values III 182
       6. Matrix elements 185
       7. Expectation values IV 186
       8. Expectation values V 187

8   Angular Momentum 189
       1. Operators and ladder operators 192
       2. Angular momentum eigenvalues 195
       3. Measurements and expectation values 197
       4. Speed of a "spinning hard sphere" electron 199
       5. Spinors 200
       6. Using Pauli matrices 200
       7. Spin ladder operators 202
       8. Electron in a B field 203
       9. Commutators 205
       10. Construct a Clebsch–Gordan table 205
       11. More Clebsch–Gordan 207
       12. Hyperfine structure of H-atom I 209
       13. Hyperfine structure of H-atom II 210
       14. Hard sphere scattering - quantum 213

9 **Indistinguishable Particles**... 217
    1. Bosons, fermions and the exchange force 218
    2. Two fermions in an SHO... 219

10 **Bound States in Three Dimensions** 223
    1. 3D L-box.. 224
    2. Density of states for a 3D L-box............................... 225
    3. Rigid rotor 226
    4. Infinitely deep spherical well 227
    5. Spherical shell well 228
    6. Classically forbidden region: H-atom 230
    7. Expectation value for H-atom.................................. 231
    8. Kramer's relation 232
    9. Bound state for unknown potential 233
    10. Relation between H-atom quantum numbers 235
    11. H-atom degeneracy... 237
    12. The quantum defect 239

11 **Approximation Methods** 243
  11.1 The WKB Approximation 243
    1. Energies of an L-box 244
    2. Energies of an SHO 245
    3. Transmission through a parabolic barrier 246
  11.2 The Variational Method 248
    1. Estimate the ground state energy of an SHO................... 249
    2. Estimate the first excited state energy of an SHO........... 251
    3. Linear potential well.. 253
    4. Quartic potential well 256
  11.3 Non-degenerate Perturbation Theory............................ 258
    1. Gravitational correction to H-atom energy 259
    2. L-box perturbed by a delta-function 260
    3. L-box perturbed by a linear function 261
    4. Relativistic correction to L-box............................. 263
    5. Charged SHO perturbed by constant electric field............. 266
    6. H-atom perturbed by delta function 270
    7. Relativistic correction to SHO 271
    8. Relativistic correction to H-atom 273
    9. Quartic perturbation of SHO.................................. 275
    10. Matrix eigenvalues 276
  11.4 The Helium Atom 280
    1. He-atom ground state energy 281
    2. He-atom: perturbation/variation 282
  11.5 Degenerate Perturbation Theory............................... 284
    1. 2-D a-box with xy perturbation 285
    2. Particle on a ring with delta-function perturbation.......... 288

           3. 2-D SHO with xy perturbation 292
           4. Select set for 3x3 matrix 294
   11.6   Time Dependent Perturbation Theory 297
           1. Gaussian perturbation applied to particle in an L-box 298
           2. E-field pulse applied to to particle in an L-box 300
           3. Exponential perturbation applied to SHO 302
           4. Pulse applied to SHO 305
           5. Exponential E-field applied to H-atom 307
           6. Fermi's Golden Rule 310
   References 311

A   **Greek Alphabet** 313

B   **Acronyms, Descriptors and Coordinates** 315
   B.1   Acronyms and Descriptors 315
   B.2   Coordinate Systems 315

C   **Units** 317

D   **Conic Sections in Polar Coordinates** 319

E   **Useful Trigonometric Identities** 323

F   **Useful Vector Relations** 325

G   **Useful Integrals** 327

H   **Useful Series** 329
   H.1   Taylor Series 329
   H.2   Binomial Expansion 330

I   **Γ-Functions** 331
   I.1   Integral Γ-Functions 331
   I.2   Half-Integral Γ-Functions 331

J   **The Dirac Delta-Function** 333

K   **Hyperbolic Functions** 335
   K.1   Manipulations of Hyperbolic Functions 335
   K.2   Relationships Between Hyperbolic and Circular Functions 337

L   **Useful Formulas** 339
   L.1   Classical Mechanics 339
   L.2   Quantum Mechanics 340
       L.2.1   One Dimension 340
       L.2.2   Three Dimensions (Central Potentials) 341

M   The Infinite Square Well 343
   M.1   The L-Box 343
   M.2   The a-Box 344

N   Operators, Eigenfunctions, and Commutators 347
   N.1   Eigenfunctions and Eigenvalues of Operators 347
   N.2   Operator Algebra; Commutators 347
   N.3   Commutator Identities 348
   N.4   Some Quantum Mechanical Commutators 348

O   The Quantum Mechanical Harmonic Oscillator 349

P   Legendre Polynomials 351
   P.1   Properties 351
   P.2   Legendre Series 351
   P.3   The Function $1/\left|r_1 - r_2\right|$ 352
   P.4   Polynomials 353

Q   Orbital Angular Momentum Operators in Spherical Coordinates ... 355

R   Spherical Harmonics 357

S   Clebsch–Gordan Tables 359

T   The Hydrogen Atom 361
   References 362

Index 363

Part I
Classical Mechanics

Chapter 1
Newtonian Physics

Most of the problems in this chapter are the type assigned in introductory courses on Classical Mechanics. Their inclusion in this volume serves two purposes. First, they provide a "warm-up" for more advanced problems. More importantly, a number of these problems are solved in subsequent chapters using more advanced methods such as Lagrangian or Hamiltonian dynamics. We are of the opinion that elementary problems solved using advanced techniques facilitate learning and better understanding of these techniques. Correct solutions obtained using the advanced methods also provide confidence in the use of these methods.

Problems

1. Solve the problem of a simple pendulum using Newtonian physics with the coordinates shown. Assume that the bob of mass m is attached to a massless rigid rod of length ℓ (Fig. 1.1).

Solution

Method 1

The force along the arc length $s = \ell\theta$ is simply $mg \sin \theta$ so according to Newton's second law, i.e. $F = ma$

$$m\ddot{s} = -mg \sin \theta \qquad (1.1)$$

© Springer International Publishing AG 2017
J.D. Kelley, J.J. Leventhal, *Problems in Classical and Quantum Mechanics*, DOI 10.1007/978-3-319-46664-4_1

Fig. 1.1 Problem 1

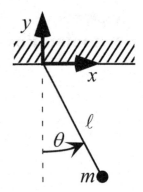

But, because s and ℓ are related according to $s = \ell\theta$ we have

$$\ddot{\theta} + \frac{g}{\ell} \sin\theta = 0 \tag{1.2}$$

In the case of small oscillations $\sin\theta \approx \theta$ so that

$$\ddot{\theta} + \frac{g}{\ell}\theta = 0 \tag{1.3}$$

and we have simple harmonic motion with frequency $\omega = \sqrt{g/\ell}$. This motion is given by

$$\theta(t) = \theta_{max} \sin(\omega t - \phi) \tag{1.4}$$

The constant ϕ is the phase angle and determines the position of the pendulum at $t = 0$. The constant θ_{max} is the amplitude, which depends upon E, the total mechanical energy (TME) as

$$E = mg\,(\ell - \ell\cos\theta_{max})$$
$$= mg\ell\,(1 - \cos\theta_{max}) \tag{1.5}$$

For small oscillations

$$E = mg\ell\left[1 - \left(1 - \frac{\theta_{max}^2}{2}\right)\right] \tag{1.6}$$

so

$$\theta_{max} = \sqrt{\frac{2E}{mg\ell}} \tag{1.7}$$

Method 2

Again we use Newton's second law, but this time in the form $\tau = I\alpha$ where τ torque, I is the moment of inertia of the bob about the pivot point and α is the angular acceleration. The torque is given by

$$\tau = r \times F = \ell mg \sin \theta \tag{1.8}$$

while the moment of inertia of the bob about the pivot point is

$$I = m\ell^2 \tag{1.9}$$

Therefore, we have

$$\ell mg \sin \theta = \left(m\ell^2\right) \ddot{\theta} \tag{1.10}$$

which produces the same differential equation, Eq. (1.2).

This simple problem is the basis for many problems in classical mechanics so it is worthwhile to work it using elementary methods.

2. A particle is dropped into a viscous fluid from rest at $y = 0$ and $t = 0$. Take the force due to fluid resistance (the "drag") to be proportional to the velocity so that $F_{\text{drag}} = kmv$, where k is a positive constant. Find the velocity v as a function of the distance y. Take y to be positive downward. Manipulate your answer to show that the answer is correct as $k \to 0$.

Fig. 1.2 Problem 2

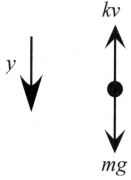

Solution

Using Fig. 1.2 and $F = ma$ we have

$$m\frac{d^2y}{dt^2} = m\frac{dv}{dt} = mg - kmv \tag{1.11}$$

Separating this differential equation we have

$$dt = \frac{1}{k}\frac{dv}{(g/k) - v} \implies t = -\frac{1}{k}\ln\left(\frac{g}{k} - v\right) + C_1 \tag{1.12}$$

Before proceeding we note that the units of k must be s^{-1}. As we proceed through the solution of his problem it is wise to check the units.

Using the initial condition to evaluate the constant of integration we arrive at

$$C_1 = \frac{1}{k}\ln\left(\frac{g}{k}\right) \tag{1.13}$$

and

$$t = -\frac{1}{k}\ln\left(\frac{g}{k} - v\right) + \frac{1}{k}\ln\left(\frac{g}{k}\right)$$

$$-kt = \ln\left(\frac{g - kv}{g}\right) = \ln\left(1 - \frac{kv}{g}\right) \tag{1.14}$$

Now solve for $v = \frac{dy}{dt}$

$$e^{-kt} = 1 - \frac{kv}{g} \implies v = \frac{g}{k}\left(1 - e^{-kt}\right) \tag{1.15}$$

so

$$dy = \frac{g}{k}\left(1 - e^{-kt}\right)dt \tag{1.16}$$

Integrating, we obtain

$$y = \frac{g}{k}\left(t + \frac{1}{k}e^{-kt}\right) + C_2 \tag{1.17}$$

Using the initial conditions to evaluate the constant of integration we find

$$0 = \frac{g}{k}\left(\frac{1}{k}\right) + C_2 \implies C_2 = -\frac{g}{k^2} \tag{1.18}$$

so

$$y = \frac{g}{k}\left(t + \frac{1}{k}e^{-kt}\right) - \frac{g}{k^2}$$

$$= \frac{g}{k}\left(t - \frac{1}{k} + \frac{1}{k}e^{-kt}\right) \tag{1.19}$$

Now we must eliminate t to obtain $y = y(v)$. To do this we solve Eq. (1.15) for t. That is

$$v = \frac{g}{k}\left(1 - e^{-kt}\right) \Longrightarrow t = -\frac{1}{k}\ln\left(1 - \frac{kv}{g}\right) \qquad (1.20)$$

so that

$$\exp(-kt) \equiv \left(1 - \frac{kv}{g}\right) \qquad (1.21)$$

Inserting Eqs. (1.20) and (1.21) into Eq. (1.19) we have

$$\begin{aligned} y &= \frac{g}{k}\left[-\frac{1}{k}\ln\left(1 - \frac{kv}{g}\right) - \frac{1}{k} + \frac{1}{k}\left(1 - \frac{kv}{g}\right)\right] \\ &= -\frac{1}{k}\left[v + \frac{g}{k}\ln\left(1 - \frac{kv}{g}\right)\right] \end{aligned} \qquad (1.22)$$

If there were no fluid resistance the result is simple to calculate using conservation of energy. Because the mass is dropped with no initial velocity the initial potential energy must equal the kinetic energy at the distance y.

$$\frac{1}{2}mv^2 = mgy \qquad (1.23)$$

As it stands Eq. (1.22) is not very illuminating; it proves useful to look at the limit as $k \to 0$. To facilitate this we use the Taylor series for $\ln(1 + x)$ given in Eq. (H.5).

$$\ln(1 + x) = x - \frac{x^2}{2} + \frac{x^3}{3} - \frac{x^4}{4} + \cdots \qquad (1.24)$$

Equation (1.22) then becomes

$$\begin{aligned} y &= -\frac{1}{k}\left\{v + \frac{g}{k}\left[\left(-\frac{kv}{g}\right) - \frac{1}{2}\left(\frac{kv}{g}\right)^2 - \frac{1}{3}\left(\frac{kv}{g}\right)^3 - \cdots\right]\right\} \\ &= -\left[-\left(\frac{1}{2g}\right)v^2 - \frac{1}{3}\left(\frac{k}{g^2}\right)v^3 - \cdots\right] \\ &= \frac{v^2}{2g} + \frac{1}{3}\left(\frac{k}{g^2}\right)v^3 + \cdots \end{aligned} \qquad (1.25)$$

The first term in Eq. (1.25) is clearly the correct answer as $k \to 0$. Moreover, the positive sign on the term involving k shows that higher values of y are required for higher values of v. Thus, with fluid resistance the particle must fall further than without resistance to attain a given velocity.

3. A mass m_2 is at rest a distance d from a brick wall. It is struck by a mass m_1 traveling with a velocity v_{1i} as shown in Fig. 1.3. The collision is elastic as is the subsequent collision of m_2 with the wall. Find the conditions under which a second collision will occur for the two cases $m_1 > m_2$ and $m_1 < m_2$. Discuss limits and find the smallest ratio for which a second collision will occur (Fig. 1.4).

Fig. 1.3 Problem 3

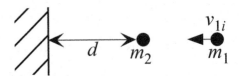

Solution

Fig. 1.4 Problem 3—
solution

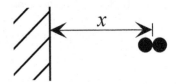

First, in terms of v_{1i}, the initial velocity of m_1 find the post-first-collision velocities of m_1 and m_2, v_{1f} and v_{2f}. All collisions are elastic so we recall an important result from introductory physics:

> In a one dimensional elastic collision the relative
> velocity of approach before collision is equal to
> the relative velocity of separation after collision.

This is a trivially derived result for the collision between m_2 and the wall (of infinite mass). Accordingly, the speed of m_2 before it hits the wall will be the same as it is after it hits the wall. Also, for the two masses

$$v_{1i} = v_{2f} - v_{1f} \tag{1.26}$$

If the student does not remember this result, it is not necessary to panic because it is readily derived, especially when m_2 is initially at rest. We digress to show this:

Conservation of momentum:

$$m_1 v_{1i} = m_1 v_{1f} + m_2 v_{2f} \implies m_1 \left(v_{1i} - v_{1f} \right) = m_2 v_{2f} \tag{1.27}$$

Conservation of TME:

$$\frac{1}{2} m_1 v_{1i}^2 = \frac{1}{2} m_1 v_{1f}^2 + \frac{1}{2} m_2 v_{2f}^2 \tag{1.28}$$

Eq. (1.28) becomes

$$m_1 \left(v_{1i} - v_{1f}\right)\left(v_{1i} + v_{1f}\right) = m_2 v_{2f}^2 \tag{1.29}$$

Substituting Eq. (1.27) for $m_1 \left(v_{1i} - v_{1f}\right)$ into Eq. (1.29) we have

$$\left(v_{1i} + v_{1f}\right) = v_{2f} \tag{1.30}$$

which is the same as Eq. (1.26).

Back to the problem at hand; using conservation of momentum with Eq. (1.30) we have

$$v_{1i} = v_{1f} + \frac{m_2}{m_1} v_{2f}$$

$$= v_{1f} + \frac{m_2}{m_1} \left(v_{1i} + v_{1f}\right) \tag{1.31}$$

Solving for v_{1f} and v_{2f} in terms of v_{1i} we have

$$v_{1f} = \left(\frac{m_1 - m_2}{m_1 + m_2}\right) v_{1i} \tag{1.32}$$

and

$$v_{2f} = \frac{2m_1}{(m_1 + m_2)} v_{1i} \tag{1.33}$$

Now, consider two different cases, $m_1 > m_2$ and $m_1 < m_2$. Let x denote the distance from the wall at which m_1 and m_2 collide for the second time. Note that this distance can be greater than d, less than d or equal to d (for the case in which $m_1 = m_2$).

Case I: $m_1 > m_2$. From Eq. (1.32) we see that for $m_1 > m_2$ v_{1f} is positive (v_{1f} is the velocity, not the speed) so $x < d$.

At the second of the collisions the masses will have traveled different distances. Let

$$z_1 = \text{distance } m_1 \text{ travels between collisions} = d - x$$

$$z_2 = \text{distance } m_2 \text{ travels between collisions} = d + x \tag{1.34}$$

The travel time for each mass t_{tr} is the same for both masses. As noted above, the speed of m_2 before it hits the wall will be the same as it is after it hits the wall, i.e. v_{2f}. We have

$$t_{tr} = \frac{z_1}{v_{1f}} = \frac{z_2}{v_{2f}} = \frac{d - x}{\left(\dfrac{m_1 - m_2}{m_1 + m_2}\right) v_{1i}} = \frac{d + x}{\dfrac{2m_1}{(m_1 + m_2)} v_{1i}} \tag{1.35}$$

Solving this equation for x we have

$$\frac{d-x}{m_1-m_2} = \frac{d+x}{2m_1} \Rightarrow x = \frac{\left(1+\dfrac{m_1}{m_2}\right)}{\left(3\dfrac{m_1}{m_2}-1\right)}d \qquad (1.36)$$

Case II: $m_1 < m_2$. Using Eq. (1.32) for this case we see that v_{1f} is negative so $x > d$. Because m_1 backs up and moves to the right we can let

$$z_1' = \text{distance } m_1 \text{ travels between collisions} = x - d = -z_1$$

$$z_2' = \text{distance } m_2 \text{ travels between collisions} = d + x = z_2$$

We require the speed to calculate the time so we must reverse the sign of v_{1f} for $m_1 < m_2$. But this reversal of sign is the same as reversing the sign of z_1' to make $t_{tr} > 0$. This makes Eq. (1.36) valid for $m_1 < m_2$.

Limits:

From Eq. (1.36) we have:

$m_1 = m_2 \Longrightarrow x = d$ (this is sensible because from Eq. (1.32) $v_{1f} = 0$)

$m_2 \to \infty \Longrightarrow x = -d$ (no second collision occurs because m_2 doesn't move)

From the denominator of Eq. (1.36) we see that if the ratio $m_1/m_2 = 1/3$ then the second collision will occur at $x = \infty$. Clearly, if $m_1/m_2 < 1/3$ a second collision cannot occur because $x < 0$ is forbidden so $m_1/m_2 = 1/3$ is the smallest ratio for which a second collision can occur.

4. A mass m_1 with initial velocity v_0 strikes a massless plate on a spring that is attached to a mass m_2 as shown in Fig. 1.5.

    The spring constant is k and the table top is frictionless.

    (a) What is x_0 the maximum compression of the spring?
    (b) What are the final velocities of m_1 and m_2?
    (c) What criterion will insure that both masses are travelling in the same direction after the collision?

Solution

(a) At maximum compression of the spring x_0 the two masses have the same velocity, call it v'. Therefore, using conservation of momentum and energy we have

Fig. 1.5 Problem 4

$$m_1 v_0 = (m_1 + m_2) \, v' \Rightarrow v' = \left(\frac{m_1}{m_1 + m_2} \right) v_0 \tag{1.37}$$

Conservation of energy gives

$$\frac{1}{2} m_1 v_0^2 = \frac{1}{2} (m_1 + m_2) \, v'^2 + \frac{1}{2} k x_0^2 \tag{1.38}$$

Eliminating v' using Eq. (1.37) we have

$$x_0^2 = \frac{1}{k} \left(\frac{m_1 m_2}{m_1 + m_2} \right) v_0^2$$

$$= \frac{1}{k} \left(\frac{1}{\frac{1}{m_1} + \frac{1}{m_2}} \right) v_0^2 \tag{1.39}$$

Clearly the maximum compression depends directly upon the square of v_0 and inversely on k. Because of the symmetry in m_1 and m_2 in the result, Eq. (1.39), these masses play an equal role in determining x_0.

(b) Before the collision there is no potential energy because the spring is neither compressed nor extended. Similarly, long after the collision it is neither compressed nor extended. Therefore, the problem is the same as if there were no spring and we have a perfectly elastic collision, the details of which were worked out in Problem 3 of this chapter. The result is the same as that of Eqs. (1.32) and (1.33)

$$v_{1f} = \left(\frac{m_1 - m_2}{m_1 + m_2} \right) v_{1i} \quad \text{and} \quad v_{2f} = \frac{2m_1}{(m_1 + m_2)} v_{1i} \tag{1.40}$$

(c) The final velocity of m_2, v_{2f}, is manifestly positive (to the left in Fig. 1.5). For both masses to be moving in the same direction after the collision v_{1f} and v_{2f} must have the same sign so we must have $v_{1f} > 0$. From Eq. (1.40) it is clear that only for $m_1 > m_2$ will $v_{1f} > 0$. This seems reasonable in the limit if one considers m_1 to be a bowling ball and m_2 to be a ping pong ball.

5. Two unequal point masses m_1 and m_2 are attached to a spring having a spring constant k and unextended length ℓ as shown in Fig. 1.6.

   The system is free to oscillate on a frictionless table. Find the frequency of oscillation for this system. It will be convenient to use the coordinates given.

Fig. 1.6 Problem 5

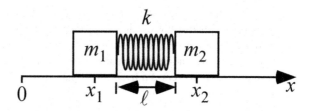

Solution

Using $F = ma$ the equations of motion of each block are

$$m_1\ddot{x}_1 = -k\left[x_1 - (x_2 - \ell)\right]$$
$$m_2\ddot{x}_2 = -k\left[x_2 - (x_1 + \ell)\right] \tag{1.41}$$

Because it is the mass separation, $x_2 - x_1$, that is important we re-write Eq. (1.41) as

$$\ddot{x}_1 = -\frac{k}{m_1}\left[x_1 - x_2 + \ell\right]$$
$$\ddot{x}_2 = -\frac{k}{m_2}\left[x_2 - x_1 - \ell\right] \tag{1.42}$$

Subtracting these equations we get

$$\ddot{x}_2 - \ddot{x}_1 = -k\left[\frac{1}{m_2}(x_2 - x_1 - \ell)\right] - k\left[\frac{1}{m_1}(x_2 - x_1 - \ell)\right]$$
$$= -k\frac{(m_1 + m_2)}{m_1 m_2}(x_2 - x_1 - \ell) \tag{1.43}$$

Now define

$$\varsigma = (x_2 - x_1 - \ell) \tag{1.44}$$

so Eq. (1.43) becomes

$$\ddot{\varsigma} + k\frac{(m_1 + m_2)}{m_1 m_2}\varsigma = 0 \tag{1.45}$$

which is the familiar equation that describes simple harmonic motion with frequency ω given by

$$\omega = \sqrt{\frac{(m_1 + m_2)\,k}{m_1 m_2}} \tag{1.46}$$

6. Imagine a hole being dug through a diameter of the earth and an object of mass m being dropped from rest at one surface. How long does it take for the object to reach the other side of the earth? What is the nature of the motion of the object? Take the radius of the earth to be R and ignore the rotation of the earth. Assume that m traverses a frictionless path (Fig. 1.7).

Fig. 1.7 Problem 6

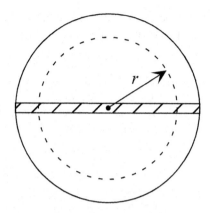

Solution

Because the gravitational force varies as $1/r^2$, Gauss's law applies. Therefore, at any distance r from the center of the earth the attractive force toward the center is proportional to the mass enclosed in a sphere of radius r. That is,

$$F(r) = -G\frac{mM(r)}{r^2} \tag{1.47}$$

where $M(r)$ is the mass contained in the sphere of radius r. Taking ρ to be the density of the earth

$$M(r) = \frac{4}{3}\pi r^3 \rho \tag{1.48}$$

Thus, the force on the mass when it is situated at r is

$$F(r) = -G\frac{m}{r^2} \cdot \frac{4}{3}\pi r^3 \rho$$
$$= -Gm\frac{4}{3}\pi r\rho \tag{1.49}$$

Using $F(r) = ma = m\ddot{r}$ we have

$$m\ddot{r} + m\left(G\frac{4}{3}\pi\rho\right)r = 0 \tag{1.50}$$

This equation can be simplified by noting that when $r = R$ in Eq. (1.47)

$$-F/m = g = \left(G\frac{4}{3}\pi\rho\right)R \tag{1.51}$$

where g is the gravitational acceleration at the surface of the earth. Equation (1.50) then becomes

$$\ddot{r} + \frac{g}{R}r = 0 \tag{1.52}$$

This is the equation for simple harmonic motion with frequency $\omega = \sqrt{g/R} = 2\pi/T$ where T is the period. Therefore, it takes one half T to arrive at the other side of the earth, i.e.

$$\frac{1}{2}T = \pi\sqrt{\frac{R}{g}} \tag{1.53}$$

Furthermore, if the mass m is not picked up when it arrives at the other side of the earth it will return to its original drop-off point and continue to execute simple harmonic motion indefinitely.

7. Imagine a hole that is dug through a chord of the earth, e.g. New York to Los Angeles, and an object of mass m is dropped from rest at one end as depicted in Fig. 1.8. How long does it take for the object to reach the other side? What is the nature of the motion of the object? Take the radius of the earth to be R; ignore friction and the rotation of the earth.

Solution

This problem is a generalization of Problem 6 of this chapter. Again Gauss's law applies so the attractive force toward the center of the earth at a radial distance r from the center is proportional to the mass enclosed in a sphere of

Fig. 1.8 Problem 7

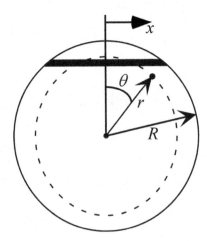

radius r. That is,

$$F(r) = G \frac{mM(r)}{r^2} \tag{1.54}$$

where $M(r)$ is the mass contained in the sphere of radius r and ρ is the density of the earth. Thus,

$$M(r) = \frac{4}{3}\pi r^3 \rho \tag{1.55}$$

As in Problem 6 of this chapter this equation can be simplified by noting that the gravitational acceleration at the surface of the earth is given by

$$g = G\frac{4}{3}\pi \rho R \tag{1.56}$$

so that

$$F(r) = mg\left(\frac{r}{R}\right) \tag{1.57}$$

In this case, however, only the component of $F(r)$ in the x-direction (see Fig. 1.8) is the accelerating force. This component is

$$F_x = F(r)\sin\theta = F(r)\frac{x}{r} = mg\left(\frac{x}{R}\right) \tag{1.58}$$

Using $F = ma$ we have

$$m\ddot{x} + \left(\frac{mg}{R}\right)x = 0 \tag{1.59}$$

and that Eq. (1.50) becomes

$$\ddot{x} + \frac{g}{R}x = 0 \tag{1.60}$$

Not surprisingly (based on the result of Problem 6) the motion is simple harmonic. Perhaps it *is* surprising though that the frequency of the oscillation is exactly the same no matter where the hole is bored, through a diameter or a chord.

8. A spherical cavity of diameter R is made in a sphere of radius R and uniform density ρ. The surfaces of the hole and the sphere coincide as shown in Fig. 1.9.

   With what force will the large sphere with the hollowed out portion attract a small sphere of mass m which lies a distance $d > R$ from the center of the large sphere on a line connecting the centers of both spheres?

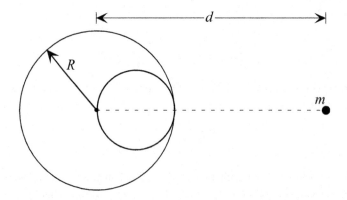

Fig. 1.9 Problem 8

Solution

This problem demonstrates the magic of superposition. We treat the sphere as if the hole doesn't exist and treat the hole as a sphere having negative mass.

Assume the mass of the sphere prior to having cut the hole in it is M and the mass of the hole is M'. These "masses" are then

$$M = \left(\frac{4}{3}\pi R^3\right)\rho \tag{1.61}$$

and

$$M' = -\rho\left[\frac{4}{3}\pi\left(\frac{R}{2}\right)^3\right] = -\frac{M}{8} \tag{1.62}$$

Therefore, using Gauss's law and ignoring the hole, the force on m due to M is

$$F_M = G\frac{mM}{d^2} \tag{1.63}$$

and the force on m from the (fictitious) "negative mass" M' is

$$F_{M'} = G\frac{mM'}{(d-R/2)^2}$$

$$= -\frac{G}{8}\frac{mM}{(d-R/2)^2} \tag{1.64}$$

There are no vectors involved so the total force on the mass m is

$$F_m = F_M + F_{M'}$$

$$= GmM \left[\frac{1}{d^2} - \frac{1}{8(d - R/2)^2} \right]$$

$$= \frac{GmM}{d^2} \left[1 - \frac{1}{8} \frac{1}{(1 - R/2d)^2} \right] \qquad (1.65)$$

Note the trivial limit that as $R \to 0$, $M \to 0$ so $F_m \to 0$ as expected. Suppose, however, that m is put at the center of the hole so $d = R/2$. The force F_m in Eq. (1.65) blows up due to the denominator in the second term in Eq. (1.65). This tells us that something is wrong. Upon reflection we recall that the fields inside and outside a uniform distribution of mass (or charge) are not the same (Gauss's Law). In short, Eq. (1.65) is not valid if $d < R$.

We have already calculated the force on a mass m that is inside the sphere in Problem 6 of this chapter. Using Gauss's law for $d < R$, the force on m due to the solid sphere (of radius d) without the hole is

$$F_M = mG \frac{4}{3} \pi \rho d \qquad (1.66)$$

and

$$F_{M'} = mG \frac{4}{3} \pi \rho \left(d - \frac{R}{2} \right) \qquad (1.67)$$

where $R/2 < d < R$. Then

$$F_m = F_M + F_{M'}$$

$$= mG \frac{4}{3} \pi \rho d - mG \frac{4}{3} \pi \rho \left(d - \frac{R}{2} \right)$$

$$= mG \frac{4}{3} \pi \rho \frac{R}{2} \qquad (1.68)$$

Replacing ρ using Eq. (1.61) we have

$$F_m = \frac{1}{2} \frac{mGM}{R^2} \qquad (1.69)$$

According to Eq. (1.69) the force on m is independent of d. In fact, the force on m is not only independent of d, it is also independent of the location of m within the hole.

Finally, let us compare the two results for F_m, one obtained for $d > R$, Eq. (1.65), and the other with $d < R$, Eq. (1.69) at the only point they have in common, $d = R$. Inserting $d = R$ into Eq. (1.65) we have

$$F_m = \frac{1}{2} \frac{mGM}{R^2} \tag{1.70}$$

which is identical with Eq. (1.69). This is an interesting result because it shows that the gravitational *field* is continuous at the common point.

9. The inclined plane of mass M shown in Fig. 1.10 rests on a frictionless surface. The block of mass m is released from rest at the top of the frictionless incline. The force on the block normal to the plane is N. Using the coordinates in Fig. 1.10 find $\ddot{X}$ the acceleration of the inclined plane.

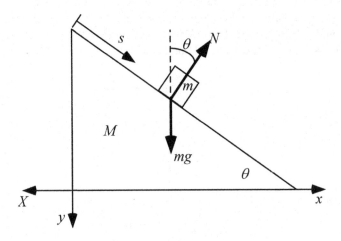

Fig. 1.10 Problem 9

Solution

Using Newton's second law:

$$m\ddot{y} = mg - N\cos\theta$$
$$m\ddot{x} = N\sin\theta$$
$$M\ddot{X} = N\sin\theta \tag{1.71}$$

Because $\ddot{x}$ and $\ddot{X}$ are antiparallel, the net horizontal of m is the sum $(\ddot{x} + \ddot{X})$. Thus, the ratio of the vertical to the horizontal acceleration is

$$\frac{\ddot{y}}{\ddot{x} + \ddot{X}} = \tan \theta$$

$$= \frac{(mg - N \cos \theta)/m}{N \sin \theta \left(\dfrac{1}{M} + \dfrac{1}{m} \right)} \tag{1.72}$$

Solving for N we have

$$N = \frac{g}{\tan \theta \sin \theta \left(\dfrac{1}{M} + \dfrac{1}{m} \right) + \dfrac{1}{m} \cos \theta} \tag{1.73}$$

Putting Eq. (1.73) into the equation for $\ddot{X}$ in Eq. (1.71) and solving for $\ddot{X}$ we have

$$\ddot{X} = \frac{g \sin \theta}{\tan \theta \sin \theta \left(1 + \dfrac{M}{m} \right) + \dfrac{M}{m} \cos \theta}$$

$$= \frac{g \sin \theta}{\dfrac{\sin^2 \theta}{\cos \theta} + \dfrac{M \sin^2 \theta}{m \cos \theta} + \dfrac{M}{m} \cos \theta \left(\dfrac{\cos \theta}{\cos \theta} \right)}$$

$$= \frac{g \sin \theta}{\left(\dfrac{m}{m} \right) \dfrac{\sin^2 \theta}{\cos \theta} + \dfrac{M}{m} \dfrac{1}{\cos \theta}}$$

$$= \frac{mg \sin \theta \cos \theta}{m \sin^2 \theta + M} \quad \text{(to the left)} \tag{1.74}$$

Notice that if $\theta = 0$ or if $\theta = \pi/2$, $\ddot{X}$ vanishes as it should. Moreover, as $M \to \infty$, $\ddot{X} \to 0$ as it should.

This problem will be worked later in this volume in a tidier fashion using Lagrangian dynamics (see Problem 14, Chap. 2).

10. A particle of mass m moves in one dimension without friction under the influence of a potential energy function

$$U(x) = U_0 \cos x \tag{1.75}$$

The motion is restricted to the range $\pi/2 < x < 3\pi/2$.

(a) Sketch the potential energy function for the region of interest.
(b) What is the period T_0 of the bounded motion for amplitudes small enough so that the motion can be considered to be simple harmonic? Explain.

Solution

(a) The potential is a cosine function as shown in Fig. 1.11.

Fig. 1.11
Problem 10—solution

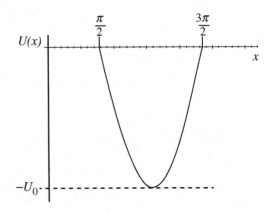

(b) For simple harmonic motion the potential energy is

$$U_{\text{shm}}(x) = -\frac{1}{2}m\omega^2 x^2 \qquad (1.76)$$

where m is the mass of the oscillating particle and ω is the angular frequency of the motion. To find the conditions for which simple harmonic motion is a good approximation to motion under $U(x)$ we must find the coefficient of the quadratic term in a series expansion of the potential energy, the well-known Taylor series given in Eq. (H.3).

$$U(x) = U_0 \cos x$$
$$= U_0 \left(1 - \frac{x^2}{2!} + \frac{x^4}{4!} - \cdots\right) \qquad (1.77)$$

Thus, $U_0/2$ is equivalent to $\frac{1}{2}m\omega^2$ and we have

$$\frac{1}{2}m\omega^2 = \frac{U_0}{2} \Rightarrow \omega^2 = \frac{U_0}{m}$$
$$\omega = \sqrt{\frac{U_0}{m}} \qquad (1.78)$$

Now, we know the period T_0 in terms of ω. It is

$$\omega = \frac{2\pi}{T_0} \Rightarrow T_0 = \frac{2\pi}{\omega} \qquad (1.79)$$

Therefore

$$T_0 = 2\pi \sqrt{\frac{m}{U_0}} \qquad (1.80)$$

This harmonic approximation is valid for displacements in x small enough that $x^4/4! << x^2/2!$.

This problem highlights the necessity of having the mathematical tools to solve a problem readily accessible. In this case the Taylor series given in Eq. (H.3) was a necessity.

11. A particle of mass m moves in one dimension without friction under the influence of a potential energy function

$$U(x) = U_0 \left(1 - e^{-\alpha^2 x^2}\right) \qquad (1.81)$$

where α is a constant having suitable dimensions.

(a) Sketch the potential energy function. Draw lines of constant TME for $E_0 < E_1 < U_0$ and $E_2 > U_0$.
(b) For a TME that gives bounded motion, what is the period T_0 of the bounded motion for amplitudes small enough for the motion to be considered simple harmonic? Use the graph to discuss qualitatively the conditions under which harmonic motion is a good approximation.

Solution

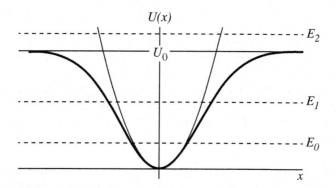

Fig. 1.12 Problem 11—solution

(a) The heavy line is $U(x)$ and the lighter line represents the leading term of the Taylor series expansion, the quadratic term, i.e. a parabola.

(b) To find the parameters for simple harmonic motion we must find the coefficient of the quadratic term in the potential energy as in Problem 10 of this chapter. We must therefore expand $U(x)$ in a Taylor' series. Using Eq. (H.3) we have

$$U(x) = U_0 \left(1 - e^{-\alpha^2 x^2}\right)$$

$$= U_0 \left[1 - \left(1 - \frac{\alpha^2 x^2}{1!} + \frac{\alpha^4 x^4}{2!} - \cdots\right)\right]$$

$$\approx U_0 \alpha^2 x^2 \tag{1.82}$$

Thus, $U_0 \alpha^2$ is equivalent to $\frac{1}{2} m\omega^2$, the coefficient of x^2 in the potential energy function for simple harmonic motion. We have

$$\frac{1}{2} m\omega^2 = U_0 \alpha^2 \Rightarrow \omega^2 = \frac{2 U_0 \alpha^2}{m}$$

$$\omega = \sqrt{\frac{2 U_0 \alpha^2}{m}} \tag{1.83}$$

Now, we know the period T_0 in terms of ω. It is

$$\omega = \frac{2\pi}{T_0} \Rightarrow T_0 = \frac{2\pi}{\omega} \tag{1.84}$$

Therefore

$$T_0 = \frac{2\pi}{\alpha} \sqrt{\frac{m}{2 U_0}} = \frac{\pi}{\alpha} \sqrt{\frac{2m}{U_0}} \tag{1.85}$$

Note that E_1 (see Fig. 1.12) would not produce approximately harmonic motion because the potential energy function is not very parabolic in that region. The quadratic term in Eq. (1.82) is responsible for simple harmonic motion so the region in which $U(x)$ most nearly approximates a parabola is the region in which harmonic motion is expected to be a good approximation, roughly that for which the TME is equal to or less than E_0 as drawn on the graph.

12. Consider a particle that approaches a hard sphere of radius R, impacts the surface and then bounces away elastically. The interaction potential is therefore

$$U(r) = \infty \quad r \leq R$$

$$= 0 \quad r > R \tag{1.86}$$

Fig. 1.13 Problem 12

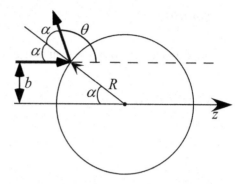

The mass of the sphere is very much greater than that of the particle (think of a small ball-bearing, a "BB," impacting a bowling ball). The relevant parameters are shown in Fig. 1.13.Take the incident velocity to be parallel to the z-axis shown; the parameter b is known as the impact parameter and is given by $b = R \sin \alpha$ for a hard sphere collision. The angles of incidence and reflection are equal and also given by α. The "scattering angle" is θ.

This problem is three-dimensional, but it has circular symmetry in the azimuthal angle ϕ around the z-axis. We define a "differential scattering cross-section" $D(\theta)$ for particles incident with impact parameter b and scattered into an infinitesimal solid angle $d\Omega$.

$$D(\theta) = \frac{d\sigma}{d\Omega} \tag{1.87}$$

where

$$d\sigma = b\,db\,d\phi$$
$$d\Omega = \sin\theta\,d\theta\,d\phi \tag{1.88}$$

From Eq. (1.88) we see that σ has units of area and the units of Ω are steradians (unitless). Now,

$$D(\theta) = \frac{b}{\sin\theta}\frac{db}{d\theta} \tag{1.89}$$

(a) Find the relationship between the scattering angle θ and the angle of incidence α, and use this to obtain the relationship between θ and b for a hard sphere. Obtain $D(\theta)$ for a hard sphere. What happens when $b > R$?
(b) Evaluate the total elastic scattering cross section σ by integrating $D(\theta)$ over all space.

Solution

(a) From the diagram, the scattering angle is

$$\theta = \pi - 2\alpha \tag{1.90}$$

so

$$b = R \sin\left(\frac{\pi}{2} - \frac{\theta}{2}\right)$$

$$= R \cos\frac{\theta}{2} \tag{1.91}$$

The latter relation immediately leads to

$$\theta = 2\cos^{-1}\left(\frac{b}{R}\right) \quad \text{if } b \leq R$$

$$= 0 \quad \text{if } b > R \tag{1.92}$$

Clearly, if $b > R$ the particle misses the sphere, so there is no scattering. Using Eq. (1.91) we can solve for $D(\theta)$.

$$D(\theta) = \frac{R\cos\frac{\theta}{2}}{\sin\theta}\frac{1}{2}R\sin\frac{\theta}{2}$$

$$= \frac{R^2}{4} \tag{1.93}$$

(b) The total cross section is

$$\sigma = \int_\Omega D(\theta)\,d\Omega$$

$$= \frac{R^2}{4}\int_\Omega d\Omega$$

$$= \pi R^2 \tag{1.94}$$

This result is (or should be) intuitively obvious. The total scattering cross section σ in the hard sphere case is just the cross sectional area of the sphere. If the particles were BBs and a large number were shot horizontally with a wide range of impact parameters, a large sheet of cardboard placed behind the sphere would be riddled with holes except for an untouched circular region of radius R. When the hard sphere is replaced by an arbitrary radial interaction potential $U(r)$ of longer range than a hard sphere, Eq. (1.89) remains valid, but the relationship between b and θ may be quite different.

Chapter 2
Lagrangian and Hamiltonian Dynamics

The Lagrangian and Hamiltonian formulations of mechanics contain no physics beyond Newtonian physics. They are simply reformulations that provide recipes to solve problems that are difficult to solve using elementary methods. Additionally, these formulations, especially Hamiltonian dynamics, play an important role in the development of quantum mechanics and in the elucidation of the deep relationships between quantum and classical physics.

The Lagrangian equations of motion are obtained from Hamilton's principle: *The actual path followed by a dynamical system is that which minimizes the time integral of the Lagrangian $\mathcal{L}$.* The Lagrangian

$$\mathcal{L} = T - U \tag{2.1}$$

is the difference between the kinetic energy T and potential energy U. The minimizing path is found using the calculus of variations [3]. In equation form this path is

$$\delta \int_{t_1}^{t_2} \mathcal{L} dt = 0 \tag{2.2}$$

This procedure leads to Lagrange's equations of motion, which are

$$\frac{\partial \mathcal{L}}{\partial q_i} - \frac{d}{dt} \left(\frac{\partial \mathcal{L}}{\partial \dot{q}_i} \right) = 0 \tag{2.3}$$

where q_i and $\dot{q}_i$ represent the coordinates and their time derivatives. It is a simple matter to remember where the d/dt must be in Eq. (2.3). We simply make sure that the units of each term are the same.

© Springer International Publishing AG 2017
J.D. Kelley, J.J. Leventhal, *Problems in Classical and Quantum Mechanics*, DOI 10.1007/978-3-319-46664-4_2

Equation (2.3) does not explicitly contain momentum, but the *generalized momentum* p_i, also termed "conjugate momentum" (conjugate to the coordinate q_i), is defined as

$$p_i = \frac{\partial \mathcal{L}}{\partial \dot{q}_i} \qquad (2.4)$$

Thus, the Lagrangian equations of motion do contain the momentum, but it is disguised. Note that if q_i happens to be a linear coordinate then the generalized momentum is the usual linear momentum. If q_i is an angle as in cylindrical or spherical coordinates, then the generalized momentum is an angular momentum. Other p_i's may not be recognizable as momenta, but they are nonetheless called generalized momenta. The generalized coordinate q_i and the generalized momentum are termed "canonically conjugate."

The Lagrangian $\mathcal{L}$ and the Hamiltonian H are related by

$$H(q_i, p_i, t) = \sum_j \dot{q}_j p_j - \mathcal{L}(q_i, p_i, t) \qquad (2.5)$$

In Eq. (2.5) we have included the possibility that the time might appear explicitly in the Lagrangian and the Hamiltonian. From Eq. (2.5) Hamilton's equations of motion are derived. They are

$$\dot{q}_i = \frac{\partial H}{\partial p_i}$$

$$-\dot{p}_i = \frac{\partial H}{\partial q_i} \qquad (2.6)$$

It is important to remember that to use these formulations the Lagrangian $\mathcal{L}$ and the Hamiltonian H must be written in terms of their proper variables. For generalized coordinates q_i the Lagrangian must be written in terms of the q_i, their time derivatives $\dot{q}_i$ and possibly the time t. In contrast, the Hamiltonian H must be written in terms of q_i and p_i. Equation (2.4) provides the link between p_i and $\dot{q}_i$.

If the Hamiltonian does not *explicitly* contain the time, then H is a conserved quantity. To show this we write the *total* derivative of H with respect to time as though it contains q, p, and t; replacing the partial derivatives with the Hamiltonian equations of motion, Eq. (2.6). This leads to

$$\frac{dH}{dt} = \frac{\partial H}{\partial q_i}\dot{q}_i + \frac{\partial H}{\partial p_i}\dot{p}_i + \frac{\partial H}{\partial t}$$

$$= -\dot{p}_i\dot{q}_i + \dot{q}_i\dot{p}_i + \frac{\partial H}{\partial t}$$

$$\equiv \frac{\partial H}{\partial t} \qquad (2.7)$$

Thus, if H does not contain the time explicitly the partial time derivative is zero as is the total derivative and H is a constant of the motion. This constant of the motion is the TME if:

- The potential U contains no velocity dependence.
- The equations connecting Cartesian and generalized coordinates are independent of time. This requires the kinetic energy to be a homogeneous quadratic function of the $\dot{q}_i$ [3].

Having noted the conditions under which the Hamiltonian (and often the TME) is a constant of the motion we turn to the generalized momenta p_i as defined in Eq. (2.4). From the second of the Hamiltonian equations of motion, Eq. (2.6), it is clear that $\dot{p}_i$ vanishes if the Hamiltonian does not explicitly contain q_i which infers conservation of p_i. In such a case the canonically conjugate coordinate q_i is referred to as a "cyclic coordinate." It is also known as an "ignorable coordinate," an unfortunate appellation.

A good way to begin solving problems using Lagrangian or Hamiltonian dynamics is to solve simple problems, problems for which you already know the answers. This will give you confidence that these seemingly abstract formulations are correct.

Problems

1. One of the first applications of the calculus of variations, upon which Lagrangian dynamics is based, was to the classic problem known as the brachistochrone. The goal is to find the path of a particle moving in a constant force field, e.g. gravity, starting from rest at some point (x_1, y_1) and ending at a lower point (x_2, y_2) such that the time to traverse this path is a minimum. It is well known that this path is a cycloid. It is convenient to make the minimum point of the cycloid at $x = 2a$. The parametric equations of the cycloid in this form are

$$x = a(1 + \cos\theta)$$
$$y = a(\theta + \sin\theta) \tag{2.8}$$

where the coordinates axes, x and y, are shown in Fig. 2.1.

What is not so well-known about the *brachistochrone* problem is that the time to traverse the cycloidal path is independent of the starting point (x_i, y_i). To determine that the path is a cycloid the particle is always placed at $x = 0$ on this coordinate system. It is a fact, however, that the time to reach the bottom of the cycloid $x = 2a$ is independent of the location of the starting point. Prove this assertion.

Fig. 2.1 The coordinates
used to derive the parametric
equations of the trajectory of
the *brachistochrone*,
Eq. (2.8). These coordinates
are also used in Problem 1 of
this chapter

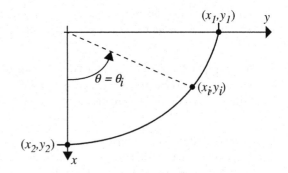

Solution

The time dt it takes the particle to traverse a distance ds along the cycloid is
simply

$$dt = \frac{ds}{v\,(x)} \tag{2.9}$$

where $v\,(x)$ is the speed of the particle at any point (x, y) along the cycloid
between (x_i, y_i) and the minimum $(x_2 = 2a, y_2 = 0)$; this speed clearly depends
only upon x.

We define as the zero of potential energy to be at (x_i, y_i). Then, because the
particle starts from rest, the TME is zero and at any point (x, y) we have

$$U\,(x) + T\,(x) = 0$$

$$mg\,(x_i - x) + \frac{1}{2}m\,[v\,(x)]^2 = 0 \tag{2.10}$$

At any point (x, y) the speed of the particle $v\,(x)$ is given by

$$v\,(x) = \sqrt{2g\,(x - x_i)} \tag{2.11}$$

The increment of path ds is

$$ds = \sqrt{dx^2 + dy^2}$$

$$= \sqrt{1 + y'^2}\,dx \tag{2.12}$$

where $y' = dy/dx$. The time to reach the minimum is

$$t = \int_{x_i}^{2a} \frac{ds}{v}$$

$$= \int_{x_i}^{2a} \sqrt{\frac{1 + y'^2}{2g\,(x - x_1)}}\,dx \tag{2.13}$$

To perform the integration we use the parametric equations for the cycloid and find that

$$y' = \frac{dy}{d\theta}\frac{d\theta}{dx} = a\,(1 + \cos\theta) \cdot \frac{1}{-a\sin\theta}$$

$$= -\frac{(1 + \cos\theta)}{\sin\theta} \qquad (2.14)$$

and

$$dx = -a\sin\theta\,d\theta \qquad (2.15)$$

Also, when $x = x_i$, $\theta = \theta_i$ and when $x = 2a$, $\theta = 0$ so we have

$$t = \frac{1}{\sqrt{2g}} \int_{\theta_1}^{0} \left[\frac{1 + \frac{1+2\cos\theta+\cos^2\theta}{\sin^2\theta}}{a + a\cos\theta - a - a\cos\theta_1} \right]^{1/2} (-a\sin\theta\,d\theta)$$

$$= \sqrt{\frac{a}{2g}} \int_{0}^{\theta_1} \left[\frac{2 + 2\cos\theta}{\cos\theta - \cos\theta_1} \right]^{1/2} d\theta$$

$$= \sqrt{\frac{a}{g}} \int_{0}^{\theta_1} \left[\frac{1 + \cos\theta}{\cos\theta - \cos\theta_1} \right]^{1/2} d\theta \qquad (2.16)$$

Using Eq. (E.9),

$$\cos\theta = 2\cos^2\frac{\theta}{2} - 1 \qquad (2.17)$$

we have

$$t = \sqrt{\frac{a}{g}} \int_{0}^{\theta_1} \left[\frac{2\cos^2\frac{\theta}{2}}{2\left(\cos^2\frac{\theta}{2} - \cos^2\frac{\theta_1}{2}\right)} \right]^{1/2} d\theta$$

$$= \sqrt{\frac{a}{g}} \int_{0}^{\theta_1} \frac{\cos\frac{\theta}{2}}{\left(\cos^2\frac{\theta}{2} - \cos^2\frac{\theta_1}{2}\right)^{1/2}} d\theta$$

$$= \sqrt{\frac{a}{g}} \int_{0}^{\theta_1} \frac{\cos\frac{\theta}{2}\,d\theta}{\sqrt{\left(1 - \sin^2\frac{\theta}{2}\right) - \left(1 - \sin^2\frac{\theta_1}{2}\right)}}$$

$$= \sqrt{\frac{a}{g}} \int_{0}^{\theta_1} \frac{\cos\frac{\theta}{2}\,d\theta}{\sqrt{\sin^2\frac{\theta_1}{2} - \sin^2\frac{\theta}{2}}} \qquad (2.18)$$

To evaluate this integral we let

$$z = \sin\frac{\theta}{2} \Rightarrow dz = \frac{1}{2}\cos\frac{\theta}{2}d\theta$$

$$\theta = 0 \rightarrow z = 0, \ \theta = \theta_1 \rightarrow z = \sin\frac{\theta_1}{2} \tag{2.19}$$

The time to reach the bottom of the cycloid, Eq. (2.18) becomes

$$t = \sqrt{\frac{a}{g}}\int_0^{\sin\frac{\theta_1}{2}} \frac{2dz}{\sqrt{\sin^2\frac{\theta_1}{2} - z^2}} \tag{2.20}$$

Applying the integral in Eq. (G.7)

$$\int \frac{dx}{\sqrt{a^2 - x^2}} = \sin^{-1}\frac{x}{a} \tag{2.21}$$

to Eq. (2.20) we have

$$t = 2\sqrt{\frac{a}{g}}\left[\sin^{-1}\left(\frac{z}{\sin\frac{\theta_1}{2}}\right)\right]_0^{\sin\frac{\theta_1}{2}}$$

$$= 2\sqrt{\frac{a}{g}}\left[\sin^{-1}(1) - \sin^{-1}(0)\right]$$

$$= \sqrt{\frac{a}{g}}\pi \tag{2.22}$$

It is surprising that no matter where the particle starts its path along the trajectory, the time to reach the bottom of the cycloidal path is always the same. This seems counterintuitive when comparing two different starting points. If, however, the particle starts from (x_1, y_1) as in Fig. 2.1, then it has gained speed when it reaches (x_i, y_i) so the path from (x_i, y_i) to (x_2, y_2) is covered in less time than if the particle had started from rest at (x_i, y_i). This argument does not prove the result, but it at least rationalizes it.

As stated above, this problem was included as an illustration of the calculus of variations which is used as the tool applied to Hamilton's principle to develop Lagrangian dynamics.

2. Use a one-dimensional formulation to show that Lagrangian dynamics is equivalent to Newton's second law. Assume that the kinetic energy is a function of only the velocity and the potential energy a function of only the position.

Solution

Because the Lagrangian is invariant under a transformation from any set of generalized coordinates to another we may use Cartesian coordinates for convenience. We write Lagrange's equation for a single particle in one dimension.

$$\frac{\partial \mathcal{L}}{\partial x} - \frac{d}{dt}\frac{\partial \mathcal{L}}{\partial \dot{x}} = 0$$

$$\frac{\partial [T(\dot{x}) - U(x)]}{\partial x} - \frac{d}{dt}\frac{\partial [T(\dot{x}) - U(x)]}{\partial \dot{x}} = 0 \qquad (2.23)$$

where we have noted explicitly that the kinetic energy T is a function of only the velocity $\dot{x}$ and the potential energy U is a function of only the position x. Carrying out the indicated differentiations we have

$$-\frac{\partial U(x)}{\partial x} = \frac{d}{dt}\frac{\partial T(\dot{x})}{\partial \dot{x}}$$

$$= \frac{d}{dt}\frac{\partial}{\partial \dot{x}}\left[\frac{1}{2}\left(m\dot{x}^2\right)\right]$$

$$= \frac{d}{dt}(m\dot{x})$$

$$= m\ddot{x} \qquad (2.24)$$

For a conservative system the negative derivative of the potential energy is the force so the last equation is

$$F = m\ddot{x} \qquad (2.25)$$

which is Newton's second law. This demonstrates that Lagrange's equations (and therefore Hamilton's equations) are nothing more than Newton's laws recast into a form that permits solution of problems that are more complex than those encountered in elementary physics courses.

3. Solve the problem of the one-dimensional frictionless simple harmonic oscillator (SHO) of mass m and spring constant k initially at rest with initial displacement x_0 using

   (a) Lagrangian dynamics
   (b) Hamiltonian dynamics

Solution

(a) Designating $x \, (= q)$ as the displacement and defining ω as the angular frequency ($\omega^2 = k/m$), the potential and kinetic energies are

$$U(x) = \frac{1}{2}kx^2 = \frac{1}{2}m\omega^2 x^2; \ T = \frac{1}{2}m\dot{x}^2 \tag{2.26}$$

which leads to a Lagrangian

$$\mathcal{L} = \frac{1}{2}m\dot{x}^2 - \frac{1}{2}m\omega^2 x^2 \tag{2.27}$$

Inserting $\mathcal{L}$ into Eq. (2.3) produces

$$\ddot{x} + \omega^2 x^2 = 0 \tag{2.28}$$

which is one of the most often encountered equations in physics. The general solution for the oscillator response $x(t)$ is well known. It is

$$x(t) = A \sin(\omega t) + B \cos(\omega t) \tag{2.29}$$

Upon inserting the boundary conditions given in the statement of the problem we find that $A = 0$ and $B = x_0$ so

$$x(t) = x_0 \cos(\omega t) \tag{2.30}$$

(b) To solve using Hamiltonian dynamics we must first find the Hamiltonian as given in Eq. (2.5). Everything in this equation is already known to us except the generalized momentum p which, according to Eq. (2.4) is

$$p = \frac{\partial \mathcal{L}}{\partial \dot{x}} = m\dot{x} \tag{2.31}$$

In this case p is the usual linear momentum because the lone coordinate x has dimensions of length. The Hamiltonian is, according to Eq. (2.5)

$$\begin{aligned} H(x, p) &= \left(\frac{p}{m}\right)p - \frac{1}{2}m\dot{x}^2 + \frac{1}{2}m\omega^2 x^2 \\ &= \frac{p^2}{2m} + \frac{1}{2}m\omega^2 x^2 \\ &= T + U \end{aligned} \tag{2.32}$$

Note that we did not have to calculate H using $\mathcal{L}$ because the potential energy is independent of the velocity and the kinetic energy is a homogeneous quadratic function of the $\dot{x}$. We could have simply written Eq. (2.32) because under these conditions the Hamiltonian *is* the TME.

Inserting the Hamiltonian into Eq. (2.6) we obtain two linear first order differential equations.

$$\dot{x} = \frac{\partial H}{\partial p} = \frac{p}{m}; \quad -\dot{p} = \frac{\partial H}{\partial x} = m\omega^2 x \qquad (2.33)$$

This is in contrast to the Lagrangian formulation which led to a single second order differential equation. We can eliminate $\dot{p}$ in the second equation by differentiating the first equation with respect to time and inserting the resulting expression for $\dot{p}$ into the second. We obtain

$$-m\ddot{x} = m\omega^2 x \qquad (2.34)$$

which is the same as Eq. (2.28).

4. Use Lagrangian dynamics to find the frequency of the simple pendulum shown in Fig. 2.2 for small vibrations.

Fig. 2.2 Problem 4

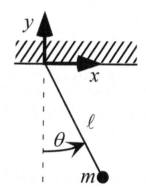

Solution

This problem is easily solved using elementary methods (see Problem 1 of Chap. 1). Nevertheless, it provides a good opportunity to practice skills with Lagrangian dynamics on a problem with a well-known solution. We begin by writing the kinetic and potential energies in Cartesian coordinates and then converting to the single coordinate θ using

$$x = \ell \sin \theta$$
$$y = -\ell \cos \theta \qquad (2.35)$$

The kinetic and potential energies are

$$T = \frac{1}{2}m\left(\dot{x}^2 + \dot{y}^2\right)$$

$$= \frac{1}{2}m\ell^2\dot{\theta}^2 \tag{2.36}$$

and

$$U = mgy = -mg\ell\cos\theta \tag{2.37}$$

so

$$\mathcal{L} = \frac{1}{2}m\ell^2\dot{\theta}^2 + mg\ell\cos\theta \tag{2.38}$$

The Lagrange equation of motion for θ is

$$\frac{d}{dt}\left(\frac{\partial \mathcal{L}}{\partial \dot{\theta}}\right) - \frac{\partial \mathcal{L}}{\partial \theta} = 0$$

$$m\ell^2\ddot{\theta} + mg\ell\sin\theta = 0 \tag{2.39}$$

which is the same equation derived using elementary methods in Problem 1 of Chap. 1 (see Eq. (1.2)).

For small angles $\sin\theta \approx \theta$ so we have

$$\ddot{\theta} + \frac{g}{\ell}\theta = 0 \tag{2.40}$$

which is the equation for undamped harmonic motion with frequency $\omega = \sqrt{g/\ell}$.

Note that in the small angle approximation the mass executes linear motion in x with this same frequency because $\theta \approx x/\ell$. In other words, the differential equation in Eq. (2.40) could have just as well have been written as

$$\ddot{x} + \frac{g}{\ell}x = 0 \tag{2.41}$$

under the small angle approximation.

There are a few things to note about this simple problem. First, let us investigate the consequences of applying the small angle approximation to the Lagrangian *before* obtaining the equations of motion. The small angle approximation is

$$\cos\theta \approx 1 - \frac{\theta^2}{2!} + \cdots \tag{2.42}$$

When the approximation $\sin\theta \approx \theta$ is employed only the first term in Eq. (2.42), i.e. unity, is taken. The Lagrangian then becomes

$$\mathcal{L} = \frac{1}{2}m\ell^2\dot{\theta}^2 + mg\ell \tag{2.43}$$

and the Lagrangian equation of motion is

$$\ddot{\theta} = 0 \tag{2.44}$$

This is clearly incorrect because we have eliminated the effect of the potential. The lesson to be learned here is that we have approximated ourselves out of a meaningful solution. Clearly we must retain (at least) the first two terms in Eq. (2.42). Doing so makes the Lagrangian

$$\mathcal{L} = \frac{1}{2}m\ell^2\dot{\theta}^2 + mg\ell - mg\ell\frac{\theta^2}{2} \tag{2.45}$$

so the Lagrangian equation of motion is

$$\frac{d}{dt}\left(\frac{\partial\mathcal{L}}{\partial\dot{\theta}}\right) - \frac{\partial\mathcal{L}}{\partial\theta} = 0$$

$$m\ell^2\ddot{\theta} + mg\ell\theta = 0 \tag{2.46}$$

or

$$\ddot{\theta} + \omega_p^2\theta = 0 \tag{2.47}$$

where

$$\omega_p = \sqrt{\frac{g}{\ell}} \tag{2.48}$$

is the natural frequency of the simple pendulum.

5. The pivot point of a simple pendulum of mass m and length ℓ moves in the y-direction according to $y = f(t)$. Using Lagrangian dynamics show that the motion of the pendulum is that of a simple pendulum in a gravitational field $g + \ddot{f}(t)$.

Fig. 2.3 Problem 5

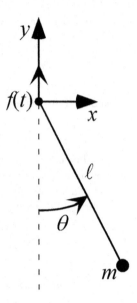

Solution

The x- and y-coordinates of the mass are

$$x = \ell \sin \theta \quad \text{and} \quad y = f(t) - \ell \cos \theta \qquad (2.49)$$

so the x- and y-coordinates of the velocities are

$$\dot{x} = \ell \dot{\theta} \cos \theta \quad \text{and} \quad \dot{y} = \dot{f}(t) + \ell \dot{\theta} \sin \theta \qquad (2.50)$$

The kinetic energy is

$$
\begin{aligned}
T &= \frac{1}{2} m \left(\dot{x}^2 + \dot{y}^2 \right) \\
&= \frac{1}{2} m \left\{ \ell^2 \dot{\theta}^2 \cos^2 \theta + \left[\dot{f}(t) \right]^2 + \ell^2 \dot{\theta}^2 \sin^2 \theta + 2 \dot{f}(t) \, \ell \dot{\theta} \sin \theta \right\} \\
&= \frac{1}{2} m \left\{ \ell^2 \dot{\theta}^2 + \left[\dot{f}(t) \right]^2 + 2 \dot{f}(t) \, \ell \dot{\theta} \sin \theta \right\}
\end{aligned}
\qquad (2.51)
$$

The potential energy is

$$U = mgy = mg \left[f(t) - \ell \cos \theta \right] \qquad (2.52)$$

The Lagrangian is

$$\mathcal{L} = \frac{1}{2} m \left\{ \ell^2 \dot{\theta}^2 + \left[\dot{f}(t) \right]^2 + 2 \dot{f}(t) \, \ell \dot{\theta} \sin \theta \right\} - mg \left[f(t) - \ell \cos \theta \right] \qquad (2.53)$$

Then

$$\frac{d}{dt}\left(\frac{\partial \mathcal{L}}{\partial \dot{\theta}}\right) = \frac{d}{dt}m\left[\ell^2 \dot{\theta} + \dot{f}(t)\,\ell \sin\theta\right]$$

$$= m\left[\ell^2 \ddot{\theta} + \ddot{f}(t)\,\ell \sin\theta + \dot{f}(t)\,\ell\dot{\theta}\cos\theta\right] \qquad (2.54)$$

and

$$\frac{\partial \mathcal{L}}{\partial \theta} = m\dot{f}(t)\,\ell\dot{\theta}\cos\theta - mg\ell\sin\theta \qquad (2.55)$$

so Lagrange's equation is

$$\left[\ell^2 \ddot{\theta} + \ddot{f}(t)\,\ell\sin\theta + \dot{f}(t)\,\ell\dot{\theta}\cos\theta\right] - \dot{f}(t)\,\ell\dot{\theta}\cos\theta + g\ell\sin\theta = 0$$

$$\ell^2 \ddot{\theta} + \ddot{f}(t)\,\ell\sin\theta + g\ell\sin\theta = 0$$

$$\ddot{\theta} + \frac{[\ddot{f}(t) + g]}{\ell}\sin\theta = 0 \qquad (2.56)$$

This is the same differential equation as that derived in Problem 1, Chap. 1 (see Eq. (1.2)) except that g has been replaced by $[\ddot{f}(t) + g]$.

It is interesting to consider different forms of $f(t)$ and their effect on the pendulum. Perhaps the most interesting is the case of free fall. In that case $\ddot{f}(t) = -g$ and the coefficient of $\sin\theta$ in Eq. (2.56) vanishes so that $\ddot{\theta} = 0$. Integration yields

$$\theta = \dot{\theta}_0 t + \theta_0 \qquad (2.57)$$

where θ_0 and $\dot{\theta}_0$ are the initial angular displacement and angular velocity of the bob of mass m. To be specific let us suppose that we have a simple pendulum in an elevator and we pull the bob out to some initial angle θ_0. We release the bob from rest so that $\dot{\theta}_0 = 0$ at the instant that the elevator cable is severed. The elevator and pendulum are therefore in free fall. The equation that describes the θ motion becomes

$$\theta = \theta_0 \qquad (2.58)$$

Thus, the bob floats in space at $\theta = \theta_0$ as the elevator drops. In effect, there is no gravity to cause the periodic motion of a simple pendulum. Free fall has negated the effect of gravity!

6. Describe the motion of the system shown in Fig. 2.4, a sliding pendulum system, for small vibrations of the pendulum. The mass M can slide without friction in either direction. It is permissible to make the small vibration

Fig. 2.4 Problem 6

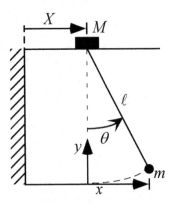

approximation in the Lagrangian before obtaining the equations of motion. Use the coordinates shown in Fig. 2.4; X is the distance of the sliding block of mass M from the wall, x is the distance of the pendulum bob of mass m from the wall and y is the vertical coordinate of the bob measured from the location of the mass m when $\theta = 0$.

Solution

There are four variables, but they are not independent. They are related according to

$$x = X + \ell \sin \theta \ \text{ and } \ y = \ell - \ell \cos \theta \tag{2.59}$$

There are only two degrees of freedom and we choose to work the problem in terms of the variables X and θ.

The kinetic energy is

$$T = \frac{1}{2} \left[M\dot{X}^2 + m \left(\dot{x}^2 + \dot{y}^2 \right) \right] \tag{2.60}$$

Eliminating x and y using Eq. (2.59) we have

$$T = \frac{1}{2}M\dot{X}^2 + \frac{1}{2}m \left[\left(\dot{X} + \ell\dot{\theta} \cos \theta \right)^2 + \ell^2\dot{\theta}^2 \sin^2 \theta \right]$$
$$= \frac{1}{2}M\dot{X}^2 + \frac{1}{2}m \left(\dot{X}^2 + 2\dot{X}\ell\dot{\theta} \cos \theta + \ell^2\dot{\theta}^2 \right) \tag{2.61}$$

Now we wish to make the small angle approximation, but we must be careful. In Problem 4 of this chapter we saw that making the approximation

$\cos\theta \approx 1$ for small angles was too much of an approximation and led to an absurd result. In this case we do not lose any terms in the kinetic or potential energies with the small-angle approximation, and it greatly simplifies the term in the parentheses in Eq. (2.61). We have

$$T \simeq \frac{1}{2}M\dot{X}^2 + \frac{1}{2}m\left(\dot{X} + \ell\dot{\theta}\right)^2 \tag{2.62}$$

The potential energy taking $U(y = 0) = 0$ is

$$U(y) = mgy$$
$$= mg\ell(1 - \cos\theta)$$
$$\approx mg\ell\frac{\theta^2}{2} \tag{2.63}$$

Notice that making the small angle approximation in Eq. (2.63) required retention of the first two terms in the Taylor series (see Appendix H) of $\cos\theta$.

With the small angle approximation the Lagrangian becomes

$$\mathcal{L} = \frac{1}{2}M\dot{X}^2 + \frac{1}{2}m\left(\dot{X} + \ell\dot{\theta}\right)^2 - mg\ell\frac{\theta^2}{2} \tag{2.64}$$

The equations of motion are

$$\ddot{X} + \ell\ddot{\theta} + g\theta = 0$$
$$(M + m)\ddot{X} + m\ell\ddot{\theta} = 0 \tag{2.65}$$

Solving the second of Eq. (2.65) for $\ddot{X}$ we have

$$\ddot{X} = -\left(\frac{m}{m + M}\right)\ell\ddot{\theta} \tag{2.66}$$

Now, substituting Eq. (2.66) for $\ddot{X}$ in the first of Eq. (2.65) eliminates X and gives

$$\left(\frac{M}{m + M}\right)\ddot{\theta} + \frac{g}{\ell}\theta = 0 \tag{2.67}$$

which we recognize to be the equation for small-angle harmonic motion in a simple pendulum with frequency

$$\omega^2 = \frac{g}{\ell}\left(\frac{M + m}{M}\right) \tag{2.68}$$

Integrating Eq. (2.66)

$$\dot{X} = -\left(\frac{m}{m+M}\right)\ell\dot{\theta} + C \qquad (2.69)$$

we see that if $\dot{X} \neq 0$, but the pendulum remains at rest with $\dot{\theta} = 0$, then the
motion in X proceeds with a constant velocity C, the constant of integration
in Eq. (2.69). If $C = 0$ and $\dot{\theta} \neq 0$, then the block and pendulum oscillate
in opposite directions. In the general case the block slides along the support
at a constant average velocity with a superimposed oscillation at frequency ω.
Suppose the mass M were very large compared with m. Then Eq. (2.68) gives
the frequency for a simple pendulum attached to a fixed point, and Eq. (2.66)
describes translation of the block undisturbed by the pendulum motion. If the
two masses were equal, ω^2 would be twice that of the simple pendulum because
the effective length of the pendulum is half that of the simple pendulum—the
block and pendulum oscillate in opposite direction about a point at $y = \ell/2$.

7. An Atwood's machine consists of two point masses connected by an inexten-
    sible massless string of length ℓ that passes over a frictionless peg as shown in
    Fig. 2.5. Take y to be the vertical displacement of m_2. For convenience assume
    that $m_2 > m_1$ and that the diameter of the peg is negligible.

    (a) Find the Lagrangian in terms of the coordinates y and $\dot{y}$.
    (b) Find the acceleration of the masses using Lagrangian dynamics.

Fig. 2.5 Problem 7

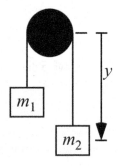

Solution

(a) The kinetic energy T is the sum of the kinetic energies of the masses m_1
    and m_2.

$$T = \frac{1}{2}(m_1 + m_2)\dot{y}^2 \qquad (2.70)$$

The potential energy measured from $y = 0$ is

$$U = -m_2 gy - m_1 g (\ell - y)$$
$$= (m_1 - m_2) gy - m_1 g\ell \qquad (2.71)$$

We are confident that this is correct because when $y = 0$ (m_2 at the top) the potential energy is due solely to m_1 and is $-m_1 g\ell$. When $y = \ell$ (m_1 at the top) the potential energy is entirely due to m_2 and is $-m_2 g\ell$.
The Lagrangian is

$$\mathcal{L} = \frac{1}{2} (m_1 + m_2) \dot{y}^2 - (m_1 - m_2) gy + m_1 g\ell \qquad (2.72)$$

(b) Lagrange's equation in terms of y and $\dot{y}$ is

$$\frac{\partial \mathcal{L}}{\partial y} - \frac{d}{dt} \frac{\partial \mathcal{L}}{\partial \dot{y}} = 0 \Rightarrow -(m_1 - m_2) g - \frac{d}{dt} [(m_1 + m_2) \dot{y}] = 0 \qquad (2.73)$$

Therefore

$$\ddot{y} = \frac{(m_2 - m_1)}{(m_2 + m_1)} g \qquad (2.74)$$

Although this simple result is often obtained in introductory physics it is satisfying to get the same results using a more advanced technique. Note that because $m_2 > m_1$, $\ddot{y}$ is manifestly positive because in this coordinate system (y positive down) g is a positive number. Moreover, the answer is correct in the limiting cases $m_2 = m_1$ and $m_1 = 0$ ($\ddot{y} = 0$ and $\ddot{y} = g$, respectively). Thus, manipulating this solution leads us to believe that it is correct because it has the correct units, the correct sign and the limiting values are correct.

8. A system consists of a pulley of mass M, radius R, and moment of inertia $I = MR^2$. Masses m_1 and m_2 are connected by an inextensible massless string that passes over the pulley as shown in Fig. 2.6. When the pulley rotates the string moves without slipping at the interface with the pulley. There is no friction between m_1 and the table. The cord is massless and $m_2 > m_1$.

(a) Find the Lagrangian in terms of y and θ.
(b) Using Lagrangian dynamics find the acceleration of the masses assuming that the system starts with m_2 at $y = 0$ with $\theta = 0$.

Solution

(a) The kinetic energy is

$$T = \frac{1}{2}(m_1 + m_2)\dot{y}^2 + \frac{1}{2}(MR^2)\dot{\theta}^2 \tag{2.75}$$

Taking the potential energy to be zero at $y = 0$ we have

$$U = -m_2 gy \tag{2.76}$$

so that

$$\mathcal{L} = \frac{1}{2}(m_1 + m_2)\dot{y}^2 + \frac{1}{2}MR^2\dot{\theta}^2 + m_2 gy \tag{2.77}$$

(b) Since we are asked for the acceleration of the masses, i.e. $\ddot{y}$, we may eliminate $\dot{\theta}$ using the relationship between y and θ (no slipping), which is

$$\theta = \frac{y}{R} \tag{2.78}$$

In terms of y and $\dot{y}$ the Lagrangian is

$$\mathcal{L}(y, \dot{y}) = \frac{1}{2}(m_1 + m_2 + M)\dot{y}^2 + m_2 gy \tag{2.79}$$

Lagrange's equation is

$$m_2 g - \frac{d}{dt}(m_1 + m_2 + M)\dot{y} = 0 \tag{2.80}$$

from which we obtain

Fig. 2.6 Problem 8

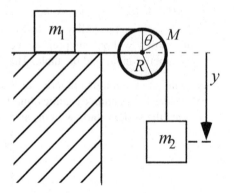

$$\ddot{y} = \frac{m_2}{(m_1 + m_2 + M)}g \tag{2.81}$$

As usual it is advisable to examine the extreme cases. If $m_2 \gg m_1$ and $m_2 \gg M$, then m_2 would essentially be in free fall so that $\ddot{y} \approx g$. On the other hand, if either m_1 or the pulley mass M dominates, virtually nothing happens or, at least, the motion is quite slow.

9. Solve Problem 8, part (b) of this chapter using Hamiltonian dynamics.

Solution

We could formally find the Hamiltonian using the relationship between it, the Lagrangian and the generalized momentum. That is

$$H(q_j, p_j) = \sum_{i=1}^{N} \dot{q}_i p_i - \mathcal{L}(q_j, \dot{q}_j) \tag{2.82}$$

where p_j is the generalized momentum defined as

$$p_j = \frac{\partial \mathcal{L}}{\partial \dot{q}_j} \tag{2.83}$$

In the present case, in terms of the coordinate y this would be

$$H = p_y \dot{y} - \mathcal{L}(y, \dot{y}) \tag{2.84}$$

We note, however, that the kinetic energy for this problem, Eq. (2.75), is a homogeneous function of the coordinates y and θ. Moreover, as in most cases, the potential energy is not a function of the velocity. These two conditions assure us that the Hamiltonian is the TME. Because we already have both the kinetic and potential energies it is simply a matter of calculating the generalized momenta in order to convert the time derivative of the coordinates in the kinetic energy into their momentum equivalents. This conversion is necessary because the derivatives taken in Hamilton's equations are those of the generalized momenta and the generalized coordinates. In terms of the generalized coordinates the Hamiltonian equations of motion are

$$\dot{q}_j = \frac{\partial H}{\partial p_j} \text{ and } -\dot{p}_j = \frac{\partial H}{\partial q_j} \tag{2.85}$$

The upshot of all of this is that we need not do much in the way of calculation other than to find the generalized momentum p_y. Because y is a linear coordinate, a Cartesian coordinate, p_y is simply a linear momentum. Using Eq. (2.79) we have

$$p_y = \frac{\partial \mathcal{L}}{\partial \dot{y}}$$

$$= (m_1 + m_2 + M) \dot{y} \qquad (2.86)$$

In terms of this momentum the TME, and thus the Hamiltonian is

$$H = \frac{p_y^2}{2 (m_1 + m_2 + M)} - m_2 g y \qquad (2.87)$$

Hamilton's equations are then

$$\dot{y} = \frac{p_y}{(m_1 + m_2 + M)} \quad \text{and} \quad \dot{p}_y = m_2 g \qquad (2.88)$$

Differentiating the first with respect to time and substituting the second for $\dot{p}_y$ we obtain

$$\ddot{y} = \frac{m_2 g}{(m_1 + m_2 + M)} \qquad (2.89)$$

which is identical with the answer obtained in part (b) of Problem 8 of this chapter, Eq. (2.81).

10. A particle of mass m is projected upward with a velocity v_0 at an angle α to the horizontal in the uniform gravitational field of the earth as shown in Fig. 2.7. Ignore air resistance and take $U (y = 0) = 0$.

Fig. 2.7 Problem 10

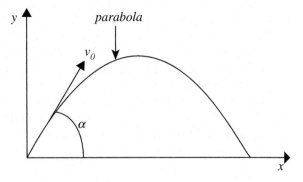

(a) Find the Lagrangian in terms of x and y and identify cyclic coordinates .
(b) Find the conjugate momenta, identify them and discuss which are conserved and why.

(c) Find the x- and y-components of the velocity as functions of time.
(d) Find the Hamiltonian.
(e) Ignoring air resistance use Hamiltonian dynamics with the coordinates shown to find the x- and y-components of the velocity as functions of time.

Solution

(a)

$$T = \frac{1}{2}m\left(\dot{x}^2 + \dot{y}^2\right); U = mgy \Longrightarrow \mathcal{L} = \frac{1}{2}m\left(\dot{x}^2 + \dot{y}^2\right) - mgy \qquad (2.90)$$

Only the x-coordinate is cyclic because it does not appear in the Lagrangian.
(b) Conjugate momenta:

$$p_x = \frac{\partial \mathcal{L}}{\partial \dot{x}} = m\dot{x}; \; p_y = \frac{\partial \mathcal{L}}{\partial \dot{y}} = m\dot{y} \qquad (2.91)$$

Because x and y are Cartesian coordinates p_x and p_y are simply the x and y components of linear momentum. From Lagrange's equation

$$\frac{\partial \mathcal{L}}{\partial q} - \frac{d}{dt}\left(\frac{\partial \mathcal{L}}{\partial \dot{q}}\right) = 0$$

$$\frac{\partial \mathcal{L}}{\partial q} - \frac{d}{dt}p_q = 0 \qquad (2.92)$$

it is clear that when the Lagrangian is cyclic in a coordinate then the momentum conjugate to that coordinate is conserved. In the current problem we have

$$\frac{dp_x}{dt} = 0 \Rightarrow p_x = \text{constant}$$

$$\frac{dp_y}{dt} = -mg \Rightarrow p_y \neq \text{constant} \qquad (2.93)$$

Because there is an outside force in the y-direction p_y is not a conserved quantity. On the other hand, there is no outside force in the x-direction so p_x is conserved.
(c) Lagrange's equations for x and y are

$$0 - m\ddot{x} = 0 \quad \text{and} \quad -mg - m\ddot{y} = 0 \qquad (2.94)$$

which lead to

$$\dot{x}(t) = \text{constant, and } \dot{y}(t) = -gt + \text{constant} \qquad (2.95)$$

Using the initial conditions we have

$$\dot{x}(t) = v_0 \cos\alpha \quad \text{and} \quad \dot{y}(t) = -gt + v_0 \sin\alpha \tag{2.96}$$

(d) The Hamiltonian is given by Eq. (2.5).

$$H(q_i, p_i, t) = \sum_j \dot{q}_j p_j - \mathcal{L}(q_i, p_i, t) \tag{2.97}$$

where $q_1 = x$ and $q_2 = y$. Using Eq. (2.91) we have

$$H = \dot{x} p_x + \dot{y} p_y - \mathcal{L}$$

$$= \frac{p_x}{m} p_x + \frac{p_y}{m} p_y - \left\{ \frac{1}{2} m \left[\left(\frac{p_x}{m}\right)^2 + \left(\frac{p_y}{m}\right)^2 \right] - mgy \right\}$$

$$= \frac{p_x^2}{2m} + \frac{p_y^2}{2m} + mgy = T + U \tag{2.98}$$

The Hamiltonian is, as expected, the TME because the potential energy is independent of the velocity.

(e) Hamilton equations of motion for x are

$$\dot{p}_x = -\frac{\partial H}{\partial x} = 0 \Longrightarrow p_x = \text{constant} = mv_0 \cos\alpha \tag{2.99}$$

and from Eq. (2.91)

$$\dot{x} = \frac{\partial H}{\partial p_x} = \frac{p_x}{m} \tag{2.100}$$

Now for the y-direction. The Hamilton equations for y are

$$\dot{p}_y = -\frac{\partial H}{\partial y} = -mg \Longrightarrow p_y = -mgt + \text{constant} \tag{2.101}$$

Using the initial conditions the constant may be evaluated.

$$\text{constant} = mv_0 \sin\alpha \tag{2.102}$$

so

$$p_y = -mgt + mv_0 \sin\alpha \tag{2.103}$$

and, again from Eq. (2.91)

$$\dot{y} = \frac{\partial H}{\partial p_y} = \frac{p_y}{m} \tag{2.104}$$

The components of momentum are then

$$p_x = m\dot{x} = mv_0 \cos\alpha; \quad p_y = m\dot{y} = -mgt + mv_0 \sin\alpha \qquad (2.105)$$

so the velocity components are

$$\dot{x} = v_0 \cos\alpha; \quad \dot{y} = -gt + v_0 \sin\alpha$$

This projectile problem is usually the first one encountered in introductory physics courses. We see, again, that simple problems may be solved easily using Lagrangian or Hamiltonian dynamics. (More importantly, we get the same answers as were obtained using elementary techniques.)

11. A massless string is wound around a disk of mass m, radius R, and moment of inertia I about the center. The disk unwinds from rest (like a yo-yo) with initial conditions $t = 0$, $y = 0$, $\theta = 0$ as shown in Fig. 2.8.

Fig. 2.8 Problem 11

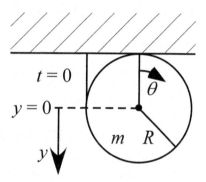

(a) Find the Lagrangian in terms of y and $\dot{y}$ only.
(b) Find p_y and the Hamiltonian and find the linear velocity of the disk in the y-direction as a function of time using Hamiltonian dynamics.

Solution

(a)

$$T = \frac{1}{2}m\dot{y}^2 + \frac{1}{2}I\dot{\theta}^2 \text{ and } \theta = \frac{y}{R}$$

$$\Rightarrow T = \frac{1}{2}m\dot{y}^2 + \frac{1}{2}I\left(\frac{\dot{y}}{R}\right)^2 \qquad (2.106)$$

and, taking $U(y = 0) = 0$ we have

$$U = -mgy \qquad (2.107)$$

Therefore

$$\mathcal{L} = \frac{1}{2}m\dot{y}^2 + \frac{1}{2}I\left(\frac{\dot{y}}{R}\right)^2 + mgy$$

$$= \frac{1}{2}\left(m + \frac{I}{R^2}\right)\dot{y}^2 + mgy \qquad (2.108)$$

(b)

$$p_y = \left(m + \frac{I}{R^2}\right)\dot{y} \Rightarrow \dot{y} = \frac{p_y}{\left(m + \frac{I}{R^2}\right)} \qquad (2.109)$$

Then, because the equations of the transformation from Cartesian to generalized coordinates do not contain the time and the potential energy does not contain the velocity the Hamiltonian is the TME, so that

$$H = T + U = \frac{1}{2}\left(m + \frac{I}{R^2}\right)\dot{y}^2 - mgy$$

$$= \frac{1}{2}\left(m + \frac{I}{R^2}\right)\frac{p_y^2}{\left(m + \frac{I}{R^2}\right)^2} - mgy$$

$$= \frac{1}{2}\frac{p_y^2}{\left(m + \frac{I}{R^2}\right)} - mgy \qquad (2.110)$$

The Hamiltonian equations of motion for y and $\dot{p}_y$ are

$$\frac{\partial H}{\partial y} = -\dot{p}_y = -mg$$

$$\frac{\partial H}{\partial p_y} = \dot{y} = \frac{p_y}{\left(m + \frac{I}{R^2}\right)} \qquad (2.111)$$

To solve these equations, we differentiate the second and substitute for $\dot{p}_y$ in the first and obtain

$$\left(m + \frac{I}{R^2}\right)\ddot{y} = mg \qquad (2.112)$$

Integrating Eq. (2.112) we get

$$\dot{y} = \frac{m}{\left(m + \frac{I}{R^2}\right)}gt + K \qquad (2.113)$$

where K is a constant. But $\dot{y}(t = 0) = 0 \Rightarrow K = 0$ because the disk was released from rest. Therefore

$$\dot{y} = \frac{m}{\left(m + \dfrac{I}{R^2}\right)} gt \tag{2.114}$$

The moment of inertia of a cylindrically symmetric object may be written $I = \beta m R^2$ where β depends upon the radial distribution of mass about the axis of rotation. For example, for a flywheel (ring) $\beta = 1$. Thus, we may write Eq. (2.114) for any such disk as

$$\dot{y} = \frac{1}{(1 + \beta)} gt \tag{2.115}$$

Note that the final speeds, both linear ($\dot{y}$) and rotational ($\dot{y}/R$), depend upon neither the mass m nor the radius of the disk R. It is the distribution of mass (β) that determines these speeds.

12. Find the linear velocity $\dot{y}$ in Problem 11 of this chapter using Lagrangian dynamics.

Solution

The Lagrangian in terms of y and $\dot{y}$ was found to be

$$\mathcal{L} = \frac{1}{2}\left(m + \frac{I}{R^2}\right)\dot{y}^2 + mgy \tag{2.116}$$

so the Lagrange equation of motion for the coordinate y is

$$\frac{\partial \mathcal{L}}{\partial y} - \frac{d}{dt}\frac{\partial \mathcal{L}}{\partial \dot{y}} = 0 \Rightarrow mg - \left(m + \frac{I}{R^2}\right)\ddot{y} = 0 \tag{2.117}$$

Integrating and applying the boundary conditions we arrive at the same answer as was obtained in Problem 11 of this chapter, Eq. (2.114).

$$\dot{y} = \frac{m}{\left(m + \frac{I}{R^2}\right)} gt \tag{2.118}$$

13. A uniform rod of mass m and length L is attached to a frictionless pivot as shown. The rod is released from rest in the horizontal position.

Fig. 2.9 Problem 13

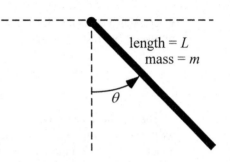

length $= L$
mass $= m$

θ

(a) Using the generalized coordinate θ shown in Fig. 2.9 write the Lagrangian.
    Take the potential energy to be zero when the rod is horizontal ($\theta = \pi/2$).
    The Lagrangian should contain only the quantities given and the gravita-
    tional acceleration.
(b) Use Lagrangian dynamics to find an expression for the angular velocity of
    the rod as a function of θ and $\dot{\theta}$.
(c) Suppose that instead of being released from the horizontal position, the rod
    were released from rest at a very small angular displacement from vertical
    $\theta_0 \ll \pi/2$. What would be the angular frequency in radians per second
    of the small oscillations of the rod?

Solution

(a)

$$T = \frac{1}{2}I_e\dot{\theta}^2 \text{ and } U = -mg\frac{L}{2}\cos\theta \qquad (2.119)$$

where I_e is the moment of inertia of the rod about one end, which can be
quickly calculated. Letting x be the coordinate along the length of the rod
we have

$$I_e = \int_0^L x^2 dm = \int_0^L x^2 \left(\frac{m}{L}\right) dx$$

$$= \frac{mL^2}{3} \qquad (2.120)$$

Using this result we find that

$$\mathcal{L} = T - U$$

$$= \frac{mL^2}{6}\dot{\theta}^2 + mg\frac{L}{2}\cos\theta \qquad (2.121)$$

(b) The Lagrangian equation of motion is

$$\frac{\partial \mathcal{L}}{\partial \theta} - \frac{d}{dt}\frac{\partial \mathcal{L}}{\partial \dot{\theta}} = 0$$

$$-mg\frac{L}{2}\sin\theta - \frac{mL^2}{3}\ddot{\theta} = 0$$

$$\ddot{\theta} + \left(\frac{3g}{2L}\right)\sin\theta = 0 \qquad (2.122)$$

Note that the coefficient of $\sin\theta$ must have the same units as $\ddot{\theta}$, that is $1/s^2$ ("radians" is unitless). Indeed, g/L has the correct units.

Equation (2.122) is a nonlinear differential equation because $\sin\theta$ is nonlinear in θ. Solution of such equations is usually non-trivial. In our case, however, we require only the solution for $\dot{\theta}$. This can be obtained using a standard "trick." We multiply through by $\dot{\theta}$ and note that the resulting equation may be written

$$\dot{\theta}\ddot{\theta} + \left(\frac{3g}{2L}\right)\dot{\theta}\sin\theta = 0$$

$$\frac{d}{dt}\left[\frac{1}{2}\dot{\theta}^2 - \left(\frac{3g}{2L}\right)\cos\theta\right] = 0 \qquad (2.123)$$

Thus

$$\left[\frac{1}{2}\dot{\theta}^2 - \left(\frac{3g}{2L}\right)\cos\theta\right] = C \qquad (2.124)$$

where C is a constant. But $\dot{\theta} = 0$ when $\theta = \pi/2$ so $C = 0$ and we have

$$\dot{\theta} = \sqrt{\frac{3g}{L}\cos\theta} \qquad (2.125)$$

This answer can be checked using elementary methods. The TME is zero because the rod is released from rest and the zero of potential energy is taken to be $\theta = \pi/2$. The TME is the Lagrangian, Eq. (2.121), with the sign of the potential energy U reversed.

$$\text{TME} = \frac{mL^2}{6}\dot{\theta}^2 - mg\frac{L}{2}\cos\theta = 0 \qquad (2.126)$$

Solving for $\dot{\theta}$ yields the same result as that obtained in Eq. (2.125).

(c) For an initial displacement θ_0 the TME is no longer zero, but this does not change the Lagrangian in Eq. (2.121). Therefore, Eq. (2.122) still describes

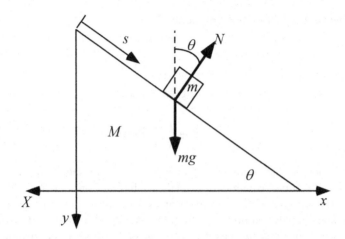

Fig. 2.10 Problem 14

the motion. Because θ_0 is small we may make the approximation $\sin \theta \simeq \theta$ (the first term in the Taylor series for $\sin \theta$). We have then

$$\ddot{\theta} + \left(\frac{3g}{2L}\right)\theta = 0 \tag{2.127}$$

This equation describes simple harmonic motion about the coordinate θ with angular frequency

$$\omega = \sqrt{\frac{3g}{2L}} \tag{2.128}$$

We note that ω has the correct units, $1/$s.

14. This problem is identical to Problem 9, Chap. 1 except now we wish to work it using Lagrangian dynamics. It will be seen that the solution is much easier.

   The inclined plane shown rests on a frictionless surface. The block of mass m is released from rest at the top of the frictionless incline. Using the coordinates in Fig. 2.10 use Lagrangian dynamics to find $\ddot{X}$ the acceleration of the inclined plane. Note that the block and plane move in opposite directions, x and X.

Solution

The total kinetic energy is

$$T = \frac{1}{2}M\dot{X}^2 + \frac{1}{2}mv^2 \tag{2.129}$$

where v is the velocity of the mass m along the coordinate s and is given, according to Fig. 2.11, by the law of cosines.

Fig. 2.11
Problem 14—solution

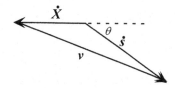

$$v^2 = \dot{X}^2 + \dot{s}^2 + 2\dot{X}\dot{s}\cos\theta \qquad (2.130)$$

Using Eq. (E.2) we see that

$$\cos(\pi \pm \theta) = \cos\pi\cos\theta$$
$$= -\cos\theta \qquad (2.131)$$

which accounts for the plus sign in the coefficient of $\cos\theta$ in Eq. (2.130). Only m contributes to the potential energy U so

$$U = -mgs\sin\theta \qquad (2.132)$$

The Lagrangian is therefore

$$\mathcal{L} = T - U$$
$$= \frac{1}{2}M\dot{X}^2 + \frac{1}{2}m\dot{X}^2 + \frac{1}{2}m\dot{s}^2 + m\dot{X}\dot{s}\cos\theta + mgs\sin\theta \qquad (2.133)$$

Lagrange's equations are then

$$\frac{d}{dt}\left(\frac{\partial\mathcal{L}}{\partial\dot{X}}\right) - \frac{\partial\mathcal{L}}{\partial X} = 0 \implies M\ddot{X} + m\ddot{X} + m\ddot{s}\cos\theta = 0 \qquad (2.134)$$

or

$$(m + M)\ddot{X} = -m\ddot{s}\cos\theta \qquad (2.135)$$

and

$$\frac{d}{dt}\left(\frac{\partial\mathcal{L}}{\partial\dot{s}}\right) - \frac{\partial\mathcal{L}}{\partial s} = 0 \implies m\ddot{s} + m\ddot{X}\cos\theta - mg\sin\theta = 0 \qquad (2.136)$$

or

$$m\ddot{s} = mg\sin\theta - m\ddot{X}\cos\theta \qquad (2.137)$$

Now substitute $m\ddot{s}$ from Eq. (2.137) into Eq. (2.135) and solve for $\ddot{X}$.

$$(m + M)\ddot{X} = -mg \sin\theta \cos\theta + m\ddot{X}\cos^2\theta$$

$$\ddot{X} = \frac{m \sin\theta \cos\theta}{m + M - m\cos^2\theta}g$$

$$= \frac{m \sin\theta \cos\theta}{M + m\sin^2\theta}g \tag{2.138}$$

which is the same answer obtained using Newton's second law directly in Problem 9 of Chap. 1.

15. A point mass m is attached to one end of a massless rigid rod of length ℓ. The other end is rotated in a *horizontal* plane with a uniform angular speed ω in a circle of radius R. Use Lagrangian dynamics to show that this configuration is identical with that of a simple pendulum in a gravitational field provided the gravitational acceleration is replaced by $R\omega^2$.

Fig. 2.12 Problem 15

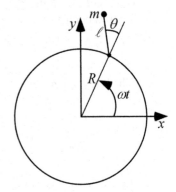

Solution

There is no potential energy so the Lagrangian is the kinetic energy T. As is frequently the case for rotating systems it is easiest to find the Cartesian coordinates of the rotating mass and then convert to the appropriate generalized coordinate, which in this case is θ. Using the axes shown we have

$$x = R \cos\omega t + \ell \cos(\omega t + \theta)$$

$$y = R \sin\omega t + \ell \sin(\omega t + \theta) \tag{2.139}$$

where x and y are the Cartesian coordinates of the rotating mass. Differentiating, we have

$$\dot{x} = -R\omega \sin \omega t - \ell \left(\omega + \dot{\theta}\right) \sin (\omega t + \theta)$$

$$\dot{y} = R\omega \cos \omega t + \ell \left(\omega + \dot{\theta}\right) \cos (\omega t + \theta) \qquad (2.140)$$

The Lagrangian is therefore

$$T = \mathcal{L} = \frac{1}{2}m\left(\dot{x}^2 + \dot{y}^2\right)$$

$$= \frac{1}{2}m[R^2\omega^2 \sin^2 \omega t + 2R\omega\ell \left(\omega + \dot{\theta}\right) \sin \omega t \sin (\omega t + \theta)$$

$$+ \ell^2 \left(\omega + \dot{\theta}\right)^2 \sin^2 (\omega t + \theta) + R^2\omega^2 \cos^2 \omega t$$

$$+ 2R\omega\ell \left(\omega + \dot{\theta}\right) \cos \omega t \cos (\omega t + \theta)$$

$$+ \ell^2 \left(\omega + \dot{\theta}\right)^2 \cos^2 (\omega t + \theta)]$$

$$= \frac{1}{2}m\left[R^2\omega^2 + 2R\omega\ell \left(\omega + \dot{\theta}\right) \cos \theta + \ell^2 \left(\omega + \dot{\theta}\right)^2\right] \qquad (2.141)$$

where we have used Eq. (E.2). Lagrange's equation is

$$\frac{d}{dt}\left(\frac{\partial \mathcal{L}}{\partial \dot{\theta}}\right) - \frac{\partial \mathcal{L}}{\partial \theta} = 0$$

$$\frac{d}{dt}\left[2R\omega\ell \cos \theta + 2\ell^2 \left(\omega + \dot{\theta}\right)\right] + 2R\omega\ell \left(\omega + \dot{\theta}\right) \sin \theta = 0$$

$$-R\omega\ell\dot{\theta} \sin \theta + \ell^2\ddot{\theta} + R\omega^2\ell \sin \theta + R\omega\ell\dot{\theta} \sin \theta = 0$$

$$\ddot{\theta} + \frac{R\omega^2}{\ell} \sin \theta = 0 \qquad (2.142)$$

This last equation reduces to that of a simple pendulum when θ is small enough to use $\sin \theta \simeq \theta$. This is derived in Problem 1, Chap. 1 (see Eq. (1.2)). In Eq. (1.2) the coefficient of $\sin \theta$ is g/ℓ, the square of the frequency for small displacements. Comparing this with the coefficient of $\sin \theta$ in Eq. (2.142) we see that the two equations are identical if g for the pendulum is replaced by $R\omega^2$.

16. Work Problem 15 of this chapter using Hamiltonian dynamics.

Solution

We have already obtained the Lagrangian for this system in Problem 15 [Eq. (2.141)] using the coordinates given in Fig. 2.12. It is

$$\mathcal{L}\left(\theta,\dot{\theta}\right) = \frac{1}{2}m\left[R^2\omega^2 + 2R\omega\ell\left(\omega + \dot{\theta}\right)\cos\theta + \ell^2\left(\omega + \dot{\theta}\right)^2\right] \quad (2.143)$$

We note that the equations of transformation connecting the Cartesian coordinates (x, y) and θ, Eq. (2.139), contain the time. Therefore, the kinetic energy cannot be a quadratic homogeneous function of the $\dot{\theta}$ as may be seen in Eq. (2.143) in which there are two linear terms in $\dot{\theta}$. As a consequence the Hamiltonian cannot be simply the TME so we must formally compute it according to the relation

$$H(\theta, p_\theta) = \dot{\theta}p_\theta - \mathcal{L}\left(\theta, \dot{\theta}\right) \quad (2.144)$$

First we must find the generalized momentum, which in this case is p_θ and is defined as

$$p_\theta = \frac{\partial \mathcal{L}}{\partial \dot{\theta}}$$

$$= mR\omega\ell\cos\theta + m\ell^2\left(\omega + \dot{\theta}\right)$$

$$= mR\omega\ell\cos\theta + m\ell^2\omega + m\ell^2\dot{\theta} \quad (2.145)$$

from which we find that

$$\dot{\theta} = \frac{p_\theta}{m\ell^2} - \frac{R\omega}{\ell}\cos\theta - \omega \quad (2.146)$$

Using Eq. (2.144) the Hamiltonian is

$$H(\theta, p_\theta) = \left(\frac{p_\theta}{m\ell^2} - \frac{R\omega}{\ell}\cos\theta - \omega\right)p_\theta - \frac{1}{2}mR^2\omega^2$$

$$- mR\omega\ell\left[\omega + \left(\frac{p_\theta}{m\ell^2} - \frac{R\omega}{\ell}\cos\theta - \omega\right)\right]\cos\theta$$

$$- \frac{1}{2}m\ell^2\left[\omega + \left(\frac{p_\theta}{m\ell^2} - \frac{R\omega}{\ell}\cos\theta - \omega\right)\right]^2 \quad (2.147)$$

which becomes

$$
\begin{aligned}
H\left(\theta, p_\theta\right) = {} & \frac{p_\theta^2}{m\ell^2} - \frac{R\omega}{\ell} p_\theta \cos\theta - \omega p_\theta - \frac{1}{2} m R^2 \omega^2 \\
& - \frac{R\omega}{\ell} p_\theta \cos\theta + m R^2 \omega^2 \cos^2\theta \\
& - \frac{p_\theta^2}{2m\ell^2} + \frac{R\omega}{\ell} p_\theta \cos\theta - \frac{1}{2} m R^2 \omega^2 \cos^2\theta
\end{aligned}
\tag{2.148}
$$

Simplifying further we have

$$
\begin{aligned}
H\left(\theta, p_\theta\right) = {} & \frac{p_\theta^2}{2m\ell^2} - \frac{R\omega}{\ell} p_\theta \cos\theta - \omega p_\theta \\
& - \frac{1}{2} m R^2 \omega^2 + \frac{1}{2} m R^2 \omega^2 \cos^2\theta
\end{aligned}
\tag{2.149}
$$

Hamilton's equations are

$$
\dot{p}_\theta = -\frac{\partial H}{\partial \theta} \quad \text{and} \quad \dot{\theta} = \frac{\partial H}{\partial p_\theta}
\tag{2.150}
$$

so we have

$$
\begin{aligned}
\dot{p}_\theta &= -\left(\frac{R\omega}{\ell} p_\theta \sin\theta - m R^2 \omega^2 \cos\theta \sin\theta \right) \\
&= -\frac{R\omega}{\ell} p_\theta \sin\theta + m R^2 \omega^2 \cos\theta \sin\theta
\end{aligned}
\tag{2.151}
$$

and

$$
\dot{\theta} = \frac{p_\theta}{m\ell^2} - \frac{R\omega}{\ell} \cos\theta - \omega
\tag{2.152}
$$

Differentiating Eq. (2.152) with respect to time we have

$$
\ddot{\theta} = \frac{\dot{p}_\theta}{m\ell^2} + \frac{R\omega}{\ell} \dot{\theta} \sin\theta
\tag{2.153}
$$

Substituting Eq. (2.151) to eliminate $\dot{p}_\theta$ in Eq. (2.153) and using Eq. (2.145) to eliminate $\dot{\theta}$ we have

$$
\begin{aligned}
\ddot{\theta} = {} & \frac{1}{m\ell^2} \left(-\frac{R\omega}{\ell} p_\theta \sin\theta + m R^2 \omega^2 \cos\theta \sin\theta \right) \\
& + \frac{R\omega}{\ell} \left(\frac{p_\theta}{m\ell^2} - \frac{R\omega}{\ell} \cos\theta - \omega \right) \sin\theta
\end{aligned}
\tag{2.154}
$$

which simplifies to

$$\ddot{\theta} + \frac{R\omega^2}{\ell} \sin\theta = 0 \qquad (2.155)$$

Equation (2.155) is identical to the differential equation obtained using Lagrangian dynamics, Eq. (2.142).

Because the Hamiltonian in this case is not the total energy there was quite a bit more labor involved in obtaining Eq. (2.155) than that required to obtain the Eq. (2.142) using Lagrangian dynamics. As always, it is comforting that we obtained the same answer using both methods.

17. This problem is the same as Problem 7 of this chapter. That is, we have an Atwood's machine consisting of two point masses connected by an inextensible massless string of length ℓ over a frictionless peg of negligible diameter. Assume that $m_2 > m_1$. In this problem we also ask for the tension in the string and the acceleration of the masses obtained by using Lagrange's method of undetermined multipliers. This method of solving the problem, as contrasted with the solution to Problem 7 of this chapter, illustrates the utility of the method of undetermined multipliers and the added information that it yields.

Solution

The undetermined multipliers are related to the forces of constraint [3]. For more detailed information the student is referred to any good textbook on classical mechanics. To include the equation of constraint we modify the generalized coordinates used in Problem 7 of this chapter (Fig. 2.13).

Fig. 2.13
Problem 17—solution

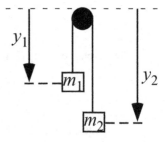

The equation of constraint is the requirement that the length of the string remain constant. Thus

$$y_1 + y_2 = \ell \Rightarrow f(y_1, y_2) = y_1 + y_2 - \ell = 0 \qquad (2.156)$$

The kinetic and potential energies are

$$T = \frac{1}{2}m_1\dot{y}_1^2 + \frac{1}{2}m_2\dot{y}_2^2; \quad U = -m_1gy_1 - m_2gy_2 \tag{2.157}$$

The Lagrange equations of motion, with λ as the undetermined multiplier, are

$$\frac{\partial \mathcal{L}}{\partial y_i} - \frac{d}{dt}\frac{\partial \mathcal{L}}{\partial \dot{y}_i} + \lambda\frac{\partial f(y_1, y_2)}{\partial y_i} = 0 \tag{2.158}$$

and the *generalized* forces of constraint are given by

$$Q_1 = \lambda\frac{\partial f(y_1, y_2)}{\partial y_1}; \quad Q_2 = \lambda\frac{\partial f(y_1, y_2)}{\partial y_2} \tag{2.159}$$

Because y_1 and y_2 are Cartesian coordinates the Qs are ordinary forces, i.e. the tension in the string [3].

Inserting the Lagrangian and the equation of constraint into Eq. (2.158) yields two uncoupled differential equations, one each in the coordinates y_1 and y_2.

$$m_1\ddot{y}_1 - m_1g - \lambda = 0 \quad \text{and} \quad m_2\ddot{y}_2 - m_2g - \lambda = 0 \tag{2.160}$$

Subtracting these equations and using the equation of constraint, Eq. (2.156), to eliminate say $\ddot{y}_1$ we immediately recover the result from Problem 7 in this chapter, namely the linear acceleration of the system

$$\ddot{y}_2 = \frac{(m_2 - m_1)}{(m_2 + m_1)}g \tag{2.161}$$

Now, substituting this result into the second of Eq. (2.160) we may solve for λ and thus the tension in the string.

$$\lambda = -\frac{2m_1m_2}{(m_2 + m_1)}g \tag{2.162}$$

Therefore

$$Q_1 = \left[-\frac{2m_1m_2}{(m_2 + m_1)}g\right] = Q_2 \tag{2.163}$$

which is the tension in the string. Notice that the tension forces are both in the same direction, up.

18. A solid sphere of mass m and radius R having moment of inertia $I = (2/5)\,mR^2$ rolls down a stationary inclined plane without slipping. Using the method of undetermined multipliers find the linear acceleration of the sphere down the plane. Also find the force and torque required to maintain the condition "rolls without slipping." The method of undetermined multipliers is necessary in this case in order to obtain the force and the torque.

Fig. 2.14 Problem 18

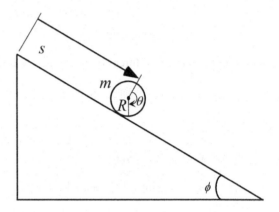

Solution

Using the generalized coordinates shown in Fig. 2.14 the kinetic energy is

$$T = \frac{1}{2}m\dot{s}^2 + \frac{1}{2}I\dot{\theta}^2 = \frac{1}{2}m\dot{s}^2 + \frac{1}{5}mR^2\dot{\theta}^2 \qquad (2.164)$$

Taking the zero of potential energy to be at $s = 0$

$$U = -mgs\sin\phi \qquad (2.165)$$

so the Lagrangian is

$$\mathcal{L} = T - U = \frac{1}{2}m\dot{s}^2 + \frac{1}{5}mR^2\dot{\theta}^2 + mgs\sin\phi \qquad (2.166)$$

The condition that defines rolling without slipping is that the distance traveled s is

$$s = R\theta \Longrightarrow f(s,\theta) = s - R\theta = 0 \qquad (2.167)$$

where $f(s,\theta)$ is the equation of constraint. We could eliminate θ from the Lagrangian using Eq. (2.167), write Lagrange's equation and solve for $\ddot{s}$, thus obtaining the linear acceleration of the disk down the plane. But, to obtain the force and the torque it is necessary to use the "rolling without slipping" constraint, Eq. (2.167), so the method of undetermined multipliers is required. We proceed as in Problem 17 in this chapter.

Using the equivalent of Eq. (2.158), Lagrange's equation with undetermined multipliers, we obtain

$$mg \sin \phi - m\ddot{s} + \lambda = 0 \quad \text{and} \quad -\frac{2}{5}mR^2\ddot{\theta} - \lambda R = 0 \qquad (2.168)$$

We may (temporarily) eliminate λ by solving the second of these equations for λ and substituting it in the first. Thus,

$$\lambda = -\frac{2}{5}mR\ddot{\theta} \qquad (2.169)$$

so that

$$mg \sin \phi - m\ddot{s} - \frac{2}{5}mR\ddot{\theta} = 0 \qquad (2.170)$$

Now, using the equation of constraint in the form $R\ddot{\theta} = \ddot{s}$, we eliminate $\ddot{\theta}$ and solve for $\ddot{s}$ the linear acceleration of the sphere down the plane

$$\ddot{s} = \frac{5g \sin \phi}{7} \qquad (2.171)$$

From the equation of constraint

$$\ddot{\theta} = \frac{\ddot{s}}{R} = \frac{5g \sin \phi}{7R} \qquad (2.172)$$

and, reintroducing λ, Eq. (2.169), and inserting Eq. (2.172) we have

$$\lambda = -\frac{2mg \sin \phi}{7} \qquad (2.173)$$

Then, from Eq. (2.159) the generalized forces are

$$Q_s = \lambda \frac{\partial f(s, \theta)}{\partial s}; \quad Q_\theta = \lambda \frac{\partial f(s, \theta)}{\partial \theta} \qquad (2.174)$$

Note that in inasmuch as s is a linear coordinate Q_s will be a force, but, because θ is an angular coordinate Q_θ is a torque. The units of these quantities, once calculated, should verify this supposition. We find that

$$Q_s = -\frac{2mg \sin \phi}{7}(1) \quad \text{and} \quad Q_\theta = -\frac{2mg \sin \phi}{7}(-R) \qquad (2.175)$$

These signs are correct inasmuch as Q_s, the force of friction that maintains the "rolling without slipping" condition is in the $-s$ direction; Q_θ is positive because θ increases as the sphere rolls. Note that Q_s and Q_θ have the units of force and torque, respectively.

19. A bead of mass m slides down a stiff straight wire without friction under the influence of gravity as shown in Fig. 2.15.

Fig. 2.15 Problem 19

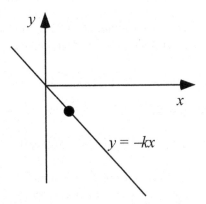

The equation of the wire, considered to be a line, is $y = -kx$ where the constant $k > 0$. Use Lagrange's method of undetermined multipliers to find the acceleration of the bead in both the x- and y-directions.

Solution

Taking the zero of potential energy to be at $y = 0$ the kinetic and potential energies are

$$T = \frac{1}{2}m\dot{x}^2 + \frac{1}{2}m\dot{y}^2; \quad U = mgy \text{ where } g > 0 \qquad (2.176)$$

and the Lagrangian is

$$\mathcal{L} = T - U = \frac{1}{2}m\dot{x}^2 + \frac{1}{2}m\dot{y}^2 - mgy \qquad (2.177)$$

The straight wire constrains the path of the bead to be

$$y = -kx \Longrightarrow f(x, y) = y + kx = 0 \qquad (2.178)$$

where $f(x, y)$ is the equation of constraint.

Lagrange's equations of motion become

$$m\ddot{x} + \lambda k = 0 \tag{2.179}$$

and

$$m\ddot{y} + mg + \lambda = 0 \tag{2.180}$$

The terms in Eqs. (2.179) and (2.180) that do not contain λ are forces. Therefore, λ must be a force (k is necessarily unitless).

To solve for $\ddot{x}$ and $\ddot{y}$ we first differentiate the equation of constraint, Eq. (2.178), twice to obtain a relation between $\ddot{x}$ and $\ddot{y}$.

$$\ddot{y} = -k\ddot{x} \tag{2.181}$$

Next, solve Eq. (2.179) for λ, insert this value and replace $\ddot{y}$ into Eq. (2.180) to obtain

$$m(-k\ddot{x}) + mg - \left(\frac{m\ddot{x}}{k}\right) = 0$$

$$\left(k + \frac{1}{k}\right)\ddot{x} = g \tag{2.182}$$

or

$$\ddot{x} = \left(\frac{k}{k^2 + 1}\right)g \tag{2.183}$$

and

$$\ddot{y} = -\left(\frac{k^2}{k^2 + 1}\right)g \tag{2.184}$$

Now let us see if Eqs. (2.183) and (2.184) give sensible answers. First examine the case for $k = 0$. This is a horizontal line so nothing happens; both $\ddot{x}$ and $\ddot{y}$ vanish if $k = 0$. So far so good. Suppose the line is coincident with the y-axis. In this case the slope of the line is infinite and we see that

$$\lim_{k \to \infty} \ddot{x} = \lim_{k \to \infty} \left(\frac{k}{k^2 + 1}\right)g$$

$$= 0 \tag{2.185}$$

This is correct because if the line is vertical $x \equiv 0$. The bead never moves left or right. On the other hand, how about $\ddot{y}$? We must examine

$$\lim_{k \to \infty} \ddot{y} = \lim_{k \to \infty} - \left(\frac{k^2}{k^2 + 1} \right) g$$

$$= -g \tag{2.186}$$

This too is correct. The bead simply falls vertically with acceleration g. Recall that $g > 0$ so the sign in Eq. (2.186) is correct.

The generalized forces are

$$Q_x = \lambda \frac{\partial f(x, y)}{\partial x}; \quad Q_y = \lambda \frac{\partial f(x, y)}{\partial y} \tag{2.187}$$

so that

$$Q_x = \lambda k; \quad Q_y = \lambda \tag{2.188}$$

First we solve for λ by inserting Eq. (2.183) into Eq. (2.179) to obtain

$$\lambda = -\frac{m}{k} \ddot{x} = - \left(\frac{1}{k^2 + 1} \right) mg \tag{2.189}$$

Therefore

$$Q_x = - \left(\frac{1}{k^2 + 1} \right) mgk; \quad Q_y = - \left(\frac{1}{k^2 + 1} \right) mg \tag{2.190}$$

Since k, the slope of the line on which the bead slides, is unitless it is clear that both Q_x and Q_y have units of force. They are the x- and y-components of the force imparted to the bead by the wire.

20. This problem is identical to Problem 5 of Chap. 1. Two unequal point masses m_1 and m_2 are attached to a spring having spring constant k and unextended length ℓ as shown in Fig. 2.16.

Fig. 2.16 Problem 20

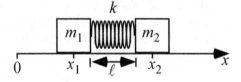

The masses are free to oscillate on a frictionless table. Use Lagrangian dynamics to find the frequency of oscillation for this system.

Solution

The kinetic and potential energies are given by

$$T = \frac{1}{2}m_1\dot{x}_1^2 + \frac{1}{2}m_2\dot{x}_2^2 \tag{2.191}$$

and the potential energy is

$$U = \frac{1}{2}k\,[x_1 - (x_2 - \ell)]^2$$
$$= \frac{1}{2}k\,(x_1 - x_2 + \ell)^2 \tag{2.192}$$

so the Lagrangian is

$$\mathcal{L} = \frac{1}{2}m_1\dot{x}_1^2 + \frac{1}{2}m_2\dot{x}_2^2 - \frac{1}{2}k\,(x_1 - x_2 + \ell)^2 \tag{2.193}$$

from which we obtain the equations of motion. They are

$$m_1\ddot{x}_1 + k\,(x_1 - x_2 + \ell) = 0 \tag{2.194}$$

and

$$m_2\ddot{x}_2 - k\,(x_1 - x_2 + \ell) = 0 \tag{2.195}$$

Adding Eqs. (2.194) and (2.195) produces

$$m_1\ddot{x}_1 + m_2\ddot{x}_2 = 0 \Rightarrow \frac{d^2}{dt^2}(m_1x_1 + m_2x_2) = 0 \tag{2.196}$$

which, upon integration twice becomes

$$m_1x_1 + m_2x_2 = c_1t + c_2 \tag{2.197}$$

where c_1 and c_2 are constants of integration. This corresponds to translational motion of the double mass system with no vibration of the spring, that is $\omega_1 = 0$.

To examine the oscillatory motion we divide Eqs. (2.194) and (2.195) by m_1 and m_2, respectively, and subtract them to obtain

$$\frac{d^2}{dt^2}(x_2 - x_1) + k\left(\frac{1}{m_1} + \frac{1}{m_2}\right)(x_2 - x_1) = k\left(\frac{1}{m_1} + \frac{1}{m_2}\right)\ell \tag{2.198}$$

which is simple harmonic motion in the coordinate $(x_2 - x_1)$ with frequency

$$\omega_2 = \sqrt{k \left(\frac{1}{m_1} + \frac{1}{m_2} \right)}$$

$$= \sqrt{\frac{(m_1 + m_2) k}{m_1 m_2}} \tag{2.199}$$

The value of ω_2 in Eq. (2.199) is identical to the answer obtained in Problem 5 of Chap. 1. Note that the grouping of masses in ω_2 is customarily referred to as μ, the reduced mass. That is

$$\mu = \frac{m_1 m_2}{(m_1 + m_2)} \tag{2.200}$$

or, as it is conveniently remembered

$$\frac{1}{\mu} = \left(\frac{1}{m_1} + \frac{1}{m_2} \right) \tag{2.201}$$

Chapter 3
Central Forces and Orbits

Central forces are of considerable importance in both classical and quantum mechanics so we briefly review their properties. In this chapter we deal with two bodies of masses m_1 and m_2. We ignore the motion of the center-of-mass so that only the relative interparticle coordinates r and θ are treated, as will be discussed below. The effective mass for motion in (r, θ) is the reduced mass $\mu = m_1 m_2 / (m_1 + m_2)$. The central forces, $F(r)$, and the potentials that produce them, $U(r)$, depend only upon the spherical coordinate r. The angular momentum vector $\mathbf{L} = \mathbf{r} \times \mathbf{p}$, where $\mathbf{r}$ is the position vector and $\mathbf{p}$ is the linear momentum vector, is conserved for *all* central potentials. This will be proven in Problem 1 of this chapter. Because $\mathbf{L}$ is conserved, the motion under the action of a central force takes place in a plane so that only two dimensions are required to describe it. Plane polar coordinates (r, θ) are almost always used in central force problems (see Appendix B.2). When written in polar coordinates, the total energy of a system under the influence of a central potential contains an angular momentum term, $\ell^2 / (2\mu r^2)$ where ℓ is the magnitude of the angular momentum $\mathbf{L}$. For constant ℓ this term acts as an additional potential, which is usually referred to as the centrifugal potential. Thus, we may define the "effective potential" as

$$U_{\text{eff}}(r) = U(r) + \frac{\ell^2}{2\mu r^2} \tag{3.1}$$

Note that $\ell = 0$ corresponds to a linear trajectory passing through the force center.

Table 3.1 contains a few especially useful equations for central potential problems using plane polar coordinates.

© Springer International Publishing AG 2017
J.D. Kelley, J.J. Leventhal, *Problems in Classical
and Quantum Mechanics*, DOI 10.1007/978-3-319-46664-4_3

Table 3.1 Useful equations for central potential problems

Quantity	Equation
Force	$F(r) = -\partial U(r)/\partial r$
Reduced mass	$\mu = m_1 m_2/(m_1 + m_2)$
Lagrangian	$\mathcal{L} = \frac{1}{2}\left(\mu\dot{r}^2 + r^2\dot{\theta}^2\right) - U(r)$
Angular momentum	$\ell = \mu r^2\dot{\theta} = \text{constant}$
Effective potential	$U_{\text{eff}}(r) = U(r) + \frac{\ell^2}{2\mu r^2}$
Total energy	$E = \frac{1}{2}\mu\dot{r}^2 + \frac{1}{2}\frac{\ell^2}{\mu r^2} + U(r)$
Orbit equation	$\frac{d^2}{d\theta^2}\left(\frac{1}{r}\right) + \frac{1}{r} = -\frac{\mu r^2}{\ell^2}F(r)$

Problems

1. The angular momentum vector $\mathbf{L}$ is given by

$$\mathbf{L} = \mathbf{r} \times \mathbf{p} \qquad (3.2)$$

where $\mathbf{r}$ is the position vector and $\mathbf{p}$ is the linear momentum vector. Show that $\mathbf{L}$ is conserved for all central potentials.

Solution

If $\mathbf{L}$ is to be conserved, then its time derivative must vanish. Using Eq. (F.4) we have

$$\frac{d\mathbf{L}}{dt} = \frac{d}{dt}(\mathbf{r} \times \mathbf{p})$$

$$= \mathbf{r} \times \frac{d\mathbf{p}}{dt} + \frac{d\mathbf{r}}{dt} \times \mathbf{p}$$

$$= \mathbf{r} \times \mathbf{F}(r) + \mathbf{v} \times \mathbf{p}$$

$$= 0 + 0 \qquad (3.3)$$

The first term vanishes because $\mathbf{F}(r)$ and $\mathbf{r}$ are colinear while the second vanishes because $\mathbf{p} = \mu\mathbf{v}$.

2. Starting with the Lagrangian in plane polar coordinates for a central potential show that momentum conjugate to the coordinate θ is conserved for any central potential. This momentum is, of course, the *angular* momentum.

Solution

For any central potential the Lagrangian is

$$\mathcal{L} = \frac{1}{2}\mu \left(\dot{r}^2 + r^2\dot{\theta}^2 \right) - U(r) \qquad (3.4)$$

Lagrange's equation for θ and $\dot{\theta}$ is

$$\frac{\partial \mathcal{L}}{\partial \theta} - \frac{d}{dt}\frac{\partial \mathcal{L}}{\partial \dot{\theta}} = 0 \qquad (3.5)$$

From Eq. (3.4) it is clear that for a central potential $\mathcal{L}$ does not contain θ explicitly (although $\dot{\theta}$ is present) so the first term in Eq. (3.5) vanishes. This means that (see Table 3.1)

$$\frac{\partial \mathcal{L}}{\partial \dot{\theta}} = \mu r^2 \dot{\theta} = \text{constant}$$

$$= |\boldsymbol{L}| = \ell \qquad (3.6)$$

Note that θ, being absent from the Lagrangian, is a *cyclic* coordinate so that the generalized momentum conjugate to this cyclic coordinate $p_\theta = \partial \mathcal{L}/\partial \dot{\theta}$ is conserved.

Theorem. *When a coordinate is cyclic the generalized momentum conjugate to this coordinate is conserved.*

Remark. Sometimes this generalized momentum is not recognizable as a momentum in the physical sense. For a central potential, however, it is the angular momentum.

3. Kepler's laws of planetary motion are

   1. *The planets travel about the sun in elliptical orbits with the sun at one focus of the ellipse.*
   2. *The radius vector from the sun to the planet sweeps out equal areas in equal times.*
   3. *The square of the period of the planet about the sun is proportional to the cube of the semi-major axis of the elliptical orbit.*

   Show that Kepler's second law is valid for any central force $F(r)$ between the sun and the planet, or more generally between any two bodies.

Solution

The central force $F(r)$ is obtained from a central potential $U(r)$ by differentiation

$$F(r) = -\frac{dU(r)}{dr} \tag{3.7}$$

The Lagrangian in spherical coordinates is

$$\mathcal{L} = T - U(r)$$
$$= \frac{1}{2}\mu\dot{r}^2 + \frac{1}{2}\mu r^2\dot{\theta}^2 - U(r) \tag{3.8}$$

where μ is the reduced mass of the sun-planet system. This Lagrangian is cyclic in the coordinate θ, i.e. it does not contain θ. Therefore, the momentum conjugate to the cyclic coordinate θ is conserved.

$$\frac{d}{dt}\left(\frac{\partial\mathcal{L}}{\partial\dot{\theta}}\right) - \frac{\partial\mathcal{L}}{\partial\theta} = 0 \Longrightarrow \left(\frac{\partial\mathcal{L}}{\partial\dot{\theta}}\right) = p_\theta = \text{constant} = \mu r^2\dot{\theta} = \ell \tag{3.9}$$

This conserved momentum is $p_\theta = \ell$, the angular momentum.

If a particle is moving under the influence of the arbitrary central potential margins $U(r)$, it sweeps out an area A in a time $t_2 - t_1$ as shown in Fig. 3.1.

Fig. 3.1 Problem 3

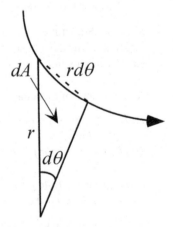

This area dA is

$$dA = \frac{1}{2}r(rd\theta) = \frac{1}{2}r^2d\theta \Rightarrow \frac{dA}{d\theta} = \frac{1}{2}r^2 \tag{3.10}$$

Using Eq. (3.9) to obtain $\dot{\theta}$ the areal velocity dA/dt is

$$\frac{dA}{dt} = \frac{dA}{d\theta}\dot{\theta}$$

$$= \frac{1}{2}r^2\left(\frac{\ell}{\mu r^2}\right) = \frac{\ell}{2\mu}$$

$$= \text{constant} \tag{3.11}$$

thus proving the validity of Kepler's second law for *any* central potential. In contrast, the first and third laws depend upon $F(r)$ being an inverse square force law.

4. Kepler deduced his laws from *observations* of the motion of the planets. His observation that the planets travel in elliptical orbits about the sun (Kepler's first law) implies that the equation of the orbit in plane polar coordinates is given by the equation of a conic section

$$\frac{\alpha}{r} = 1 + \epsilon \cos\theta \tag{3.12}$$

where α and ϵ are constants (see Appendix D). If the eccentricity ϵ is in the range $0 < \epsilon < 1$, then the orbit described by Eq. (3.12) is a closed ellipse. Newton had the advantage of knowing Kepler's laws of planetary motion and used them to obtain his famous $1/r^2$ law of gravitation.

Use the first two of Kepler's laws to deduce Newton's law of gravitation.

Solution

From Kepler's second law Newton knew that the force must be a central force. Recall that Kepler's second law (equal areas in equal times) is the result of conservation of angular momentum. The angular analog of Newton's second law $F = \dot{p}$ (where p is the linear momentum) is a relation between the torque T and the angular momentum L.

$$T = \dot{L} \tag{3.13}$$

Conservation of angular momentum dictates that $\dot{L} = 0$ so there is no torque on the particle to change its angular position. Therefore, the force is directed along the radial coordinate, a central force.

Now that we have shown that the gravitational force is central, how do we deduce the exact mathematical form of the law? We use the equation of the ellipse in Eq. (3.12) together with the orbit equation for any central force. From Table 3.1 we know the orbit equation is

$$\frac{d^2}{d\theta^2}\left(\frac{1}{r}\right) + \frac{1}{r} = -\frac{\mu r^2}{\ell^2}F(r) \tag{3.14}$$

Solving Eq. (3.12) for $1/r$ and inserting it into Eq. (3.14) we have

$$\frac{d^2}{d\theta^2}\left(\frac{1+\epsilon\cos\theta}{\alpha}\right)+\frac{1+\epsilon\cos\theta}{\alpha}=-\frac{\mu r^2}{\ell^2}F(r)$$

$$-\frac{\epsilon}{\alpha}\cos\theta+\frac{1}{\alpha}+\frac{\epsilon}{\alpha}\cos\theta=-\frac{\mu r^2}{\ell^2}F(r)\qquad(3.15)$$

so Newton's law of gravitation is, indeed, the famous $1/r^2$ law

$$F(r)=-\frac{\ell^2}{\alpha\mu r^2}\qquad(3.16)$$

This problem shows the symbiotic relationship between theory and experiment. In short, Newton didn't guess the $1/r^2$ force law. He had observational data at his disposal that led to his theory. Had Bertrand's theorem [2] existed in the latter part of the seventeenth century when Newton formulated his theory of gravitation he might have correctly guessed $1/r^2$ from Kepler's first law. Bertrand's theorem states that the only central forces that produce closed repeating orbits are those that are proportional to $1/r^2$ and those that are proportional to r^2. Unfortunately, this theorem was formulated by Joseph Bertrand and published in 1873, nearly 200 years after Newton; fortunately, Newton did not need it.

5. Express the total energy for a central potential in terms of only the radial coordinate and its derivatives.

Solution

From Problem 3 of this chapter we know that the angular momentum is conserved for any central potential so the motion is in a plane. In plane polar coordinates the total energy is (see Table 3.1)

$$E=T+U(r)$$

$$=\frac{1}{2}\mu\left(\dot{r}^2+r^2\dot{\theta}^2\right)+U(r)\qquad(3.17)$$

Also, the angular momentum is

$$\ell=\mu r^2\dot{\theta}\implies\dot{\theta}=\frac{\ell}{\mu}\cdot\frac{1}{r^2}\qquad(3.18)$$

Substituting this in Eq. (3.17) we obtain the desired result

$$E=\frac{1}{2}\mu\dot{r}^2+\frac{\ell^2}{2\mu r^2}+U(r)\qquad(3.19)$$

6. The equation of the orbit for a central potential is given by (see Table 3.1)

$$\frac{d^2}{d\theta^2}\left(\frac{1}{r}\right) + \left(\frac{1}{r}\right) = -\frac{\mu^2 r^2}{\ell^2}F(r) \tag{3.20}$$

This equation is a valuable tool when seeking the force $F(r)$ when the equation of the orbit $r(\theta)$ is known. It may also be used to determine $r(\theta)$ when $F(r)$ is known. Starting from the Lagrangian for a central potential derive Eq. (3.20) using the substitution $u = 1/r$ so that

$$\frac{d^2 u}{d\theta^2} + u = -\frac{\mu}{\ell^2}\left(\frac{1}{u^2}\right)F\left(\frac{1}{u}\right) \tag{3.21}$$

Solution

The Lagrangian for a central potential is [see Eq. (3.8)]

$$\mathcal{L} = \frac{1}{2}\mu \dot{r}^2 + \frac{1}{2}\mu r^2 \dot{\theta}^2 - U(r) \tag{3.22}$$

so the equation of motion for the coordinate r is

$$\frac{d}{dt}\frac{\partial \mathcal{L}}{\partial \dot{r}} - \frac{\partial \mathcal{L}}{\partial r} = 0$$

$$\mu \ddot{r} - \mu r \dot{\theta}^2 + \frac{dU}{dr} = 0 \tag{3.23}$$

Comparing Eq. (3.23) with Eq. (3.20) we see that we must eliminate $\dot{\theta}$ and $\ddot{r}$ from Eq. (3.23). This is easily done because angular momentum is conserved for any central potential. Thus

$$\ell = \mu r^2 \dot{\theta} \implies \dot{\theta} = \frac{\ell}{\mu r^2} \tag{3.24}$$

To eliminate $\ddot{r}$ we use the chain rule on $u = 1/r$. We have

$$\frac{du}{d\theta} = \frac{du}{dr} \cdot \frac{dr}{d\theta}$$

$$= -\frac{1}{r^2}\frac{dr}{dt} \cdot \frac{dt}{d\theta} = -\frac{1}{r^2}\frac{\dot{r}}{\dot{\theta}} \tag{3.25}$$

$$= -\frac{\mu}{\ell}\dot{r}$$

where we have used conservation of angular momentum, Eq. (3.51). Differentiating $du/d\theta$ with respect to θ we have

$$\frac{d^2u}{d\theta^2} = -\frac{\mu}{\ell}\frac{d\dot{r}}{d\theta} = -\frac{\mu}{\ell}\frac{d\dot{r}}{dt}\frac{dt}{d\theta}$$

$$= -\frac{\mu}{\ell}\frac{\ddot{r}}{\dot{\theta}}$$

$$= -\frac{\mu^2}{\ell^2}r^2\ddot{r} \qquad (3.26)$$

where we have again used conservation of angular momentum, Eq. (3.24), to eliminate $\dot{\theta}$. Solving for $\ddot{r}$ we have

$$\ddot{r} = -\frac{\ell^2}{\mu^2}\frac{1}{r^2}\frac{d^2u}{d\theta^2} \qquad (3.27)$$

Substituting Eq. (3.27) for $\ddot{r}$, Eq. (3.24) for $\dot{\theta}$ and noting that $F(r) = -dU/dr$ Eq. (3.23) becomes

$$-\frac{\ell^2}{\mu}\frac{1}{r^2}\frac{d^2u}{d\theta^2} - \mu r\left(\frac{\ell}{\mu r^2}\right)^2 - F(r) = 0$$

$$\frac{\ell^2}{\mu}u^2\frac{d^2u}{d\theta^2} + \frac{\ell^2}{\mu}u^3 + F(r) = 0 \qquad (3.28)$$

which is indeed the equation that we sought, Eq. (3.20), with $r = 1/u$. In the form most often seen it is

$$\frac{d^2u}{d\theta^2} + u = -\frac{\mu}{\ell^2}\left(\frac{1}{u^2}\right)F\left(\frac{1}{u}\right) \qquad (3.29)$$

or

$$\frac{d^2}{d\theta^2}\left(\frac{1}{r}\right) + \frac{1}{r} = -\frac{\mu}{\ell^2}r^2F(r) \qquad (3.30)$$

7. A particle of mass m is under the influence of a gravitational force of a mass M located at the origin O. Assume $M >> m$ so the center-of-mass is located at the origin. The particle of mass m has a speed of v_p at the point P, which is a distance r_0 from the origin (Fig. 3.2).

   (a) What is E, the TME of the system for an arbitrary orbit having $OP = r_0$?
   (b) Define $\alpha = v_p/v_0$ where v_0 would be the speed of the particle m in a circular orbit of radius r_0. Show that

$$E = E_0 + \frac{1}{2}mv_0^2\left(\alpha^2 - 1\right) \qquad (3.31)$$

Fig. 3.2 Problem 7

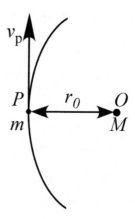

where E_0 is the energy for a circular orbit. That is,

$$E_0 = \frac{1}{2}mv_0^2 - \frac{GmM}{r_0} \tag{3.32}$$

(c) Show that the particle will be bound to M if $\alpha^2 < 2$.

Solution

(a) The TME is

$$E = \frac{1}{2}mv_p^2 - \frac{GmM}{r_0} \tag{3.33}$$

(b) Adding and subtracting the kinetic energy for a circular orbit to Eq. (3.33) we have

$$\begin{aligned}
E &= \frac{1}{2}m\alpha^2 v_0^2 - \frac{GmM}{r_0} + \left(\frac{1}{2}mv_0^2 - \frac{1}{2}mv_0^2\right) \\
&= \frac{1}{2}m\alpha^2 v_0^2 + E_0 - \frac{1}{2}mv_0^2 \\
&= E_0 + \frac{1}{2}mv_0^2 \left(\alpha^2 - 1\right)
\end{aligned} \tag{3.34}$$

where E_0 is the TME for a circular orbit.

(c) For a bound orbit the TME must be negative, $E < 0$. Eliminate r_0 from the total energy, Eq. (3.33), by finding it in terms of v_0. This is accomplished by equating the centripetal force to the gravitational force for a circular orbit

$$\frac{mv_0^2}{r_0} = \frac{GmM}{r_0^2} \Rightarrow \frac{GM}{r_0} = v_0^2 \tag{3.35}$$

Substituting this into Eq. (3.33) we have

$$E = \frac{1}{2}m\alpha^2 v_0^2 - mv_0^2$$

$$= \frac{1}{2}mv_0^2 \left(\alpha^2 - 2\right) \tag{3.36}$$

The condition for $E < 0$ is $\alpha^2 < 2$.

8. A particle moves in a spiral orbit given by the equation

$$r(\theta) = K\theta^2 \tag{3.37}$$

where K is a real constant.

(a) Use Eq. (3.19) derived in Problem 5 of this chapter to deduce the force law $F(r)$ that produces this orbit.
(b) Use Eq. (3.20) derived in Problem 6 of this chapter to deduce the force law $F(r)$ that produces this orbit.

Solution

(a) The total energy as given by Eq. (3.19) is

$$E = \frac{1}{2}\mu\dot{r}^2 + \frac{\ell^2}{2\mu r^2} + U(r) \tag{3.38}$$

Differentiating $r(\theta)$ and using conservation of angular momentum $\ell = \mu r^2\dot{\theta}$

$$\dot{r} = 2\theta\dot{\theta}K = 2K \cdot \left(\frac{r}{K}\right)^{1/2} \cdot \left(\frac{\ell}{\mu r^2}\right) \tag{3.39}$$

Inserting this result into Eq. (3.38) and solving for $U(r)$ we have

$$U(r) = E - \frac{2K\ell^2}{\mu r^3} - \frac{\ell^2}{2\mu r^2}$$

$$= E - \frac{\ell^2}{\mu}\left(\frac{2K}{r^3} + \frac{1}{2r^2}\right) \tag{3.40}$$

Then the force law is

$$F(r) = -\frac{dU(r)}{dr} \tag{3.41}$$

$$= -\frac{\ell^2}{\mu}\left(\frac{6K}{r^4} + \frac{1}{r^3}\right) \tag{3.42}$$

(b) The equation of the orbit as given by Eq. (3.20) is

$$\frac{d^2}{d\theta^2}\left(\frac{1}{r}\right) + \frac{1}{r} = -\frac{\mu}{\ell^2}r^2 F(r) \tag{3.43}$$

Now

$$\frac{1}{r} = \frac{1}{K}\theta^{-2} \tag{3.44}$$

so

$$\frac{d}{d\theta}\left(\frac{1}{r}\right) = -\frac{2}{K}\theta^{-3} \tag{3.45}$$

and

$$\frac{d^2}{d\theta^2}\left(\frac{1}{r}\right) = \frac{6}{K}\theta^{-4}$$

$$= \frac{6K}{r^2} \tag{3.46}$$

Inserting the result of Eq. (3.46) into Eq. (3.43) and solving for $F(r)$ we have

$$\frac{6K}{r^2} + \frac{1}{r} = -\frac{\mu}{\ell^2}r^2 F(r)$$

$$F(r) = -\frac{\ell^2}{\mu}\left(\frac{6K}{r^4} + \frac{1}{r^3}\right) \tag{3.47}$$

We obtain the same answer using both methods. This is not surprising because: (a) there is only one correct answer, and (b) the equation of the orbit was derived using Lagrangian dynamics based on the kinetic and potential energies so the methods are equivalent. Often, however, the equation of the orbit is a more direct way of obtaining the force law when the function $r = r(\theta)$ is known.

9. Using the equation of the orbit, see Eq. (3.20) and Table 3.1, show that there are three different types of orbits that can result from a force law of the form

$$F(r) = -\frac{C}{r^3} \tag{3.48}$$

where C is a real positive constant.

Solution

We use the equation of the orbit in the form

$$\frac{d^2u}{d\theta^2} + u = -\frac{\mu}{\ell^2}\frac{1}{u^2}F(u) \tag{3.49}$$

where $u = 1/r$ and

$$F(u) = -Cu^3 \tag{3.50}$$

so we have

$$\frac{d^2u}{d\theta^2} + u = \frac{C\mu}{\ell^2}u$$

$$\frac{d^2u}{d\theta^2} + \left(1 - \frac{C\mu}{\ell^2}\right)u = 0 \tag{3.51}$$

This differential equation is the same as that describing simple harmonic motion. By analogy, it is solved by assuming a solution of the form

$$u = e^{m\theta} \tag{3.52}$$

from which we arrive at the indicial equation

$$m^2 + K^2 = 0 \text{ where } K^2 = \left(1 - \frac{C\mu}{\ell^2}\right) \tag{3.53}$$

The general solution is

$$u = \frac{1}{r} = Ae^{K\theta} + Be^{-K\theta} \tag{3.54}$$

where A and B are constants of integration to be determined by the boundary conditions. The nature of the orbits will, however, differ greatly depending upon the sign of K^2. That is, the nature of the orbit will depend upon the relative magnitudes of the constant C and the ratio μ/ℓ^2. Examine the three cases separately.

$K^2 > 0$: In this case the exponents are real and the orbit is a spiral.

$K^2 = 0$: The differential equation, Eq. (3.51), becomes

$$\frac{d^2u}{d\theta^2} = 0 \tag{3.55}$$

the solution to which is

$$u = A\theta + B \Longrightarrow r = \frac{1}{A\theta + B} \tag{3.56}$$

This too is a spiral as shown in Fig. 3.3.

Fig. 3.3 Problem 9

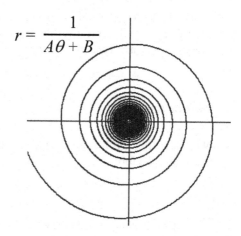

$$r = \frac{1}{A\theta + B}$$

$K^2 < 0$: The exponents are imaginary and the solutions for u are therefore sinusoidal. Thus, we may write

$$r = \frac{1}{A\cos(\theta + \delta)} \tag{3.57}$$

which is also a spiral.

The three spirals that result from the attractive inverse cube central force law are known as Cotes' spirals. They are conveniently summarized by the three different equations for their orbits $r = r(\theta)$

$$r = \frac{A}{\cos(K\theta + \delta)} \qquad C < \frac{\ell^2}{\mu}$$

$$= \frac{A}{\cosh(K\theta + \delta)} \qquad C > \frac{\ell^2}{\mu}$$

$$= \frac{1}{A\theta + B} \qquad C = \frac{\ell^2}{\mu} \tag{3.58}$$

The equation of the orbit was extremely useful in this problem because it allowed us to expeditiously arrive at the solution.

10. A force law, the same as in Problem 9 of this chapter, is given by

$$F(r) = -\frac{C}{r^3} \quad \text{where} \quad C = \frac{\ell^2}{\mu} \tag{3.59}$$

The points $(r, \theta) = (1/4, 0)$ and $(r, \theta) = (1/8, 4)$ are known to lie on the orbit of a particle subject to this force law. Also, it is known that at $t = 0$, $r = r_0$. Find r as a function of time.

Solution

We know from Problem 9 of this chapter that, because $C = \ell^2/\mu$, the orbit must have the form

$$r = \frac{1}{A\theta + B} \tag{3.60}$$

Using the two points on the orbit that are given we can evaluate A and B.

$$\frac{1}{4} = \frac{1}{A \cdot 0 + B} \Longrightarrow B = 4 \tag{3.61}$$

and

$$\frac{1}{8} = \frac{1}{A \cdot 4 + 4} \Longrightarrow A = 1 \tag{3.62}$$

Therefore, the equation of the orbit $r = r(\theta)$ is

$$r = \frac{1}{\theta + 4} \tag{3.63}$$

At this point it is worth performing a consistency check on the units. From Eq. (3.63) it might seem as though the units are not consistent because r has units of length and everything on the right-hand side of this equation appears to be unitless. Note, however, that A and B in Eq. (3.60) have units of r^{-1}, so r has the correct units.

Using conservation of angular momentum

$$\ell = \mu r^2 \dot{\theta} \tag{3.64}$$

we have

$$r^2 d\theta = \frac{\ell}{\mu} dt \tag{3.65}$$

We can eliminate $d\theta$ using Eq. (3.63)

$$\theta = \frac{1}{r} - 4 \Longrightarrow d\theta = -\frac{1}{r^2} dr \tag{3.66}$$

So Eq. (3.65) becomes

$$-dr = \frac{\ell}{\mu}dt \tag{3.67}$$

which, upon integration is

$$r(t) = -\frac{\ell}{\mu}t + D \tag{3.68}$$

where D is a constant of integration. At $t = 0$ the constant of integration $D = r_0$ so the final answer is

$$r(t) = -\frac{\ell}{\mu}t + r_0 \tag{3.69}$$

This contrived problem illustrates the manipulations that can yield the time-dependent orbit $r = r(t)$ from the equation of the orbit in polar coordinates $r = r(\theta)$. As usual in central force problems, conservation of angular momentum is central (no pun intended) to the solution.

11. A particle of mass μ having angular momentum ℓ is bound by a central force $F(r) = -kr$, where k is a positive constant. This force causes the particle to oscillate as an isotropic harmonic oscillator, "isotropic" because the spring constant k is independent of direction.

(a) Sketch the potential energy $U(r)$ and the effective potential $U_{\text{eff}}(r)$.
(b) Find the radius r_c of a circular orbit for this effective potential. Will the circular orbit be stable or unstable? Explain.

Solution

(a) This force law for an isotropic harmonic oscillator is derived from the corresponding potential energy, given by

$$F(r) = -\frac{dU(r)}{dr} \implies U(r) = \frac{1}{2}kr^2 \tag{3.70}$$

Because angular momentum is conserved for central potentials, the motion takes place in a plane and

$$U_{\text{eff}}(r, \ell) = U(r) + \frac{\ell^2}{2\mu r^2} \tag{3.71}$$

the graph of which is shown in Fig. 3.4.

Fig. 3.4
Problem 11—solution

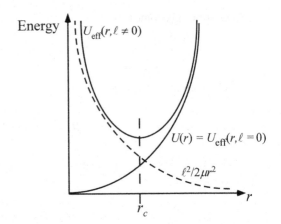

Note that the curve that represents $U_{\text{eff}}(r, \ell = 0)$ is half a parabola (r cannot be negative) while $U_{\text{eff}}(r, \ell \neq 0)$ must approach the ordinate asymptotically. Thus, the orbiting particle can never pass through the force center at the origin.

(b) The radius r_c of a stable circular orbit is the value of r at the minimum of $U_{\text{eff}}(r, \ell \neq 0)$. Setting $dU_{\text{eff}}(r)/dr = 0$ to find r_c we have

$$\left. \frac{dU_{\text{eff}}(r)}{dr} \right|_{r=r_c} = \left. kr - \frac{\ell^2}{\mu r^3} \right|_{r=r_c} = 0 \Longrightarrow r_c^4 = \frac{\ell^2}{\mu k} \qquad (3.72)$$

For r_c to be a minimum $d^2 U_{\text{eff}}(r)/dr^2 \big|_{r=r_c} > 0$ which is easily shown to be the case. If the particle in this example were moving with constant r_c, a slight perturbation would produce small oscillations about r_c. If r_c had been a maximum point, the particle would be at the top of the potential U_{eff} and a perturbation would produce a large displacement in r.

12. Describe the possible orbits of the particle of mass m under the influence of an isotropic harmonic oscillator potential (see Problem 11 of this chapter).
    [Hint: It is recommended that Cartesian coordinates be used in conjunction with $F = ma$ rather than the equation of the orbit in polar coordinates.]

Solution

Using $F = ma$ for Cartesian coordinates we have

$$F(x, y, z) = m\ddot{r} = m\left(\ddot{x}\hat{\imath} + \ddot{y}\hat{\jmath} + \ddot{z}\hat{k}\right)$$

$$= -kx\hat{\imath} - ky\hat{\jmath} - kz\hat{k} \qquad (3.73)$$

Equation (3.73) represents three equations, one for each Cartesian component. They are of the form

$$\ddot{q} + \omega^2 q = 0 \text{ where } q = x, y \text{ or } z \text{ and } \omega^2 = k/m \qquad (3.74)$$

Because the force is a central force we know that the motion takes place in a plane so we may as well let $z = 0$. The solutions for x and y are

$$x = x_m \cos(\omega t - \xi_x) \text{ and } y = y_m \cos(\omega t - \xi_y) \qquad (3.75)$$

where x_m, y_m, ξ_x and ξ_y are constants (ξ_x and ξ_y are phase angles). The momenta p_x and p_y are

$$p_x = -m\omega x_m \sin(\omega t - \xi_x) \text{ and } p_y = -m\omega y_m \sin(\omega t - \xi_y) \qquad (3.76)$$

The coordinate equation (3.75) are the well-known parametric equations of an ellipse. The phase difference $(\xi_y - \xi_x) = \delta$ determines the orientation of the ellipse with respect to the xy-coordinate system. The phase difference δ also determines the direction of motion of the particle on the ellipse. For $\delta = 0, \pm\pi$, the ellipse is a straight line with particle oscillating back and forth; for $-\pi < \delta < 0$, the particle moves clockwise on the ellipse, and for $0 < \delta < \pi$ it moves counterclockwise.

The total angular momentum, L_z is, from Eq. (3.2)

$$
\begin{aligned}
L_z &= x p_y - y p_x \\
&= -x_m \cos(\omega t - \xi_x) m\omega y_m \sin(\omega t - \xi_y) \\
&\quad + y_m \cos(\omega t - \xi_y) m\omega x_m \sin(\omega t - \xi_x) \\
&= -m\omega (x_m y_m) \bullet [\sin(\omega t - \xi_y)\cos(\omega t - \xi_x) \\
&\quad - \cos(\omega t - \xi_y)\sin(\omega t - \xi_x)] \\
&= -m\omega (x_m y_m) \sin[(\omega t - \xi_y) - (\omega t - \xi_x)] \\
&= m\omega (x_m y_m) \sin \delta \qquad (3.77)
\end{aligned}
$$

L_z is positive or negative for counterclockwise or clockwise motion, respectively, and it is zero for rectilinear motion, $\delta = 0, \pm\pi$. For a given value of the product $x_m y_m$, $|L_z|$ is a maximum for $\delta = \pm\pi/2$. If $x_m = y_m$, the ellipse is a circle.

Note that these elliptical orbits are closed paths that produce periodic motion. In Problem 4 of this chapter it was noted that there are only two central potentials for which the paths are closed, the isotropic oscillator and the Kepler potential.

13. A particle of mass μ subjected to an attractive force law $F(r) = -k/r^2$ where k is a positive constant. This is also referred to as the "Kepler problem" because planetary motion is derived from this force law, as we have already discussed.

   (a) Sketch the effective potential $U_{\text{eff}}(r, \ell)$ for three different values of angular momenta, $\ell_1 = 0$ and $0 < \ell_2 < \ell_3$.
   (b) Find r_c the radius of a circular orbit for $\ell \neq 0$. Is the circular orbit stable?

Solution

   (a) The potential is obtained from $F(r)$

$$F(r) = -k/r^2 \Longrightarrow U(r) = -k/r \tag{3.78}$$

   so the effective potential is

$$U_{\text{eff}}(r) = -k/r + \frac{\ell^2}{2\mu r^2} \tag{3.79}$$

   the graph of which is (Fig. 3.5)

Fig. 3.5
Problem 13a—solution. The
radius of a circular orbit r_c for
$\ell \neq 0$ is shown for $U_{\text{eff}}(r, \ell_2)$

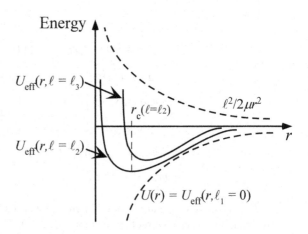

   (b) Find the minimum of the effective potential $U_{\text{eff}}(r, \ell \neq 0)$ by setting the derivative equal to zero.

$$\frac{dU_{\text{eff}}(r)}{dr} = \frac{k}{r_c^2} - \frac{\ell^2}{\mu r_c^3} = 0 \tag{3.80}$$

so

$$kr_c = \frac{\ell^2}{\mu} \implies r_c = \frac{\ell^2}{k\mu} \tag{3.81}$$

The circular orbit is stable because r_c occurs at a minimum in $U_{\text{eff}}(r)$.

14. For the gravitational force, $F(r) = -k/r^2$ (k = positive constant), the possible orbits are conic sections, the standard form of which is

$$\frac{\alpha}{r} = 1 + \epsilon \cos \theta \tag{3.82}$$

See Appendix D for details. Choose one of the constants of integration such that the polar angle θ is measured from the axis that coincides with the minimum value of r.

(a) Prove that the equations of the orbits for an attractive inverse square force (the Kepler problem) are conic sections (see Appendix D).
(b) Find the values of ϵ and α in terms of the parameters ℓ, μ, k, and E (the TME).
(c) Show that the TME for a bound orbit is

$$E = -\frac{k}{2a} \tag{3.83}$$

where a is the semi-major axis of the elliptical orbit.

Solution

(a) We begin with the equation of the orbit, Eq. (3.20) in the form

$$\frac{d^2u}{d\theta^2} + u = -\frac{\mu}{\ell^2} \frac{1}{u^2} F(u) \tag{3.84}$$

with $u = 1/r$ so

$$F(u) = -ku^2 \tag{3.85}$$

Thus, we have an inhomogeneous differential equation

$$\frac{d^2u}{d\theta^2} + u = \frac{k\mu}{\ell^2} \tag{3.86}$$

the solution of which is the sum of the solution to the homogeneous equation and the particular solution. The solution to the homogeneous equation is $B \cos (\theta - \theta_0)$ where B and θ_0 are constants of integration [see the discussion surrounding Eq. (3.51)]. The particular solution is $u = \text{constant} = \mu k / \ell^2$. Thus

$$u = \frac{1}{r} = \frac{\mu k}{\ell^2} + B \cos (\theta - \theta_0) \tag{3.87}$$

From Appendix D the standard form of a conic section with the origin at one focus is

$$\frac{\alpha}{r} = 1 + \epsilon \cos \theta \tag{3.88}$$

where α and ϵ are constants given by

$$\epsilon = \sqrt{1 + \frac{2E\ell^2}{\mu k^2}} = \text{eccentricity}$$

$$\alpha = \frac{\ell^2}{\mu k} = \text{latus rectum} \tag{3.89}$$

The shape of the conic section depends upon the value of ϵ.

$$\epsilon = 1 \quad \text{parabola}$$

$$\epsilon < 1 \quad \text{ellipse}$$

$$\epsilon > 1 \quad \text{hyperbola} \tag{3.90}$$

Equation (3.87) may be re-written

$$\frac{(\ell^2/\mu k)}{r} = 1 + \left(\frac{B\ell^2}{\mu k} \right) \cos (\theta - \theta_0) \tag{3.91}$$

from which it is clear that

$$\alpha = \frac{\ell^2}{\mu k} \tag{3.92}$$

and

$$\epsilon = B\alpha \tag{3.93}$$

It is customary to take $\theta_0 = 0$, a choice that forces r to have its minimum value $r_{\min}$ at $\theta = 0$. This convention conforms to the usual notation for the

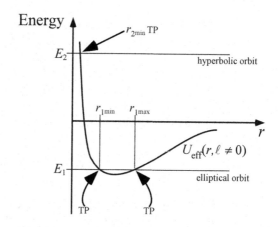

Fig. 3.6
Problem 14—solution. The energies E_1 and E_2 represent the TME for a bound and unbound orbit. The *points* labelled TP are the turning points for a particle having these energies

"standard form" of the equation of the conic section, Eq. (3.88) and we are left with

$$\frac{\alpha}{r} = 1 + B\alpha \cos\theta \qquad (3.94)$$

(b) To evaluate the constant of integration B so that we can find ϵ in terms of the known parameters we use Fig. 3.6 as a guide. We note that whether the TME is positive or negative the conic section must have at least one turning point (TP in the figure) as long as $\ell \neq 0$. Only an ellipse has two turning points, one at $r_{1\text{min}}$ and one at $r_{1\text{man}}$ (see Appendix D). From Eq. (3.94) we see that r_{min} is given in terms of B by

$$\frac{\alpha}{r_{\text{min}}} = 1 + B\alpha \qquad (3.95)$$

We may write another expression for r_{min} by noting that the effective potential must equal the TME at both $r_{1\text{min}}$ and $r_{1\text{max}}$ because the kinetic energy at these values of r is zero for $E < 0$. For $E > 0$ the kinetic energy is zero at the single turning point $r_{2\text{min}}$, but we limit our discussion to the elliptical orbits ($E < 0$). Setting $U_{\text{eff}}(r) = E$ we have

$$U_{\text{eff}}(r) = -k/r + \frac{\ell^2}{2\mu r^2} = E \qquad (3.96)$$

or

$$\frac{1}{r^2} - \frac{2\mu k}{\ell^2}\frac{1}{r} - \frac{2\mu E}{\ell^2} = 0 \qquad (3.97)$$

which is a quadratic equation in the variable $1/r$. The solutions to this equation are

$$\frac{1}{r} = \frac{\mu k}{\ell^2} \pm \sqrt{\left(\frac{\mu k}{\ell^2}\right)^2 + \left(\frac{2\mu E}{\ell^2}\right)}$$

$$= \frac{1}{\alpha} \pm \sqrt{\frac{1}{\alpha^2} + \left(\frac{1}{\alpha}\right)\frac{2E}{k}} \qquad (3.98)$$

The plus sign yields $r = r_{min}$ so we have

$$\frac{\alpha}{r_{min}} = 1 + \sqrt{1 + \frac{2E\alpha}{k}} \qquad (3.99)$$

We now have two expressions for α/r_{min}, Eqs. (3.95) and (3.99), so we may equate them to evaluate B and the eccentricity ϵ. Using Eq. (3.93) we have

$$B\alpha = \epsilon = \sqrt{1 + \frac{2E\alpha}{k}} \qquad (3.100)$$

$$= \sqrt{1 + \frac{2E\ell^2}{\mu k^2}} \qquad (3.101)$$

Note that for the bound elliptical orbit the total energy E must be negative and $\epsilon < 1$ as prescribed by Eq. (3.90). Similarly, the eccentricities of the parabolic and hyperbolic orbits follow from their properly inserted energies (see the Fig. 3.6). According to Eq. (3.88) the minimum value of $r = r_{min}$ occurs when $\theta = 0$ and for an elliptical orbit the maximum occurs when $\theta = \pi$. Thus, the polar angle θ is measured from the "short side" of the ellipse, the pericenter, as shown in Fig. 3.7. The long side, $r = r_{max}$, is referred to as the apocenter.

Fig. 3.7
Problem 14—solution

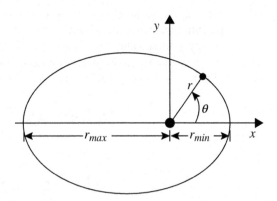

(c) From Eq. (3.95) it is clear that

$$r_{\min} = \frac{\alpha}{1+\epsilon} \quad \text{and} \quad r_{\max} = \frac{\alpha}{1-\epsilon} \tag{3.102}$$

so the semi-major axis is

$$
\begin{aligned}
a &= \frac{1}{2}(r_{\min} + r_{\max}) = \left(\frac{\alpha}{1-\epsilon^2}\right) \\
&= \left(\frac{\ell^2}{\mu k}\right)\left(\frac{\mu k^2}{2\,|E|\,\ell^2}\right) \\
&= -\frac{k}{2E} \tag{3.103}
\end{aligned}
$$

Solving for E we have

$$E = -\frac{k}{2a} \tag{3.104}$$

where we have let $|E| \rightarrow -E$ because the energy must be negative for a bound orbit.

Using the equation of the orbit we have derived the properties of the conic sections (orbits) that result from an inverse square attractive force. There are other ways to obtain this result, most of them equivalent to the method used here. One of the most intriguing aspects of this problem is that, in contrast to virtually all other central force orbits, the bound orbit for the Kepler problem, the ellipse, closes on itself and is fixed in space (Bertrand's theorem again). This aspect of the problem is seldom stressed at the introductory level. It is a consequence of a subtle symmetry of this force law that produces an additional constant of the motion, a vector along the semi-major axis that keeps the ellipse fixed in space. The perceptive student is no doubt asking "what about the precession of the perihelion of Mercury?" This motion of Mercury's elliptical orbit is the result of several perturbations to the pure Keplerian potential, the most famous being a consequence of general relativity, which to first order adds a $1/r^3$ term to the potential (see page 511 of [2]).

15. It was mentioned near the end of Problem 14 of this chapter that, in addition to energy and angular momentum , the Kepler problem has an additional constant of the motion [1]. Although it goes by other names [2], this constant of the motion is commonly called the Runge–Lenz vector. It is designated by A and defined as

$$A = p \times L - \mu k \frac{r}{r} \tag{3.105}$$

where the angular momentum vector is

$$L = r \times p \tag{3.106}$$

and the other quantities in Eq. (3.105) have their customary meanings in this chapter.

(a) Show that A is a constant of the motion.
(b) Show that $A \bullet L = 0$ so that A lies in the plane of the motion.
(c) By evaluating the scalar product $A \bullet r$ and comparing the result with Eq. (3.88) show that A points toward the pericenter (the short end) of the semi-major axis of the elliptical orbit. Show also that the magnitude of A is proportional to the eccentricity of the orbit and is given by

$$A = \mu k \sqrt{1 + \frac{2E\ell^2}{\mu k^2}} \tag{3.107}$$

Solution

(a) To show that A is a constant of the motion we must show that its time derivative vanishes.

$$\frac{dA}{dt} = \dot{p} \times L + p \times \dot{L} - \mu k \frac{d}{dt}\left(\frac{r}{r}\right) \tag{3.108}$$

For a central potential angular momentum is conserved so $\dot{L} = 0$ and

$$\frac{dA}{dt} = \dot{p} \times L - \mu k \frac{d}{dt}\left(\frac{r}{r}\right) \tag{3.109}$$

For the Kepler problem

$$\dot{p} = F(r) = -\frac{k}{r^2}\hat{a}_r = -\frac{k}{r^3}r \tag{3.110}$$

Then, using the vector identity for the triple cross product, Eq. (F.2)

$$\dot{p} \times L = -\frac{k}{r^3}[r \times (r \times \mu\dot{r})]$$

$$= -\frac{\mu k}{r^3}[r(r \bullet \dot{r}) - r^2\dot{r}] \tag{3.111}$$

Noting that

$$r \bullet \dot{r} = \frac{1}{2}\frac{d}{dt}(r \bullet r) = r\dot{r} \tag{3.112}$$

we have

$$\dot{p} \times L = -\frac{\mu k}{r^3} \left[r \dot{r} \mathbf{r} - r^2 \dot{\mathbf{r}} \right]$$

$$= -\frac{\mu k}{r^2} \left[\dot{r} \mathbf{r} - r \dot{\mathbf{r}} \right]$$

$$= \mu k \left[\frac{\dot{\mathbf{r}}}{r} - \frac{\dot{r} \mathbf{r}}{r^2} \right]$$

$$= \mu k \frac{d}{dt} \left(\frac{\mathbf{r}}{r} \right) \qquad (3.113)$$

Inserting this result into Eq. (3.109) shows that $\dot{A}$ vanishes so A is a constant of the motion for the Kepler problem.

(b) We have

$$A \bullet L = \left(p \times L - \mu k \frac{\mathbf{r}}{r} \right) \bullet L$$

$$= -\mu k \frac{\mathbf{r}}{r} \bullet L$$

$$= -\mu k \frac{\mathbf{r}}{r} \bullet (\mathbf{r} \times p)$$

$$= 0 \qquad (3.114)$$

because L is perpendicular to $(p \times L)$ and $\mathbf{r}$ is perpendicular to $(\mathbf{r} \times p)$.

(c) Taking the scalar product $A \bullet \mathbf{r}$ as directed we have

$$A \bullet \mathbf{r} = \mathbf{r} \bullet \left[(p \times L) - \mu k \frac{\mathbf{r}}{r} \right]$$

$$= L \bullet (\mathbf{r} \times p) - \mu k r$$

$$= \ell^2 - \mu k r \qquad (3.115)$$

where we have used the vector identity in Eq. (F.1).

By definition the scalar product $A \bullet \mathbf{r}$ is also given by

$$A \bullet \mathbf{r} = A r \cos \phi \qquad (3.116)$$

where ϕ is the angle between A and $\mathbf{r}$. Equating these two expressions for $A \bullet \mathbf{r}$ we see that the resulting equation is that of a conic section

$$A r \cos \phi = \ell^2 - \mu k r$$

$$r = \frac{\ell^2}{(A \cos \phi + \mu k)} \qquad (3.117)$$

or

$$\frac{(\ell^2/\mu k)}{r} = 1 + \frac{A}{\mu k}\cos\phi \qquad (3.118)$$

Comparison of Eq. (3.118) with the equation of a conic section, Eq. (3.88), shows that they are identical and $\theta = \phi$. Thus, the Runge–Lenz vector points in the direction of the pericenter, that is, in the $+x$ direction ($\theta = 0$).

We can determine the magnitude of A in terms of the Keplerian orbit parameters by comparing Eq. (3.118) with Eqs. (3.88) and (3.89). Accordingly

$$\epsilon = \frac{A}{\mu k} \qquad (3.119)$$

so that

$$A = \mu k\sqrt{1 + \frac{2E\ell^2}{\mu k^2}} \qquad (3.120)$$

16. This problem shows why conservation of the Runge–Lenz vector demands closed orbits for the Kepler problem.

   (a) Express the Runge–Lenz vector A as defined in Eq. (3.105) in terms of r and p only by eliminating L.
   (b) Show that

$$A = \mu r_{\max}\left(2E + \frac{k}{r_{\max}}\right)$$

$$= \mu r_{\min}\left(2E + \frac{k}{r_{\min}}\right) \qquad (3.121)$$

   where E is the total energy, $r = r_{\min}$ is the pericenter of the elliptical orbit, and $r = r_{\max}$ is the apocenter.
   (c) Show that A is parallel to $r_{\min}$ and antiparallel to $r_{\max}$ and that for a circular orbit $A = 0$.

Solution

(a) A is defined as

$$A = p \times L - \mu k\frac{r}{r} \qquad (3.122)$$

so, using the vector identity for the triple cross product, Eq. (F.2) and the definition of angular momentum

$$A = p \times (r \times p) - \mu k \frac{r}{r}$$

$$= p^2 r - (p \bullet r) p - \mu k \frac{r}{r} \qquad (3.123)$$

(b) First eliminate p^2 from Eq. (3.123) using the total energy

$$E = \frac{p^2}{2\mu} - \frac{k}{r} \Longrightarrow p^2 = 2\mu E + \frac{2\mu k}{r} \qquad (3.124)$$

Fig. 3.8
Problem 15b—solution

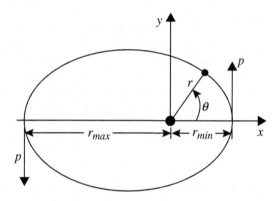

then

$$A = p^2 r - (p \bullet r) p - \mu k \frac{r}{r}$$

$$= 2\mu E r + \frac{2\mu k}{r} r - (p \bullet r) p - \frac{\mu k}{r} r$$

$$= \mu \left(2E + \frac{k}{r} \right) r - (p \bullet r) p \qquad (3.125)$$

Now, when $r = r_{\min}$ or $r = r_{\max}$, $p \perp r$ as illustrated in Fig. 3.8 so $p \bullet r = 0$. Thus, at pericenter and apocenter we have

$$A = \mu r_{\min} \left(2E + \frac{k}{r_{\min}} \right) = \mu r_{\max} \left(2E + \frac{k}{r_{\max}} \right) \qquad (3.126)$$

(c) From Problem 14 of this chapter we know that the total energy of the elliptical Kepler orbit is

$$E = -\frac{k}{2a} \tag{3.127}$$

so Eq. (3.126) may be re-written

$$A = \mu k r_{min}\left(-\frac{1}{a} + \frac{1}{r_{min}}\right)$$

$$= \mu k r_{max}\left(-\frac{1}{a} + \frac{1}{r_{max}}\right) \tag{3.128}$$

Moreover, we know that $r_{min} < a < r_{max}$ so in the first of Eq. (3.128) the quantity in parenthesis is positive; in the second, it is negative. This establishes the directions of A:

$$A \propto r_{min} \propto -r_{max} \tag{3.129}$$

For a circular orbit $r_{min} = r_{max} = a$ so, from Eq. (3.128), $A = 0$.

Although quantum mechanics has not yet been discussed here, it is worthwhile to compare the nature of Keplerian orbits to the quantum mechanical results for the hydrogen atom (H-atom) because both systems are subject to the same $1/r^2$ force law. As listed in Table T.1 the energy levels of the quantum mechanical H-atom are specified by a single quantum number n, the principal quantum number.

$$E \propto 1/n^2 \tag{3.130}$$

These allowed energies do not depend upon the angular momentum quantum number. For Keplerian elliptical orbits the energy depends only upon the value of the semi-major axis a, Eq. (3.127), and not on the semi-minor axis b which depends upon the angular momentum ℓ according to

$$b = \frac{\ell}{\sqrt{2\mu |E|}} = \ell\sqrt{\frac{a}{\mu k}} \tag{3.131}$$

Thus, elliptical Keplerian orbits that have the same a, but different values of b, all have the same energy. This is an example of a classical degeneracy. Clearly, the classical degeneracy of the Keplerian orbits is related to those of the H-atom.

Chapter 4
Normal Modes and Coordinates

Before we present the problems in this chapter we briefly review the properties of normal modes and coordinates by stating three propositions about them:

Proposition 1. *Using the normal coordinates to describe a multi-oscillator system decouples the Lagrangian equations of motion. Each of the differential equations of motion contains only a single variable, the normal coordinate, and its time derivatives.*

Proposition 2. *Motion in the normal coordinates corresponds to sinusoidal oscillation with a single well-defined frequency, a normal mode.*

Proposition 3. *When T and U are expressed in terms of the normal coordinates they contain only quadratic functions of these coordinates and their time derivatives; there are no cross terms.*

Problems

1. The masses m of two identical pendulums of length ℓ are connected by a spring having spring constant k. The motion is constrained to a plane and the angles θ_1 and θ_2, each of which is measured counterclockwise from the vertical, are assumed small. Assume also that the unstretched length of the spring is d, the same as the separation between the supports.

© Springer International Publishing AG 2017
J.D. Kelley, J.J. Leventhal, *Problems in Classical and Quantum Mechanics*, DOI 10.1007/978-3-319-46664-4_4

Fig. 4.1 Problem 1

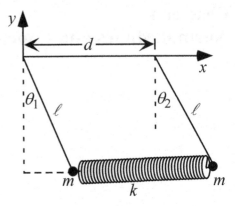

(a) Set up Lagrange's equations using the generalized coordinates θ_1 and θ_2. Use the small angle approximation.

(b) Find the normal mode frequencies and the normal coordinates. Describe the motion for each normal mode.

Solution

(a) It is easier to find the kinetic energy T and potential energy U in terms of Cartesian coordinates and then transform the result to θ_1 and θ_2.

$$x_1 = \ell \sin \theta_1; \; y_1 = -\ell \cos \theta_1$$
$$x_2 = d + \ell \sin \theta_2; \; y_2 = -\ell \cos \theta_2 \qquad (4.1)$$

so

$$T = \frac{1}{2}m \left(\dot{x}_1^2 + \dot{y}_1^2 + \dot{x}_2^2 + \dot{y}_2^2 \right)$$
$$= \frac{1}{2}m\ell^2 \left(\dot{\theta}_1^2 \sin^2 \theta_1 + \dot{\theta}_1^2 \cos^2 \theta_1 + \dot{\theta}_2^2 \sin^2 \theta_2 + \dot{\theta}_2^2 \cos^2 \theta_2 \right)$$
$$= \frac{1}{2}m\ell^2 \left(\dot{\theta}_1^2 + \dot{\theta}_2^2 \right) \qquad (4.2)$$

The potential energy due to gravity measured from the equilibrium position of the mass is

$$U_{\text{gravity}} = mgy_1 + mgy_2$$
$$= mg\ell \left[(1 - \cos \theta_1) + (1 - \cos \theta_2) \right]$$

$$\approx mg\ell \left[1 - \left(1 - \frac{\theta_1^2}{2!}\right) + 1 - \left(1 - \frac{\theta_2^2}{2!}\right) \right]$$

$$= \frac{1}{2} mg\ell \left(\theta_1^2 + \theta_2^2 \right) \tag{4.3}$$

where we have, in the spirit of the small angle approximation, retained only the first two terms of the Taylor expansion for $\cos \theta$.

The potential energy due to the spring (again using the small angle approximation) is

$$U_{spring} = \frac{1}{2} k \left[(x_2 - x_1) - d \right]^2$$

$$= \frac{1}{2} k\ell^2 \left[(d + \sin \theta_2 - \sin \theta_1) - d \right]^2$$

$$\approx \frac{1}{2} k\ell^2 (\theta_2 - \theta_1)^2 \tag{4.4}$$

so that

$$U = \frac{1}{2} mg\ell \left(\theta_1^2 + \theta_2^2 \right) + \frac{1}{2} k\ell^2 (\theta_2 - \theta_1)^2 \tag{4.5}$$

The Lagrangian is

$$\mathcal{L} = T - U$$

$$= \frac{1}{2} m\ell^2 \left(\dot{\theta}_1^2 + \dot{\theta}_2^2 \right) - \frac{1}{2} mg\ell \left(\theta_1^2 + \theta_2^2 \right) - \frac{1}{2} k\ell^2 (\theta_2 - \theta_1)^2 \tag{4.6}$$

and the Lagrangian equations of motion are

$$\frac{d}{dt} \left(\frac{\partial \mathcal{L}}{\partial \dot{\theta}_i} \right) - \left(\frac{\partial \mathcal{L}}{\partial \theta_i} \right) = 0 \tag{4.7}$$

so, using the notation

$$\omega_p^2 = \frac{g}{\ell} = \text{frequency of simple pendulum}$$

$$\omega_o^2 = \frac{k}{(m/2)} = \text{frequency of SHO} \tag{4.8}$$

where $m/2$ is the reduced mass (see Problem 4 of Chap. 1 and Table 3.1). The Lagrange equations of motion become

$$m\ell^2 \ddot{\theta}_1 + mg\ell\theta_1 - k\ell^2 (\theta_2 - \theta_1) = 0 \tag{4.9}$$

or

$$\ddot{\theta}_1 + \omega_p^2 \theta_1 - \frac{\omega_o^2}{2} (\theta_2 - \theta_1) = 0 \tag{4.10}$$

and

$$ml^2\ddot{\theta}_2 + mg\ell\theta_2 + k\ell^2(\theta_2 - \theta_1) = 0 \tag{4.11}$$

or

$$\ddot{\theta}_2 + \omega_p^2\theta_2 + \frac{\omega_o^2}{2}(\theta_2 - \theta_1) = 0 \tag{4.12}$$

Notice that these equations of motion, Eqs. (4.10) and (4.12), are coupled in the sense that each equation contains both θ_1 and θ_2.

(b) Sometimes, as in this problem, it is possible to find the normal coordinate by inspection. Let us add and subtract Eqs. (4.10) and (4.12) to cast them in a different form. Adding them we have

$$\left(\ddot{\theta}_1 + \ddot{\theta}_2\right) + \omega_p^2(\theta_1 + \theta_2) = 0 \tag{4.13}$$

Subtracting them gives

$$\left(\ddot{\theta}_1 - \ddot{\theta}_2\right) + \omega_p^2(\theta_1 - \theta_2) + \omega_o^2(\theta_1 - \theta_2) = 0 \tag{4.14}$$

Inspection of Eqs. (4.13) and (4.14) shows that θ_1 and θ_2 occur only in the forms $(\theta_1 + \theta_2)$ and $(\theta_1 - \theta_2)$. This suggests that we make the substitutions

$$\eta = \frac{1}{2}(\theta_1 + \theta_2)\,; \ \ \xi = \frac{1}{2}(\theta_1 - \theta_2) \tag{4.15}$$

which lead to

$$\ddot{\eta} + \omega_p^2\eta = 0$$

$$\ddot{\xi} + \left(\omega_p^2 + \omega_o^2\right)\xi = 0 \tag{4.16}$$

In terms of the new variables, η and ξ, Eqs. (4.10) and (4.12) are now "decoupled." Moreover, Eq. (4.16) are the differential equations of simple harmonic motion in the variables η and ξ so the system oscillates in η and ξ at the frequencies

$$\omega_\eta = \omega_p\,; \ \ \omega_\xi = \sqrt{\omega_p^2 + \omega_o^2} \tag{4.17}$$

which are the normal mode frequencies (see Proposition 2).

To check Proposition 3 we must solve Eq. (4.15) for θ_1 and θ_2 in terms of η and ξ and then find T [Eq. (4.2)] and U [Eq. (4.5)] in terms of η and ξ. We have

$$\theta_1 = \eta + \xi \ \ \text{and} \ \ \theta_2 = \eta - \xi \tag{4.18}$$

From Eq. (4.2)

$$T = \frac{1}{2}m\ell^2\left(\dot{\theta}_1^2 + \dot{\theta}_2^2\right)$$

$$= \frac{1}{2}m\ell^2\left[\left(\dot{\eta}^2 + 2\dot{\eta}\dot{\xi} + \dot{\xi}^2\right) + \left(\dot{\eta}^2 - 2\dot{\eta}\dot{\xi} + \dot{\xi}^2\right)\right]$$

$$= m\ell^2\left(\dot{\eta}^2 + \dot{\xi}^2\right) \tag{4.19}$$

and from Eq. (4.5)

$$U = \frac{1}{2}mg\ell\left(\theta_1^2 + \theta_2^2\right) + \frac{1}{2}k\ell^2(\theta_2 - \theta_1)^2$$

$$= \frac{1}{2}mg\ell\left[\left(\eta^2 + 2\eta\xi + \xi^2\right) + \left(\eta^2 - 2\eta\xi + \xi^2\right)\right]$$

$$+ \frac{1}{2}k\ell^2\left[(\eta + \xi) - (\eta - \xi)\right]^2$$

$$= m\ell^2\left[\omega_p^2\eta^2 + \left(\omega_p^2 + \omega_0^2\right)\xi^2\right] \tag{4.20}$$

Thus, we see that the kinetic and potential energies, T and U are quadratic functions of the normal coordinates η and ξ and their first time derivatives, consistent with Proposition 3.

Examination of Eq. (4.17) provides further insight into the motion of the system. In the symmetric mode, $\omega_\eta = \omega_p$, we see that the characteristics of the spring have no effect on the motion and the two pendulums sway in unison with frequency ω_p as in Fig. 4.1. In the antisymmetric mode the parameters of both the pendulum and the spring determine the frequency of the motion [see Eq. (4.17)]. In this mode the spring is alternately compressed and stretched, as shown in Fig. 4.2, and the frequency of the motion depends upon the spring constant.

Fig. 4.2
Problem 1b–solution

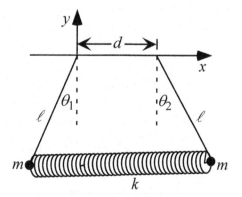

There is a more general way to obtain the normal coordinates than the method of inspection employed above. We know that the solutions will be sinusoidal so we try a solution of the form

$$\theta_i = A_i \sin \omega t \tag{4.21}$$

and attempt to solve for ω. Inserting Eq. (4.21) into Eqs. (4.10) and (4.12) we obtain

$$- A_1 \omega^2 + \omega_p^2 A_1 - \frac{\omega_o^2}{2}(A_2 - A_1) = 0$$

$$-A_2 \omega^2 + \omega_p^2 A_2 + \frac{\omega_o^2}{2}(A_2 - A_1) = 0 \tag{4.22}$$

Taking A_1 and A_2 as the independent variables we write two simultaneous equations in the more familiar form

$$\left(\omega_p^2 + \frac{\omega_o^2}{2} - \omega^2\right) A_1 - \frac{\omega_o^2}{2} A_2 = 0$$

$$-\frac{\omega_o^2}{2} A_1 + \left(\omega_p^2 + \frac{\omega_o^2}{2} - \omega^2\right) A_2 = 0 \tag{4.23}$$

These simultaneous *homogeneous* equations for A_1 and A_2 may be written in matrix form

$$\begin{pmatrix} \omega_p^2 + \frac{\omega_o^2}{2} - \omega^2 & -\frac{\omega_o^2}{2} \\ -\frac{\omega_o^2}{2} & \omega_p^2 + \frac{\omega_o^2}{2} - \omega^2 \end{pmatrix} \begin{pmatrix} A_1 \\ A_2 \end{pmatrix} = 0 \tag{4.24}$$

The only way Eq. (4.23) can have a non-trivial solution is if the determinant of the coefficients vanishes. This condition yields the secular equation, which for Eq. (4.24) is

$$\left(\omega_p^2 + \frac{\omega_o^2}{2} - \omega^2\right)^2 - \frac{\omega_o^4}{4} = 0 \tag{4.25}$$

or

$$\omega_p^4 + \omega_p^2 \omega_o^2 - 2\omega_p^2 \omega^2 - \omega_o^2 \omega^2 + \frac{\omega_o^4}{4} + \omega^4 - \frac{\omega_o^4}{4} = 0 \tag{4.26}$$

so that

$$\omega^4 - \left(2\omega_p^2 + \omega_o^2\right)\omega^2 + \left(\omega_p^4 + \omega_p^2 \omega_o^2\right) = 0 \tag{4.27}$$

which is quadratic in ω^2. Using the quadratic formula to solve for ω^2 we have

$$2\omega^2 = \left(2\omega_p^2 + \omega_o^2\right) \pm \sqrt{\left(2\omega_p^2 + \omega_o^2\right)^2 - 4\left(\omega_p^4 + \omega_p^2\omega_o^2\right)}$$
$$= \left(2\omega_p^2 + \omega_o^2\right) \pm \omega_o^2 \tag{4.28}$$

so that

$$\omega^2 = \left(\omega_p^2 + \frac{\omega_o^2}{2}\right) \pm \frac{\omega_o^2}{2} \tag{4.29}$$

These are the same two frequencies that we obtained using the less general method [see Eq. (4.17)]. Clearly $\omega_\eta = \omega_p$ corresponds to the minus sign while $\omega_\xi = \sqrt{\omega_p^2 + \omega_o^2}$ corresponds to the plus sign. To recover η and ξ we first substitute ω_η into the first of Eq. (4.23) to obtain a relation between A_1 and A_2 for $\omega = \omega_\eta$

$$\left(\omega_p^2 + \omega_o^2 - \omega_p^2\right) A_1 - \omega_o^2 A_2 = 0 \Longrightarrow A_1 = A_2 = A/2 \tag{4.30}$$

so that for $\omega = \omega_\eta$

$$\theta_1 = A \sin \omega_\eta t$$
$$\theta_2 = A \sin \omega_\eta t \tag{4.31}$$

Thus, $\theta_1 = \theta_2$ and we have recovered the symmetric solution. The time dependent motion $\eta(t)$ is given by

$$\eta(t) = A \sin \omega_\eta t \tag{4.32}$$

which is the solution to Eq. (4.13).

Substituting $\omega = \omega_\xi$ into the second of Equations

$$\left(\omega_p^2 + \omega_o^2 - \omega_p^2 - 2\omega_o^2\right) A_1 - \omega_o^2 A_2 = 0 \Longrightarrow A_1 = -A_2 = A/2 \tag{4.33}$$

so that for $\omega = \omega_\xi$

$$\theta_1 = A \sin \omega_\xi t$$
$$\theta_2 = -A \sin \omega_\xi t \tag{4.34}$$

and $\theta_1 = -\theta_2$ and we have recovered the antisymmetric solution. The time dependent motion $\xi(t)$ is given by

$$\xi(t) = A \sin \omega_\xi t \tag{4.35}$$

We see again that the sum and difference of θ_1 and θ_2 as given in Eq. (4.15) provide the sinusoidal response required of the solution to the uncoupled differential equations in Eq. (4.13). They are the "eigenvectors" corresponding to the eigenvalues ω_η and ω_ξ. They are, mathematically speaking, analogous to the unit vectors $\hat{\imath}$ and $\hat{\jmath}$ in Cartesian coordinates in the sense that any motion of the system may be considered to be a linear combination of η and ξ just as any two-dimensional vector may be considered to be a linear combination of $\hat{\imath}$ and $\hat{\jmath}$.

2. Find the frequencies and normal mode coordinates for the triple spring system shown in Fig. 4.3.

Fig. 4.3 Problem 2

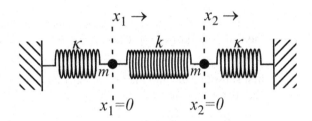

Solution

This problem is mathematically identical with Problem 1 of this chapter. For small pendulum displacements the coupled pendulums execute linear harmonic vibrations. We can immediately write the normal frequencies appropriate to this problem by simply substituting the frequency ω of an SHO with spring constant κ, that is $\omega = \sqrt{\kappa/m}$, for the frequency of the pendulum, $\omega = \sqrt{g/\ell}$ in the symmetric and anti-symmetric frequencies of vibration ω_s and ω_a, Eq. (4.17). Therefore

$$\omega_\eta = \sqrt{\frac{g}{\ell}} \rightarrow \omega_s = \sqrt{\frac{\kappa}{m}}; \quad \omega_\xi = \sqrt{\frac{g}{\ell} + 2\frac{k}{m}} \rightarrow \omega_a = \sqrt{\frac{\kappa}{m} + 2\frac{k}{m}} \qquad (4.36)$$

Now we justify the analogy and work out the details. We begin by writing Largrange's equations.

$$T = \frac{1}{2}m\dot{x}_1^2 + \frac{1}{2}m\dot{x}_2^2 \qquad (4.37)$$

and

$$U = \frac{1}{2}\kappa x_1^2 + \frac{1}{2}k(x_2 - x_1)^2 + \frac{1}{2}\kappa x_2^2 \qquad (4.38)$$

so

$$\mathcal{L} = \frac{1}{2}m\dot{x}_1^2 + \frac{1}{2}m\dot{x}_2^2 - \frac{1}{2}\kappa x_1^2 - \frac{1}{2}k(x_2 - x_1)^2 - \frac{1}{2}\kappa x_2^2 \tag{4.39}$$

Lagrange's equations are then

$$m\ddot{x}_1 + \kappa x_1 - k(x_2 - x_1) = 0$$
$$m\ddot{x}_2 + \kappa x_2 + k(x_2 - x_1) = 0 \tag{4.40}$$

As in Problem 1 of this chapter we could take the sum and difference of Eq. (4.40), but here we take the more general approach letting

$$x_i = A_i \sin \omega t \tag{4.41}$$

Since we are solving this problem using the general method we write the simultaneous equations for A_1 and A_2 in matrix form.

$$\begin{pmatrix} \dfrac{\kappa}{m} + \dfrac{k}{m} - \omega^2 & -\dfrac{k}{m} \\ -\dfrac{k}{m} & \dfrac{\kappa}{m} + \dfrac{k}{m} - \omega^2 \end{pmatrix} \begin{pmatrix} A_1 \\ A_2 \end{pmatrix} = 0 \tag{4.42}$$

Solving the secular equation we obtain a quadratic equation in ω^2 the solutions to which are

$$\omega_+^2 = \omega_s^2 = \frac{\kappa}{m} \quad \text{and} \quad \omega_-^2 = \omega_a^2 = \frac{\kappa}{m} + 2\frac{k}{m} \tag{4.43}$$

where the $\pm$ subscripts represent the solutions obtained using the $\pm$ signs in the discriminant of the quadratic formula. We see that these frequencies are identical with those obtained in Eq. (4.36) using the results of Problem 1. Not surprisingly, the normal coordinates are $\eta = x_1 + x_2$ and $\xi = x_1 - x_2$ as may be seen by adding and subtracting or by solving Eq. (4.42) for A_1 and A_2 using the two different values of ω.

3. Find the frequencies and normal mode coordinates for small displacements of the double pendulum system shown. Do not make the small displacement approximation until after the Lagrange equations of motion have been derived. Is it possible for the system to swing as a single pendulum?

Fig. 4.4 Problem 3

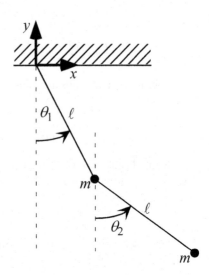

Check that the normal mode coordinates that you obtain are correct by verifying Proposition 3. For this part of the problem care should be taken in making the small angle approximation in the Lagrangian. The degree of approximation is different than that required in the equations of motion.

Solution

To find the Lagrangian it is perhaps easiest to first write the Cartesian coordinates in terms of the coordinates θ_1 and θ_2 and then find the kinetic and potential energies. Letting 1 and 2 denote the coordinates of the upper and lower masses, respectively, we have

$$x_1 = \ell \sin \theta_1; \quad y_1 = -\ell \cos \theta_1$$
$$\dot{x}_1 = \ell \dot{\theta}_1 \cos \theta_1; \quad \dot{y}_1 = \ell \dot{\theta}_1 \sin \theta_1 \tag{4.44}$$

For the lower mass we have

$$x_2 = \ell \sin \theta_1 + \ell \sin \theta_2; \quad y_2 = -\ell \cos \theta_1 - \ell \cos \theta_2$$
$$\dot{x}_2 = \ell \dot{\theta}_1 \cos \theta_1 + \ell \dot{\theta}_2 \cos \theta_2; \quad \dot{y}_2 = \ell \dot{\theta}_1 \sin \theta_1 + \ell \dot{\theta}_2 \sin \theta_2 \tag{4.45}$$

Then, the kinetic energy is

$$T = \frac{1}{2}m\left(\dot{x}_1^2 + \dot{y}_1^2\right) + \frac{1}{2}m\left(\dot{x}_2^2 + \dot{y}_2^2\right)$$

$$= \left[\frac{1}{2}m\ell^2\dot{\theta}_1^2\right] + \frac{1}{2}m\ell^2\dot{\theta}_1^2 + \frac{1}{2}m\ell^2\dot{\theta}_2^2$$

$$+ m\ell^2\dot{\theta}_1\dot{\theta}_2\left(\cos\theta_1\cos\theta_2 + \sin\theta_1\sin\theta_2\right) \tag{4.46}$$

This may be re-written with the help of the trigonometric identity, Eq. (E.2):

$$\cos(A - B) = \cos A\cos B + \sin A\sin B \tag{4.47}$$

as

$$T = m\ell^2\left[\dot{\theta}_1^2 + \frac{1}{2}\dot{\theta}_2^2 + \dot{\theta}_1\dot{\theta}_2\cos(\theta_1 - \theta_2)\right] \tag{4.48}$$

The potential energy is

$$U = mgy_1 + mgy_2$$

$$= mg\ell\left(-2\cos\theta_1 - \cos\theta_2\right) \tag{4.49}$$

The Lagrangian is therefore

$$\mathcal{L} = m\ell^2\left[\dot{\theta}_1^2 + \frac{1}{2}\dot{\theta}_2^2 + \dot{\theta}_1\dot{\theta}_2\cos(\theta_1 - \theta_2)\right] + mg\ell\left(2\cos\theta_1 + \cos\theta_2\right) \tag{4.50}$$

Lagrange's equations are

$$2\ell\ddot{\theta}_1 + \ell\ddot{\theta}_2\cos(\theta_1 - \theta_2) + \ell\dot{\theta}_2^2\sin(\theta_1 - \theta_2) + 2g\sin\theta_1 = 0$$

$$\ell\ddot{\theta}_2 + \ell\ddot{\theta}_1\cos(\theta_1 - \theta_2) - \ell\dot{\theta}_1^2\sin(\theta_1 - \theta_2) + g\sin\theta_2 = 0 \tag{4.51}$$

These are two coupled nonlinear differential equations the solutions to which are not only difficult, but probably not very enlightening. As directed in the statement of the problem we make the small angle approximation to simplify and retain only the leading terms. Retaining only the leading terms eliminates the nonlinearity of the equations because we ignore the $\dot{\theta}_1^2$ and $\dot{\theta}_2^2$ terms. Also, letting

$$\sin\theta_i \approx \theta_i \quad \text{and} \quad \cos(\theta_1 - \theta_2) \approx 1 \tag{4.52}$$

we have

$$2\ell\ddot{\theta}_1 + \ell\ddot{\theta}_2 + 2g\theta_1 = 0$$
$$\ell\ddot{\theta}_2 + \ell\ddot{\theta}_1 + g\theta_2 = 0 \tag{4.53}$$

Noting that the frequency of a single simple pendulum is given by

$$\omega_p^2 = \frac{g}{\ell} \tag{4.54}$$

we re-write Eq. (4.53) as

$$2\ddot{\theta}_1 + 2\omega_p^2\theta_1 + \ddot{\theta}_2 = 0$$
$$\ddot{\theta}_2 + \ddot{\theta}_1 + \omega_p^2\theta_2 = 0 \tag{4.55}$$

Letting

$$\theta_j = A_j \sin \omega t \tag{4.56}$$

we obtain

$$-2A_1\omega^2 + 2\omega_p^2 A_1 - A_2\omega^2 = 0$$
$$-A_2\omega^2 - A_1\omega^2 + \omega_p^2 A_2 = 0 \tag{4.57}$$

or

$$-2\left(\omega^2 - 2\omega_p^2\right) A_1 - \omega^2 A_2 = 0$$
$$-\omega^2 A_1 + \left(-\omega^2 + \omega_p^2\right) A_2 = 0 \tag{4.58}$$

The secular equation (see Problem 1 of this chapter) is therefore

$$\begin{vmatrix} -2\left(\omega^2 - \omega_p^2\right) & -\omega^2 \\ -\omega^2 & -\left(\omega^2 - \omega_p^2\right) \end{vmatrix} = 0 \tag{4.59}$$

or

$$2\left(\omega^2 - \omega_p^2\right)^2 - \omega^4 = 0 \tag{4.60}$$

which is a quadratic equation in ω^2. The two frequencies are

$$\omega_\pm^2 = \frac{1}{2}\left(4 \pm \sqrt{16 - 4\cdot 1\cdot 2}\right)\omega_p^2$$
$$= \left(2 \pm \sqrt{2}\right)\omega_p^2 \tag{4.61}$$

If we put the trial solution given in Eq. (4.56) into the first of Eq. (4.55), we obtain

$$\left(-2\omega^2 + 2\omega_p^2\right)\theta_1 - \omega^2\theta_2 = 0 \tag{4.62}$$

Substituting each of the values of the frequencies in Eq. (4.61) into Eq. (4.62) leads to a relationship between θ_1 and θ_2. We obtain

$$\theta_1 = -\frac{1}{\sqrt{2}}\theta_2 \quad \omega = \omega_+$$

$$\theta_1 = \frac{1}{\sqrt{2}}\theta_2 \quad \omega = \omega_- \tag{4.63}$$

We see that the second of the relations in Eq. (4.63), in which θ_1 and θ_2 are in the same direction, is represented by Fig. 4.4. This is commonly known as the symmetric mode so it is best to change the notation and let $\omega_- = \omega_s$. The first of the relations in Eq. (4.63) represents the antisymmetric mode so $\omega_+ = \omega_a$ and is shown in Fig. 4.5.

Fig. 4.5 Problem 3–solution

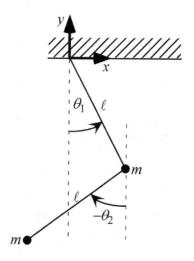

The frequency of the antisymmetric mode is considerably higher than that of the symmetric mode, roughly 2.4 times higher. For the system to swing as a single pendulum we must have $\theta_1 = \theta_2$, but Eq. (4.63) show that this is not the case. Therefore, the double pendulum cannot swing as a single pendulum.

To find the normal coordinates, we seek combinations of θ_1 and θ_2 that vary sinusoidally with each of the normal mode frequencies, ω_s and ω_a. To do this we write θ_1 and θ_2 as (arbitrary) linear combinations of the sine functions in Eq. (4.56). First we write

$$\theta_1 = B_1 \sin\omega_a t + B_2 \sin\omega_s t \tag{4.64}$$

where, from Eq. (4.61),

$$\omega_s^2 = \left(2 - \sqrt{2}\right)\omega_p^2$$

$$\omega_a^2 = \left(2 + \sqrt{2}\right)\omega_p^2 \tag{4.65}$$

Using the relations between θ_1 and θ_2 in Eq. (4.63) we write θ_2 as a linear combination of the two possible relations between it and θ_1.

$$\theta_2 = -\sqrt{2}B_1 \sin \omega_a t + \sqrt{2}B_2 \sin \omega_s t \tag{4.66}$$

Multiplying Eq. (4.64) by $\sqrt{2}$, adding and subtracting it to and from Eq. (4.66), we arrive at

$$\frac{1}{2}\left(\theta_1 + \frac{\theta_2}{\sqrt{2}}\right) = B_2 \sin \omega_s t$$

$$\frac{1}{2}\left(\theta_1 - \frac{\theta_2}{\sqrt{2}}\right) = B_1 \sin \omega_a t \tag{4.67}$$

Thus, the normal coordinates are

$$\eta = \frac{1}{2}\left(\theta_1 + \frac{\theta_2}{\sqrt{2}}\right) \quad \text{(antisymmetric)}$$

$$\xi = \frac{1}{2}\left(\theta_1 - \frac{\theta_2}{\sqrt{2}}\right) \quad \text{(symmetric)} \tag{4.68}$$

because it is these combinations of θ_1 and θ_2 that oscillate with one of the single normal mode frequencies, ω_s (η) or ω_a (ξ).

4. This problem is similar to Problem 6 of Chap. 2 except that the mass at the pivot point of the pendulum is attached to a spring. A plane pendulum of length ℓ and mass m is attached to another equal mass m that is attached to a spring of constant k that slides on a frictionless surface as shown in Fig. 4.6. Find the normal mode frequencies using x and θ as the generalized coordinates.

Fig. 4.6 Problem 4

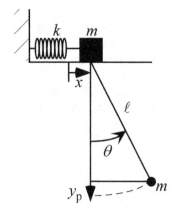

The coordinate x is the distance from the equilibrium position of the mass that is attached to the spring. Assume small oscillations after obtaining the equations of motion.

Solution

The kinetic energy and the potential energy (measured from $y = 0$) of the mass on the end of the spring are

$$T_s = \frac{1}{2}m\dot{x}^2 \text{ and } U_s = \frac{1}{2}kx^2 \tag{4.69}$$

The x and y positions of the pendulum bob are

$$x_p = x + \ell \sin\theta \text{ and } y_p = \ell \cos\theta \tag{4.70}$$

The kinetic energy of the pendulum bob is

$$T_p = \frac{1}{2}m\left(\dot{x}_p^2 + \dot{y}_p^2\right)$$
$$= \frac{1}{2}m\left[\dot{x}^2 + 2\dot{x}\dot{\theta}\ell\cos\theta + \left(\ell\dot{\theta}\right)^2\right] \tag{4.71}$$

and the potential energy of the pendulum bob is

$$U_p = -mgy_p$$
$$= -mg\ell\cos\theta \tag{4.72}$$

The Lagrangian is therefore

$$\mathcal{L} = T_s + T_p - U_s - U_p$$
$$= m\dot{x}^2 + m\dot{x}\dot{\theta}\ell\cos\theta + \frac{1}{2}m\left(\ell\dot{\theta}\right)^2$$
$$- \frac{1}{2}kx^2 + mg\ell\cos\theta \tag{4.73}$$

so the Lagrange equation for θ is

$$\frac{d}{dt}\left(\frac{\partial\mathcal{L}}{\partial\dot{\theta}}\right) - \frac{\partial\mathcal{L}}{\partial\theta} = 0$$
$$\ddot{x}\cos\theta - \dot{x}\dot{\theta}\sin\theta + \ell\ddot{\theta} + g\sin\theta = 0 \tag{4.74}$$

For small oscillations we may drop the term $\dot{x}\dot{\theta}\sin\theta$. Also, we take $\sin\theta \approx \theta$ and $\cos\theta \approx 1$ and arrive at

$$\ddot{x} + \ell\ddot{\theta} + g\theta = 0 \tag{4.75}$$

The Lagrange equation for x is obtained as follows.

$$2m\ddot{x} - m\ell\dot{\theta}^2\sin\theta + m\ell\ddot{\theta}\cos\theta + kx = 0 \tag{4.76}$$

For small oscillations we drop the $\dot{\theta}^2$ term and again take $\cos\theta \approx 1$ obtaining

$$2m\ddot{x} + m\ell\ddot{\theta} + kx = 0 \tag{4.77}$$

Now we assume a sinusoidal solution and, because x and θ do not have the same dimensions, we assume the solution to be of the form

$$x = A_1\sin\omega t \text{ and } \theta = \frac{A_2}{\ell}\sin\omega t \tag{4.78}$$

as was done in Problem 6 of Chap. 2. Inserting these trial solutions into the equations of motion, Eqs. (4.75) and (4.77) we obtain

$$-\omega^2 A_1 + \left(\frac{g}{\ell} - \omega^2\right) A_2 = 0$$
$$\left(k - 2m\omega^2\right) A_1 - m\omega^2 A_2 = 0 \tag{4.79}$$

which may be written in terms of the natural frequencies of the oscillator and the simple pendulum.

$$\omega_o^2 = \frac{k}{m} \text{ and } \omega_p^2 = \frac{g}{\ell} \tag{4.80}$$

Therefore

$$-\omega^2 A_1 + \left(\omega_p^2 - \omega^2\right) A_2 = 0$$
$$\left(\omega_o^2 - 2\omega^2\right) A_1 - \omega^2 A_2 = 0 \tag{4.81}$$

To solve for ω we solve the secular equation.

$$\omega^4 - \left(\omega_p^2 - \omega^2\right)\left(\omega_o^2 - 2\omega^2\right) = 0$$
$$\omega^4 - \left(2\omega_p^2 + \omega_o^2\right)\omega^2 + \omega_p^2\omega_o^2 = 0 \tag{4.82}$$

Solving the quadratic equation for ω^2 we obtain the normal mode frequencies which we denote by $\omega_\pm$.

$$\omega_\pm^2 = \frac{1}{2}\left[\left(2\omega_p^2 + \omega_o^2\right) \pm \sqrt{\left(2\omega_p^2 + \omega_o^2\right)^2 - 4\omega_p^2\omega_o^2}\right] \tag{4.83}$$

Now we can examine special cases. If, for example, we set $\omega_o^2 = 2\omega_p^2$, then the solution becomes

$$\omega^2 = \omega_o^2 \left(1 \pm \frac{1}{\sqrt{2}}\right) \tag{4.84}$$

Replacing ω^2 with $\omega_0^2(1 + 1/\sqrt{2})$ in the first of Eq. (4.81) and using $\omega_p^2 = \omega_o^2/2$, we get

$$\frac{A_1}{A_2} = -\frac{(1/2 + 1/\sqrt{2})}{(1 + 1/\sqrt{2})} \tag{4.85}$$

so the oscillator and pendulum move in opposite directions. Using $\omega_0(1-1/\sqrt{2})$ in the first of Eq. (4.81) yields the ratio

$$\frac{A_1}{A_2} = \frac{(1/\sqrt{2} - 1/2)}{(1 - 1/\sqrt{2})} \tag{4.86}$$

which is positive so the pendulum and spring oscillate in the same direction.

Suppose we let the spring constant $k \to 0$ in Eq. (4.80). The two roots for this case become

$$\omega^2 = 0, \ 2\omega_p^2 \tag{4.87}$$

Setting $k = 0$ allows the block to become freely sliding, and the $\omega^2 = 0$ root represents the constant velocity solution obtained in Problem 6 Chap. 2. The $\omega^2 = 2\omega_p^2$ root represents the pendulum motion obtained in that problem for the equal-mass case.

5. A very thin uniform rod of length ℓ and mass m hangs from two identical massless springs with spring constant k (Fig. 4.7). The coordinate y is the vertical displacement of the rod center from its equilibrium position. The displacements of the rod ends are $y_1 = -(\ell/2) \sin \theta$ and $y_2 = (\ell/2) \sin \theta$. Consider only vertical motion of the rod-spring system and assume small displacements.

Fig. 4.7 Problem 5

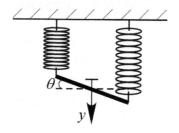

(a) Describe the system using coordinates (y_1, y_2) where each represents the coordinate of an end of the rod. Find the normal mode coordinates and frequencies.

(b) Use the (y, θ) coordinates to find the normal mode coordinates and frequencies.

Solution

(a) Using Newton's second law the force is given by

$$-k(y_1 + y_2) - mg = \frac{m}{2}(\ddot{y}_1 + \ddot{y}_2) \tag{4.88}$$

The torque about the center of the rod, $\tau = I\alpha$ (α=angular acceleration, I =moment of inertia), is

$$\left(\frac{I}{\ell}\right)(\ddot{y}_2 - \ddot{y}_1) = -\left(\frac{k\ell}{2}\right)(y_2 - y_1) \tag{4.89}$$

where we have used

$$\alpha = \frac{1}{\ell}(\ddot{y}_2 - \ddot{y}_1) \tag{4.90}$$

consistent with the small-displacement assumption. The moment of inertia of the thin rod about its center of mass is

$$I = \frac{1}{12}m\ell^2 \tag{4.91}$$

Equations (4.88) and (4.89) are easily separable if we define new coordinates

$$\eta = y_1 + y_2 + \frac{mg}{k}$$

$$\xi = (y_2 - y_1) \tag{4.92}$$

In the new coordinates Eqs. (4.88) and (4.89) become

$$\ddot{\eta} + \left(\frac{2k}{m}\right)\eta = 0 \tag{4.93}$$

and

$$\ddot{\xi} + \left(\frac{6k}{m}\right)\xi = 0 \tag{4.94}$$

$$\omega_\eta = \sqrt{\frac{2k}{m}} \tag{4.95}$$

Equations (4.93) and (4.94) describe harmonic motion in η with frequencies

$$\omega_\eta = \sqrt{\frac{2k}{m}} \text{ and } \omega_\xi = \sqrt{\frac{6k}{m}} \tag{4.96}$$

respectively.

The coordinates η and ξ are therefore the normal coordinates in the (y_1, y_2) system. (The constant term mg/k in the equation of transformation for η, Eq. (4.92), has no effect on the frequency.)

(b) The kinetic energy for the system in the (y, θ) system, after substituting for I, is

$$
\begin{aligned}
T &= \frac{1}{2}I\dot{\theta}^2 + \frac{1}{2}m\dot{y}^2 \\
&= \frac{1}{24}m\ell^2\dot{\theta}^2 + \frac{1}{2}m\dot{y}^2
\end{aligned}
\tag{4.97}
$$

The potential energy is

$$
\begin{aligned}
U &= -mgy + \frac{1}{2}k\left(y + \frac{\ell}{2}\sin\theta\right)^2 + \frac{1}{2}k\left(y - \frac{\ell}{2}\sin\theta\right)^2 \\
&\approx -mgy + ky^2 + k\frac{\ell^2}{4}\theta^2
\end{aligned}
\tag{4.98}
$$

We note that T contains only the squares of the time derivatives of the coordinates and that a simple translation of the y coordinate would make U dependent upon only the squares of both coordinates. Thus, from Proposition 3 we know that θ is a normal mode coordinate and that a simple translation of y will produce the other normal coordinate.

The Lagrangian is

$$
\begin{aligned}
\mathcal{L} &= T - U \\
&= \frac{1}{24}m\ell^2\dot{\theta}^2 + \frac{1}{2}m\dot{y}^2 + mgy - ky^2 - k\frac{\ell^2}{4}\theta^2
\end{aligned}
\tag{4.99}
$$

and the Lagrange equations of motion are

$$
\ddot{y} + 2\frac{k}{m}y - g = 0
$$

$$
\ddot{\theta} + 6\frac{k}{m}\theta = 0 \tag{4.100}
$$

Equation (4.100) are seen to represent harmonic motion in y and θ with frequencies

$$\omega_y = \sqrt{6\frac{k}{m}} \text{ and } \omega_\theta = \sqrt{6\frac{k}{m}} \qquad (4.101)$$

As in part (a) the constant term in the y-equation does not affect the calculated frequency. The frequencies are identical to those obtained in part (a) as they must be. This result is not a surprise, because with the small-angle approximation y and θ are proportional to η and ξ, respectively, so they are essentially the same coordinates.

6. Three particles are arranged along a line as shown in Fig. 4.8. The two end particles are identical, each with mass M, and the middle particle has mass m. The particles are connected by identical springs each having a spring constant k, and they are constrained to move along the x-axis so that there is no bending motion. This system can be thought of as representing the *linear* vibrations a linear triatomic molecule such as CO_2. The coordinates x are displacement coordinates, and $x_1 = x_2 = x_3 = 0$ when the system is at rest.

Fig. 4.8 Problem 6

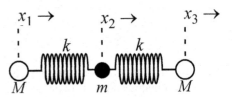

(a) Construct the Lagrangian in terms of the x_i and find the Lagrangian equations of motion.
(b) Assume sinusoidal motion so that $x_i = A_i \sin \omega t$ where A_i is the amplitude of the oscillations in the x_i coordinate. Show that substituting this expression for x_i into each of the three Lagrange equations leads to three simultaneous equations for ω^2 and solve for each value of ω^2. These are the normal mode frequencies.
(c) Find the relationships between the x_i that describe the normal mode motion and discuss this motion.

Solution

(a) The kinetic energy for each displacement coordinate is

$$T = \frac{1}{2}M\dot{x}_1^2 + \frac{1}{2}m\dot{x}_2^2 + \frac{1}{2}M\dot{x}_3^2 \qquad (4.102)$$

and the potential energy is

$$U = \frac{1}{2}k\left(x_2 - x_1\right)^2 + \frac{1}{2}k\left(x_3 - x_2\right)^2 \tag{4.103}$$

so the Lagrangian is

$$\begin{aligned}\mathcal{L} &= T - U \\ &= \frac{1}{2}M\dot{x}_1^2 + \frac{1}{2}m\dot{x}_2^2 + \frac{1}{2}M\dot{x}_3^2 - \frac{1}{2}k\left(x_2 - x_1\right)^2 - \frac{1}{2}k\left(x_3 - x_2\right)^2 \end{aligned} \tag{4.104}$$

The Lagrangian equations of motion are given by

$$\frac{d}{dt}\frac{\partial \mathcal{L}}{\partial \dot{x}_i} - \frac{\partial \mathcal{L}}{\partial x_i} = 0 \tag{4.105}$$

which yield

$$\begin{aligned} M\ddot{x}_1 - k\left(x_2 - x_1\right) &= 0 \\ m\ddot{x}_2 + k\left(x_2 - x_1\right) - k\left(x_3 - x_2\right) &= 0 \\ M\ddot{x}_3 + k\left(x_3 - x_2\right) &= 0 \end{aligned} \tag{4.106}$$

(b) Substituting

$$x_i = A_i \sin \omega t \tag{4.107}$$

into the Lagrange equations of motion, Eq. (4.106), we obtain

$$\begin{aligned} \left(\omega^2 - \frac{k}{M}\right)A_1 + \frac{k}{M}A_2 &= 0 \\ \frac{k}{m}A_1 + \left(\omega^2 - 2\frac{k}{m}\right)A_2 + \frac{k}{m}A_3 &= 0 \\ \frac{k}{M}A_2 + \left(\omega^2 - \frac{k}{M}\right)A_3 &= 0 \end{aligned} \tag{4.108}$$

The only way these homogeneous equations can have a non-trivial solution is if the determinant of the coefficients vanishes. The secular equation is

$$\begin{vmatrix} \left(\omega^2 - \frac{k}{M}\right) & \frac{k}{M} & 0 \\ \frac{k}{m} & \left(\omega^2 - 2\frac{k}{m}\right) & \frac{k}{m} \\ 0 & \frac{k}{M} & \left(\omega^2 - \frac{k}{M}\right) \end{vmatrix} = 0 \tag{4.109}$$

or

$$\left(\omega^2 - \frac{k}{M}\right)\left[\left(\omega^2 - 2\frac{k}{m}\right)\left(\omega^2 - \frac{k}{M}\right) - \frac{k^2}{mM}\right]$$

$$- \frac{k}{M}\left[\frac{k}{m}\left(\omega^2 - \frac{k}{M}\right)\right] = 0 \qquad (4.110)$$

which reduces to

$$0 = \left(\omega^2 - \frac{k}{M}\right)\left(\omega^4 - \frac{k}{M}\omega^2 - 2\frac{k}{m}\omega^2 + 2\frac{k^2}{mM} - \frac{k^2}{mM} - \frac{k^2}{mM}\right)$$

$$0 = \left(\omega^2 - \frac{k}{M}\right)\left(\omega^4 - \frac{k}{M}\omega^2 - 2\frac{k}{m}\omega^2\right)$$

$$0 = \omega^2\left(\omega^2 - \frac{k}{M}\right)\left[\omega^2 - \left(\frac{k}{M} + 2\frac{k}{m}\right)\right] \qquad (4.111)$$

There are three different values of ω^2 that are solutions to this equation. They are

$$\omega_1^2 = 0$$

$$\omega_2^2 = \frac{k}{M}$$

$$\omega_3^2 = \left(\frac{m + 2M}{mM}\right)k \qquad (4.112)$$

(c) To describe the motion of each of the three normal modes we must find the relations among the mode amplitudes for each frequency listed in Eq. (4.112). We must therefore replace ω^2 in Eq. (4.108) with ω_1, ω_2 and ω_3 and examine the motion for each of these normal modes.

- $\omega^2 = \omega_1^2 = 0$. From the first of Eq. (4.108) we see that $A_1 = A_2$. From the third of these equations we obtain $A_3 = A_2$. There is therefore no relative motion of the masses. In this normal mode each atom in the molecule is simply shifted along the axis by the same amount, so that the entire molecule moves (translates) without compressing or stretching the springs. This is referred to as "free" motion.

- $\omega^2 = \omega_2^2 = k/M$. From the first (or third) of Eq. (4.108) we find that $A_2 = 0$. Inserting $A_2 = 0$ into the second equation we obtain $A_1 = -A_3$. For this mode of vibration the center mass m is stationary while the outside masses vibrate in opposite directions, each with the same amplitude. This is referred to as the "symmetric stretch" mode.

- $\omega^2 = \omega_3^2 = [(m + 2M)/mM]k$. Inserting ω_3^2 into the first of Eq. (4.108) we find that

$$\left(\frac{k}{M}+2\frac{k}{m}-\frac{k}{M}\right)A_1+\frac{k}{M}A_2=0 \Rightarrow A_1=-\frac{1}{2}\frac{m}{M}A_2 \qquad (4.113)$$

Now we eliminate A_2 in the second of Eq. (4.108)

$$\frac{k}{m}A_1-\left(\frac{(m+2M)}{mM}k-2\frac{k}{m}\right)2\frac{M}{m}A_1+\frac{k}{m}A_3=0$$

$$\frac{k}{m}A_1-\left(\frac{k}{M}+2\frac{k}{m}-2\frac{k}{m}\right)2\frac{M}{m}A_1+\frac{k}{m}A_3=0$$

$$\frac{k}{m}A_1-2\frac{k}{m}A_1+\frac{k}{m}A_3=0$$

$$A_1=A_3 \qquad (4.114)$$

In this mode we see that the two outer masses vibrate in the same direction with the same amplitude while the center mass vibrates in the opposite direction with a different amplitude. This is called the "asymmetric stretch" mode.

References

1. Burkhardt CE, Leventhal JJ (2006) Topics in atomic physics. Springer, New York
2. Goldstein H (1980) Classical mechanics, 2nd edn. Addison-Wesley, Reading.
3. Thornton ST, Marion JB (2004) Classical dynamics of particles and systems, 5th edn. Harcourt Brace Jovanovich, New York

Part II
Quantum Mechanics

Chapter 5
Introductory Concepts

The important relationships and notation are in the appendices.

Problems

1. In creating the Bohr theory of the atom Niels Bohr assumed that the electron travels around the proton in a series of circular orbits. He set forth two postulates.

   I. *An atom exists in a series of energy states such that the accelerating electron does not radiate energy when in these states. These states are called stationary states.*

   II. *Radiation is absorbed or emitted during a transition between two stationary states. The frequency of the absorbed or emitted radiation is given by Planck's theory.*

   In many books a third postulate is added that has the form

   III. *The angular momentum of the orbits of the allowed states is quantized in units of $\hbar$.*

   In fact, Bohr made no such postulate as the last one. Why would he? Did he have an epiphany? Ask yourself why would he choose $\hbar$ and not, for example h or $2\pi h$ or some other constant. Using Postulate III in a derivation of the Bohr model of the atom minimizes Bohr's contribution. Bohr did have a third postulate, known today as the correspondence principle, but he did not state it as a postulate. The correspondence principle states that when quantum numbers become large the system behavior tends toward that of comparable classical systems. Bohr used his correspondence principle together with the observations of the hydrogen atom (H-atom) spectrum made by the spectroscopist Johann Balmer to deduce the correct energy levels. Using "Postulate III" above, the correct energy levels are obtained without appealing to the correspondence

© Springer International Publishing AG 2017
J.D. Kelley, J.J. Leventhal, *Problems in Classical
and Quantum Mechanics*, DOI 10.1007/978-3-319-46664-4_5

principle. This principle is then presented as a consequence of the Bohr model of the atom, but this was not Bohr's line of reasoning.

In addition to the postulates, Bohr drew heavily on previous work by others, particularly Planck, Einstein, and Balmer. Planck and Einstein showed that there is a relation between the frequency of emitted light and the energy of the photon emitted.

$$E = h\nu = \hbar\omega \tag{5.1}$$

where ν, the frequency in hertz, is equal to $\omega/2\pi$. Balmer had shown that the wavelength of the lines in the visible spectrum of hydrogen were given by an empirical formula

$$\frac{1}{\lambda_n} = R_H \left(\frac{1}{2^2} - \frac{1}{n^2} \right) \tag{5.2}$$

where n is an integer and R_H is a constant known as the Rydberg constant because of earlier work on spectroscopy by Johannes Rydberg. In fact, he generalized Eq. (5.2) by replacing the 2 in the denominator with another integer.

Now, the problem: From the Bohr model [1], using the correspondence principle, the radius of the nth Bohr orbit r_n is given by

$$r_n = n^2 a_0 \tag{5.3}$$

where a_0 is the radius of the first Bohr orbit

$$a_0 = \frac{(4\pi\epsilon_0)\,\hbar^2}{m_e e^2} \Rightarrow \frac{e^2}{4\pi\epsilon_0} = \frac{\hbar^2}{m_e a_0}$$

$$= \frac{\hbar}{m_e c \alpha} \tag{5.4}$$

The constant α in Eq. (5.4) is called the fine structure constant and is given by

$$\alpha = \left[\frac{e^2}{(4\pi\epsilon_0)\,\hbar c} \right]$$

$$\simeq \frac{1}{137} \tag{5.5}$$

This constant α is a fundamental and unitless quantity.

Beginning with Eq. (5.3) derive Postulate III given above. Show also that the speed of the electron in the nth Bohr orbit is $v_n = \alpha c/n \simeq (1/137)\,c/n$.

Solution

In the construction of the Bohr model of the atom the centripetal force is equated to the Coulomb force giving

$$\frac{m_e v^2}{r} = \left(\frac{e^2}{4\pi\epsilon_0}\right)\frac{1}{r^2} \tag{5.6}$$

which is a relationship between the orbital radii r and the orbital velocity v. Assuming quantized orbits (Postulate I above), attaching subscripts to the parameters of these orbits and solving Eq. (5.6) for v_n we obtain

$$v_n = \sqrt{\frac{1}{m_e}\left(\frac{e^2}{4\pi\epsilon_0}\right)\frac{1}{n^2 a_0}} \tag{5.7}$$

Using Eq. (5.4), Eq. (5.7) may be rewritten as

$$v_n = \sqrt{\frac{1}{m_e}\left(\frac{\hbar^2}{m_e a_0}\right)\frac{1}{n^2 a_0}}$$

$$= \frac{1}{m_e}\frac{\hbar}{n a_0} \tag{5.8}$$

The angular momentum in the nth orbit is

$$L_n = m_e v_n r_n = m_e \left(\frac{1}{m_e}\frac{\hbar}{n a_0}\right)(n^2 a_0)$$

$$= n\hbar \tag{5.9}$$

so Eq. (5.9) shows that the angular momentum is quantized in units of $\hbar$. This result was obtained using Postulates I and II and the correspondence principle.

To find v in terms of the speed of light and the fine structure constant we solve Eq. (5.9) for v_n and use Eqs. (5.4) and (5.5) to obtain

$$v_n = \frac{n\hbar}{m_e r_n}$$

$$= \frac{n\hbar}{m_e}\frac{1}{n^2 a_0}$$

$$= \left(\frac{\hbar}{n m_e}\right)\left(\frac{m_e c\alpha}{\hbar}\right)$$

$$= \frac{\alpha}{n}c \simeq \frac{1}{137}\left(\frac{c}{n}\right) \tag{5.10}$$

This is an important relation because it shows that the orbital speed v_n of the electron is much smaller than c. Moreover, the electronic speed decreases as n increases, thus justifying a nonrelativistic treatment of the quantum mechanical H-atom (ignoring spin which is an inherently relativistic phenomenon). One way of describing the fine structure constant is that it is the ratio of the electron velocity in the first Bohr orbit to the speed of light.

Some of the concepts that Bohr employed to construct his model are incorrect. For example, the idea of a well-defined orbit for the electron is incorrect. Additionally, the result derived in this problem that the angular momentum is equal to $n\hbar$, Eq. (5.9), is not entirely correct because $n = 1, 2, 3 \ldots$ and we now know that the angular momentum of an electron in an atom can be zero. Nonetheless, Bohr's application of empirical facts and his assertion of the correspondence principle were the stepping stones to modern quantum physics. Although physicists are well aware of the wave nature of matter and the consequent position and momentum distributions for an electron bound to a nucleus, most occasionally find themselves thinking in terms of the Bohr model if they are not careful.

An important property of the Bohr model is that it gives the correct order of magnitude for atomic parameters. The Bohr model, in most cases, gives the correct dependence of the atomic parameters on the principal quantum number. Perhaps the most dramatic consequence of the Bohr model is that it gives the correct quantized energies of the H-atom.

2. Find the expression for the radius of the first Bohr orbit ignoring the electro-magnetic interaction between the electron and the proton and assuming that the only force that binds them is the gravitational force.

Solution

The Coulomb attractive force between the proton and the electron in the Bohr atom is

$$F\left(r = a_0\right) = -\left(\frac{e^2}{4\pi\epsilon_0}\right)\frac{1}{a_0^2} \tag{5.11}$$

Denoting the radius of the first Bohr orbit that is due to only the gravitational force by a_0^G the gravitational force between them at a_0^G is

$$F\left(r = a_0^G\right) = -\left(G m_e m_p\right)\frac{1}{\left(a_0^G\right)^2} \tag{5.12}$$

where G is the gravitational constant; m_e and m_p are the masses of the electron and proton, respectively. Comparing the two expressions for the forces in

Eqs. (5.11) and (5.12) we see that the quantities in parentheses are analogous. The expression for a_0, the "real" Bohr radius [see Eq. (5.4)] is

$$a_0 = \left(\frac{4\pi\epsilon_0}{e^2}\right)\frac{\hbar^2}{m_e}$$

$$\approx 5.3 \times 10^{-11}\,\text{m} \tag{5.13}$$

so we simply replace $4\pi\epsilon_0/e^2$ by $1/Gm_e m_p$ to obtain a_0^G. This gives

$$a_0^G = \frac{\hbar^2}{Gm_p m_e^2} \approx 1.1 \times 10^{29}\,\text{m} \tag{5.14}$$

A currently accepted value for the diameter of the observable universe is the order of 10^{27} m, roughly one hundred times smaller than a_0^G which suggests that gravity plays little role in determining interatomic forces.

3. Show that the de Broglie wavelength of an electron in the nth Bohr orbit is constant and is $1/n$ times the circumference of the nth Bohr orbit. This shows that de Broglie waves "fit" into the Bohr orbits thus justifying the designation of the Bohr states as "stationary states."

Solution

Equating the centripetal force to the Coulomb force that binds the electron to the proton in the nth Bohr orbit we have

$$\frac{m_e v_n^2}{r_n} = \left(\frac{e^2}{4\pi\epsilon_0}\right)\frac{1}{r_n^2} \Rightarrow p_n^2 = m_e\left(\frac{e^2}{4\pi\epsilon_0}\right)\frac{1}{r_n} \tag{5.15}$$

Bohr's quantized radii are given by [1]

$$r_n = n^2 a_0 \text{ where } a_0 = \left(\frac{4\pi\epsilon_0}{e^2}\right)\frac{\hbar^2}{m_e} \Rightarrow \left(\frac{e^2}{4\pi\epsilon_0}\right) = \frac{\hbar^2}{a_0 m_e} \tag{5.16}$$

so the linear momentum in the nth Bohr orbit is

$$p_n^2 = m_e\left(\frac{e^2}{4\pi\epsilon_0}\right)\frac{1}{n^2 a_0}$$

$$= m_e\left(\frac{\hbar^2}{a_0 m_e}\right)\frac{1}{n^2 a_0}$$

$$= \frac{h^2}{4\pi^2 n^2 a_0^2} \Longrightarrow p_n = \frac{h}{2\pi n a_0} \tag{5.17}$$

The de Broglie wavelength in the nth orbit is [see Eq. (L.6)]

$$\lambda_n = \frac{h}{p_n} = 2\pi n a_0$$

$$= \frac{1}{n}\left[2\pi\left(n^2 a_0\right)\right] \qquad (5.18)$$

or

$$n\lambda_n = (2\pi r_n) = \text{circumference of the } n\text{th Bohr orbit} \qquad (5.19)$$

It is interesting to note that the de Broglie wavelength for a Bohr orbit (a *stationary state*) is consistent with the correspondence principle. The angular momentum in the nth Bohr orbit is

$$L_n = p_n r_n \qquad (5.20)$$

so that

$$L_n = \left(\frac{h}{2\pi n a_0}\right)\left(n^2 a_0\right) \equiv n\hbar \qquad (5.21)$$

4. Use the uncertainty principle to estimate the ground state energy of the H-atom.

Solution

The uncertainty principle is

$$\Delta x \Delta p \geq \hbar/2 \qquad (5.22)$$

Now, one often sees these "back of the envelope" calculations with the uncertainty principle performed in such a way that the answer conforms with the exact (known) answer. We take the approach that we do not know the answer until the end and then compare the computed and exact results. To that end we take the uncertainty in x for the ground state of the H-atom to be some fraction β of the orbital radius r of the ground state orbit. Thus, $\Delta x \approx \beta r$. Because we know nothing about the momentum we take the uncertainty in the momentum Δp to be the momentum itself so $\Delta p \approx p$. Inserting these estimates into the uncertainty principle we have

$$\beta r p \approx \hbar/2 \qquad (5.23)$$

Using Eq. (5.23) total energy is

$$E = \frac{p^2}{2m_e} - \frac{e^2}{4\pi\epsilon_0 r}$$

$$= \frac{\hbar^2}{8\beta^2 m_e r^2} - \frac{e^2}{4\pi\epsilon_0 r} \qquad (5.24)$$

where m_e is the rest mass of the electron.

To estimate the ground state energy we must minimize E with respect to r to obtain the corresponding value, r_{min}.

$$\frac{dE}{dr} = -\frac{\hbar^2}{4\beta^2 m_e r_{min}^3} + \frac{e^2}{4\pi\epsilon_0 r_{min}^2} = 0 \qquad (5.25)$$

Multiplying through by r_{min}^3 we have

$$\frac{\hbar^2}{4\beta^2 m_e} = \frac{e^2 r_{min}}{4\pi\epsilon_0} \implies r_{min} = \left(\frac{1}{4\beta^2}\right) \cdot \frac{4\pi\epsilon_0 \hbar^2}{m_e e^2} \qquad (5.26)$$

Now, the Bohr radius of the atom, which is the radius of the first Bohr orbit, is

$$a_0 = \frac{(4\pi\epsilon_0)\,\hbar^2}{m_e e^2} \qquad (5.27)$$

If we had chosen $\beta = \frac{1}{2}$, then we would have obtained the exact value of a_0 that leads to the exact energy of the ground state of the H-atom. The exact value of β and the uncertainties Δx and Δp that are chosen are not important because this is an *estimate*. For example, if the reasonable values

$$\beta = 1 \; ; \; \Delta x \approx r \; ; \; \Delta p \approx p \; ; \; \Delta x \Delta p = \hbar \qquad (5.28)$$

are chosen, then the exact answer ($r_{min} = a_0$) is obtained, but this is a designed coincidence. What is important is that the order of magnitude is correct.

In terms of the fine structure constant α [see Eq. (C.1)] the energy of the H-atom ground state is given by

$$E = -\frac{1}{2}\alpha^2 m_e c^2 \qquad (5.29)$$

where c is the speed of light and the fine structure constant is

$$\alpha = \left[\frac{e^2}{(4\pi\epsilon_0)\,\hbar c}\right]$$

$$\simeq \frac{1}{137} \qquad (5.30)$$

It is easy to calculate the ground state ionization energy of the H-atom using Eq. (5.29). Given the electron's rest mass (which should be known by all physicists) is $m_e c^2 = 0.51\,\text{MeV}$ and $\alpha = 1/137$, one obtains $-E = 13.6\,\text{eV}$.

5. Equate the electron's rest energy to the electrostatic energy of the electron regarded as a charge e that is uniformly distributed throughout a sphere of radius r_e. Use this relation to estimate r_e. This radius is known as the *classical radius of the electron*. Put your answer in terms of the fine structure constant α and the Bohr radius a_0 (see Problem 4 of this chapter). The electrostatic energy of a uniform distribution of charge e throughout the sphere is given by [3]

$$E_{\text{electrostatic}} = \left(\frac{3}{5}\right)\frac{e^2}{4\pi\epsilon_0 r_e} \tag{5.31}$$

but, because this approach gives a very rough approximation to r_e, the $3/5$ is usually omitted. Comment on the magnitude of the answer obtained.

Solution

Following the directions given in the statement of the problem we have

$$m_e c^2 = \frac{e^2}{4\pi\epsilon_0 r_e} \tag{5.32}$$

so that

$$r_e = \frac{e^2}{4\pi\epsilon_0 m_e c^2} \tag{5.33}$$

Using Eqs. (5.27) and (5.30) as guides we write

$$r_e = \frac{e^2}{4\pi\epsilon_0 m_e c^2}\left\{\left[\frac{(4\pi\epsilon_0)\hbar^2}{m_e e^2}\right]\frac{1}{(4\pi\epsilon_0)}\cdot\frac{m_e e^2}{\hbar^2}\right\} \tag{5.34}$$

where the expression in curly brackets is unity. Regrouping we have

$$r_e = \left(\frac{e^2}{4\pi\epsilon_0\hbar c^2}\right)^2\left[\frac{(4\pi\epsilon_0)\hbar^2}{m_e e^2}\right]$$

$$= \alpha^2 a_0 \approx \left(\frac{1}{137}\right)^2 a_0 \tag{5.35}$$

Thus, from this point of view, the electron's "size" is the order of 10^{-5} the orbital radius of the electron.

6. The magnetic moment μ of a uniformly charged sphere of radius R is spinning with an angular frequency ω and carrying a total charge Q is given by

$$\mu = \frac{Q\omega R^2}{5} \tag{5.36}$$

Use this result to find the velocity v_s of a point on the surface of a uniform spherical charge distribution, the radius of which is the classical radius of the electron r_e as given by Eq. (5.35). Show that this point on the surface would be moving at several hundred times the speed of light if the magnetic dipole moment due to the spinning electron is taken to be the measured value, one Bohr magneton μ_B.

$$\mu_B = \frac{e\hbar}{2m_e} \tag{5.37}$$

Solution

In this problem we may insert into Eq. (5.36) $R \rightarrow r_e$, $Q \rightarrow e$, $\omega \rightarrow v_s/r_e$ where v_s is the speed of a point on the surface of the assumed spherical electron. Equating this magnetic moment to μ_B in Eq. (5.37) we have

$$\mu_B = \frac{ev_s r_e}{5} = \frac{e\hbar}{2m_e} \tag{5.38}$$

so that

$$v_s = \left(\frac{5}{r_e}\right)\left(\frac{\hbar}{2m_e}\right) \tag{5.39}$$

At this point we must put in values for the quantities in Eq. (5.39) to determine the order of magnitude of v_s relative to c the speed of light. It is much easier if we use atomic units (a.u.) (see Table C.1 of Appendix C) for these quantities because all quantities in Eq. (5.39) except r_e are unity in that system. From Eq. (5.35) we can write r_e in terms of a_0 (which is unity in a.u.), i.e. $r_e = \alpha^2 a_0$ where α is the fine structure constant [see Eq. (C.1)]. We have

$$v_s = \left(\frac{5}{2\alpha^2}\right) = \frac{5}{2} \cdot 137^2 \approx 340c \tag{5.40}$$

because the speed of light in a.u. is 137. Therefore, in this model a point on the surface of the sphere would be moving about 340 times the speed of light. The crux of this exercise is to show that the term "spin," which was invented to satisfy the physicists concept of reality, is not to be taken literally. Remember that we are working with nonrelativistic quantum mechanics in which particles are *point* particles. The idea that a point is spinning is, of course, nonsense. Nonetheless, most physicists think of a spinning electron even if they will not admit it.

7. Use the uncertainty principle to estimate the ground state energy of an SHO. Use the information in Appendix O that the ground state position and momentum eigenfunctions are Gaussians. Moreover, Gaussians produce the minimum uncertainty wave packet so that

$$\Delta x \Delta p = \hbar/2 \tag{5.41}$$

Solution

We take the uncertainties to be the quantities themselves, a reasonable assumption (because we know we will obtain the correct answer). That is

$$\Delta x = x \quad \text{and} \quad \Delta p = p \tag{5.42}$$

so

$$p = \frac{\hbar}{2x} \tag{5.43}$$

The total energy is then

$$\begin{aligned} E &= \frac{p^2}{2m} + \frac{1}{2}m\omega^2 x^2 \\ &= \frac{\hbar^2}{8mx^2} + \frac{1}{2}m\omega^2 x^2 \end{aligned} \tag{5.44}$$

Minimizing E with respect to x to obtain x_{min} we have

$$\left.\frac{dE}{dx}\right|_{x=x_{min}} = -\frac{\hbar^2}{4mx_{min}^3} + m\omega^2 x_{min} = 0$$

$$\implies x_{min} = \sqrt{\frac{\hbar}{2m\omega}} \tag{5.45}$$

Substituting this value for x_{min} back into the expression for the total energy, Eq. (5.44), we obtain the estimated ground state energy E_0

$$\begin{aligned} E_0 &= \frac{\hbar^2}{8mx_{min}^2} + \frac{1}{2}m\omega^2 x_{min}^2 \\ &= \frac{\hbar^2}{8m}\left(\frac{2m\omega}{\hbar}\right) + \frac{1}{2}m\omega^2\left(\frac{\hbar}{2m\omega}\right) \end{aligned}$$

$$= \frac{1}{4}\hbar\omega + \frac{1}{4}\hbar\omega$$

$$= \frac{1}{2}\hbar\omega \tag{5.46}$$

That we obtained the exact value in this case is not surprising in view of the fact that we knew the correct answer a priori. The most suspicious move that we employed was that of taking the uncertainties to be equal to the quantities themselves.

8. The normalized wave function at $t = 0$ for a particle of mass m is given by

$$\Psi(x, 0) = \left[\frac{\sqrt{2}}{4}\psi_1(x) + \frac{1}{2}\psi_2(x) + \frac{\sqrt{10}}{4}\psi_3(x) \right] \tag{5.47}$$

where the $\psi_n(x)$ are the orthonormal eigenfunctions of the Hamiltonian $\hat{H}$. Each $\psi_n(x)$ has energy eigenvalue $E_n - (1/n)E_0$ where E_0 is a positive number.

(a) If an energy measurement is made, what are the possible results of the measurement?
(b) What is the probability of measuring each of these energies?
(c) What is the expectation value of the energy?
(d) Suppose there is another physical quantity that may be measured, a quantity that is represented mathematically by Q. Assume that the $\psi_n(x)$ are also eigenfunctions of the operator that represents this quantity, $\hat{Q}$, with eigenvalues nQ_0. That is

$$\hat{Q}\psi_n(x) = nQ_0\psi_n(x) \tag{5.48}$$

If the energy is measured first and found to be $E_3 = -(1/3)E_0$ and then a measurement of Q is made, what will be the value of Q that is measured?

Solution

(a) The only constituents of $\Psi(x, 0)$ are those representing $n = 1, 2$, and 3. Therefore the only values of the energy that can be measured are

$$-\frac{E_0}{1}; -\frac{E_0}{2}; -\frac{E_0}{3} \tag{5.49}$$

(b) Because $\Psi(x, 0)$ is normalized, the sum of the squared coefficients of the $\psi_n(x)$ is unity and the probability of measuring each of the energies is (in order)

$$\frac{2}{16}; \frac{4}{16}; \frac{10}{16} \tag{5.50}$$

(c) Remembering that the $\psi_n(x)$ are orthonormal we have

$$\langle E \rangle = \int_{-\infty}^{\infty} \Psi^*(x,0)\, \hat{H} \Psi(x,0)\, dx$$

$$= \frac{2}{16} \cdot \left(-\frac{E_0}{1} \right) + \frac{4}{16} \cdot \left(-\frac{E_0}{2} \right) + \frac{10}{16} \cdot \left(-\frac{E_0}{3} \right)$$

$$= -\frac{1}{16}\,(2 + 2 + 10/3)\, E_0 = -0.46 E_0 \tag{5.51}$$

(d) This part of the problem is truly *fundamental* quantum mechanics. Any measurement of a quantum mechanical system forces the system into an eigenstate of the operator corresponding to the quantity that was measured. Therefore, because the energy was measured to be E_3 the measurement "collapses the wave function" into the $n = 3$ state. This means that the wave function after the measurement is no longer given by Eq. (5.47). It is now simply the eigenfunction $\psi_3(x)$. Thus, after the measurement at $t = 0$ the collapsed wave function $\Psi_{\text{collapsed}}(x,0)$ is simply

$$\Psi_{\text{collapsed}}(x,0) = \psi_3(x) \tag{5.52}$$

From Eq. (5.48) we know that $\psi_3(x)$ is an eigenfunction of $\hat{Q}$ such that

$$\hat{Q}\psi_3(x) = 3Q_0\psi_3(x) \tag{5.53}$$

so the measurement of Q yields $3Q_0$.

9. The wave function at $t = 0$ for a one-dimensional harmonic oscillator (SHO) is given by

$$\psi(x) = \frac{1}{\sqrt{5}}\psi_1(x) + \frac{2}{\sqrt{5}}\psi_2(x) \tag{5.54}$$

where the $\psi_n(x)$ are eigenfunctions of the time independent harmonic oscillator Hamiltonian.

$$\psi_1(x) = \sqrt{\frac{\alpha}{2\sqrt{\pi}}}\, 2\,[(\alpha x)]\, e^{-\alpha^2 x^2/2}$$

$$\psi_2(x) = \sqrt{\frac{\alpha}{2\sqrt{\pi}}}\, \left[2\,(\alpha x)^2 - 1 \right] e^{-\alpha^2 x^2/2} \tag{5.55}$$

(a) Find $\Psi(x,t)$, the wave function for $t > 0$.
(b) Find the expectation value of the energy for $t > 0$.
(c) Find $\langle x(t) \rangle$, the expectation value of the position for $t > 0$.

Solution

(a) Simply multiply each term in the expansion by the time dependent part of the eigenfunction. Therefore

$$\Psi(x,t) = \frac{1}{\sqrt{5}}\psi_1(x)\exp\left[-\frac{3}{2}i\omega t\right] + \frac{2}{\sqrt{5}}\psi_2(x)\exp\left[-\frac{5}{2}i\omega t\right] \qquad (5.56)$$

(b)

$$\langle E \rangle = \int_{-\infty}^{\infty} \Psi^*(x,t)\hat{H}\Psi(x,t)\,dx$$

$$= \int_{-\infty}^{\infty} \Psi^*(x,t)\left\{\frac{1}{\sqrt{5}}E_1\psi_1(x)\exp\left[-\frac{3}{2}i\omega t\right]\right.$$

$$\left. + \frac{2}{\sqrt{5}}E_2\psi_2(x)\exp\left[-\frac{5}{2}i\omega t\right]\right\}dx$$

$$= \frac{1}{5}E_1 + \frac{4}{5}E_2$$

$$= \frac{1}{5}\cdot\frac{3}{2}\hbar\omega + \frac{4}{5}\cdot\frac{5}{2}\hbar\omega = \frac{23}{10}\hbar\omega \qquad (5.57)$$

where we have omitted writing the components of $\Psi^*(x,t)$ in the integrand. Note that the cross terms in Eq. (5.57) vanish because the $\psi_n(x)$ are orthonormal. As a result $\langle E \rangle$ is independent of time. Also, because $\psi(x)$ is heavily weighted toward $\psi_2(x)$ (by a factor of 2) the average energy is skewed toward $E_2 = (25/10)\hbar\omega$ as is clear from Eq. (5.57).

(c)

$$\langle x(t) \rangle = \langle \Psi(t)|\hat{x}|\Psi(t)\rangle$$

$$= \int_{-\infty}^{\infty} \Psi^*(x,t)x\Psi(x,t)\,dx \qquad (5.58)$$

Comparing this integral to that in part (b) we see that the cross terms do not vanish here because the components of $\Psi(x,t)$ are not eigenfunctions of x as they were of $\hat{H}$. Therefore, there will be a time dependence in the expectation value.

$$\langle x(t) \rangle = \int_{-\infty}^{\infty} \Psi^*(x,t)x\Psi(x,t)\,dx$$

$$= \frac{1}{5}\int_{-\infty}^{\infty}|\psi_1(x)|^2\,xdx + \frac{4}{5}\int_{-\infty}^{\infty}|\psi_2(x)|^2\,xdx$$

$$+ \frac{2}{5}e^{i\omega t}\int_{-\infty}^{\infty}\psi_1^*(x)x\psi_2(x)\,dx + \frac{2}{5}e^{-i\omega t}\int_{-\infty}^{\infty}\psi_2^*(x)x\psi_1(x)\,dx$$

$$(5.59)$$

The first two integrals vanish because they have odd integrands. The second two must, however, be evaluated. The eigenfunctions $\psi_n(x)$ are real so these two integrals are identical. We have

$$\langle \hat{x}(t) \rangle = \frac{2}{5}\left(e^{i\omega t} + e^{-i\omega t}\right) \int_{-\infty}^{\infty} \psi_1^*(x)\, x\psi_2(x)\, dx$$

$$= \frac{4}{5}\cos\omega t \int_{-\infty}^{\infty} \psi_1^* x\psi_2(x)\, dx \tag{5.60}$$

The integral is

$$I = \int_{-\infty}^{\infty} \psi_1^* x\psi_2(x)\, dx$$

$$= \frac{\alpha}{2\sqrt{\pi}} \int_{-\infty}^{\infty} \left\{[2(\alpha x)]\cdot x \cdot \left[2(\alpha x)^2 - 1\right] e^{-\alpha^2 x^2}\right\} dx$$

$$= \frac{\alpha}{\sqrt{\pi}} \int_{-\infty}^{\infty} \left(2\alpha^3 x^4 e^{-\alpha^2 x^2} - \alpha x^2 e^{-\alpha^2 x^2}\right) dx$$

$$= \frac{2\alpha^4}{\sqrt{\pi}} \int_{-\infty}^{\infty} x^4 e^{-\alpha^2 x^2}\, dx - \frac{\alpha^2}{\sqrt{\pi}} \int_{-\infty}^{\infty} x^2 e^{-\alpha^2 x^2}\, dx \tag{5.61}$$

To evaluate the integral in Eq. (5.61) we use Eq. (G.4) which is

$$\int_0^{\infty} x^m e^{-\beta x^2}\, dx = \frac{\Gamma[(m+1)/2]}{2\beta^{(m+1)/2}} \tag{5.62}$$

where $\Gamma[(m+1)/2]$ is a Γ-function (see Appendix I). Changing the limits of integration in Eq. (5.61) to match those in Eq. (5.62) we have

$$I/2 = \left\{\frac{2\alpha^4}{\sqrt{\pi}}\frac{\Gamma[5/2]}{2\alpha^5} - \frac{\alpha^2}{\sqrt{\pi}}\frac{\Gamma[3/2]}{2\alpha^3}\right\} \tag{5.63}$$

From Appendix I we have

$$\Gamma[3/2] = \frac{\sqrt{\pi}}{2} \quad\text{and}\quad \Gamma[5/2] = \frac{3\sqrt{\pi}}{4} \tag{5.64}$$

so, using Eq. (G.4)

$$I/2 = \frac{2\alpha^4}{\sqrt{\pi}}\frac{1}{2\alpha^5}\cdot\frac{3\sqrt{\pi}}{4} - \frac{\alpha^2}{\sqrt{\pi}}\frac{1}{2\alpha^3}\cdot\frac{\sqrt{\pi}}{2}$$

$$= \frac{3}{4\alpha} - \frac{1}{4\alpha} = \frac{1}{2\alpha} \Rightarrow I = \frac{1}{\alpha} \tag{5.65}$$

and

$$\langle \hat{x}(t) \rangle = \frac{4}{5\alpha} \cos \omega t$$

10. For an L-box with potential energy

$$U(x) = 0 \quad 0 \le x \le L$$
$$= \infty \quad -\infty < x < 0 \,;\, L < x < \infty \qquad (5.66)$$

the eigenfunctions are given in Eq. (M.2). They are

$$\psi_n(x) = A \sin\left(\frac{n\pi x}{L}\right) \quad 0 \le x \le L \quad n = 1, 2, 3\ldots$$
$$= \infty \quad -\infty < x < 0 \,;\, L < x < \infty \qquad (5.67)$$

(a) Calculate the classical probability density of finding the particle in an increment of space Δx within the box.
(b) Find the classical average values $\langle x \rangle_{\text{classical}}$ and $\langle x^2 \rangle_{\text{classical}}$ and show that the quantum mechanical values $\langle x \rangle$ and $\langle x^2 \rangle$ approach the classical values as the quantum number $n \to \infty$.

Solution

(a) The speed of the particle in the box v is constant because the potential energy within the box is constant (zero). The probability of finding the particle in a given interval Δx must then be constant. Therefore, the classical probability of finding the particle in an increment Δx is simply

$$P_{cl}(x) = \frac{\Delta x}{L} \qquad (5.68)$$

(b) The average value of any function $f(x)$ over an interval (x_1, x_2) is

$$\langle f(x) \rangle = \frac{1}{(x_2 - x_1)} \int_{x_1}^{x_2} f(x)\, dx \qquad (5.69)$$

so

$$\langle x \rangle_{\text{classical}} = \frac{1}{L} \int_0^L x\, dx$$
$$= \frac{L}{2} \qquad (5.70)$$

and

$$\langle x^2\rangle_{\text{classical}} = \frac{1}{L}\int_0^L x^2 dx$$

$$= \frac{L^2}{3} \tag{5.71}$$

Quantum mechanically

$$\langle x\rangle_n = \frac{2}{L}\int_0^L \sin\left(\frac{n\pi x}{L}\right) x \sin\left(\frac{n\pi x}{L}\right) dx$$

$$= \frac{2}{L}\int_0^L x \sin^2\left(\frac{n\pi x}{L}\right) dx \tag{5.72}$$

where we have used the eigenfunctions for an L-box as given in Eq. (5.67). Let

$$y = \frac{n\pi x}{L} \tag{5.73}$$

so that

$$\langle x\rangle_n = \frac{2}{L}\left(\frac{L}{n\pi}\right)^2\int_0^{n\pi} y\sin^2 y\,dy$$

$$= \frac{2L}{n^2\pi^2}\left[\frac{y^2}{4} - \frac{y\sin(2y)}{4} - \frac{\cos(2y)}{8}\right]_0^{n\pi}$$

$$= \frac{L}{2} \tag{5.74}$$

where we have used the integration formula in Eq. (G.9). Clearly the classical and quantum mechanical values of $\langle x\rangle$ are identical. Note that the quantum mechanical value is independent of the quantum number n.

The quantum mechanical expectation value of x^2 in the nth eigenstate is

$$\langle x^2\rangle_n = \frac{2}{L}\int_0^L x^2 \sin^2\left(\frac{n\pi x}{L}\right) dx \tag{5.75}$$

Using the same substitution as above, Eq. (5.73), we have

$$\langle x\rangle_n = \frac{2}{L}\left(\frac{L}{n\pi}\right)^3\int_0^{n\pi} y^2\sin^2 y\,dy$$

$$= \frac{2L^2}{n^3\pi^3}\left[\frac{y^3}{6} - \left(\frac{y^2}{4}-\frac{1}{8}\right)\sin(2y) - \frac{y\cos(2y)}{4}\right]_0^{n\pi}$$

$$= \frac{2L^2}{n^3\pi^3}\left[\frac{n^3\pi^3}{6} - \frac{n\pi}{4}\right] = \frac{L^2}{n^3\pi^3}\left[\frac{n^3\pi^3}{3} - \frac{n\pi}{2}\right] \tag{5.76}$$

$$= \frac{L^2}{3}\left[1 - \frac{3}{2n^2\pi^2}\right] \tag{5.77}$$

where we have used the integration formula in Eq. (G.10).

While the quantum mechanical value of $\langle x^2\rangle_n$ differs from the classical value we see that the quantum mechanical value approaches the classical value as $n \rightarrow \infty$. This is a manifestation of Bohr's correspondence principle.

11. A rectangular potential barrier of length L and height U_0 has a beam of particles each of mass m and kinetic energy $E < U_0$ incident on it from the left as shown in Fig. 5.1.The potential energy is

$$U(x) = 0 \,; x < 0$$

$$= U_0 \,; 0 < x < L$$

$$= 0 \,; x > L \tag{5.78}$$

Find the transmission coefficient T and the reflection coefficient R for this barrier.

Fig. 5.1 Problem 11. The rectangular potential barrier

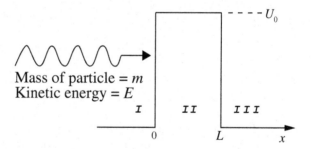

Mass of particle $= m$
Kinetic energy $= E$

Solution

Classically, because $E < U_0$ the particles would simply bounce off the barrier. Quantum mechanically, however, this total reflection is not the case. The occurrence of transmission can be understood if we first write the wave functions in each of the three regions of space. Because the potential energies in each of these regions is constant the wave functions are

$$\Psi_I(x, t) = \left(Ae^{ikx} + Be^{-ikx}\right)e^{-i\omega t}$$

$$\Psi_{II}(x, t) = \left(Ce^{\kappa x} + De^{-\kappa x}\right)e^{-i\omega t}$$

$$\Psi_{III}(x, t) = \left(Fe^{ikx} + Ge^{-ikx}\right)e^{-i\omega t} \tag{5.79}$$

where

$$k^2 = 2m\left(E - U_0\right)/\hbar$$

$$\kappa^2 = 2m\left(U_0 - E\right)/\hbar \tag{5.80}$$

and

$$\omega = E/\hbar \tag{5.81}$$

Quantum mechanics permits the incoming beam to penetrate the classically forbidden region (inside the barrier). If the wave function is non-zero at $x = L$, then particles can emerge from the barrier into region III. These particles have "tunneled" through the barrier.

To solve the problem we must find the coefficients in Eq. (5.79) using the usual requirements of continuity of the wave function and its first derivative at the boundaries. We need not consider the time component of the wave functions in Eq. (5.79) because ω, the energy corresponding to the spatial part of the wave function is the same in all three regions of space.

We immediately set the constant $G = 0$ in Eq. (5.79) because e^{-ikx} represents a plane wave traveling in the $-x$ direction. While it is possible that such a wave will be present in region I (reflection), there cannot be any particles travelling in the $-x$ direction in region III (see Problem 13 of this chapter). Because a plane wave traveling in the $-x$ direction is possible in region I we must assume that $B \neq 0$. There is also a wave function in region II, but, $E < U_0$ so the exponents are real. The wave functions without the time dependence are

$$\psi_I(x) = Ae^{ikx} + Be^{-ikx} \quad x < 0$$

$$\psi_{II}(x) = Ce^{\kappa x} + De^{-\kappa x} \quad 0 < x < L$$

$$\psi_{III}(x) = Fe^{ikx} \quad L < x \tag{5.82}$$

To better understand the boundary conditions and to solve for the constants in each region, we employ the equation for the probability current density $j(x)$, which must be conserved and continuous. The probability current density is defined as [see Eq. (L.12)]

$$j(x) = \frac{\hbar}{2im}\left[\psi^*(x)\frac{\partial\psi(x)}{\partial x} - \psi(x)\frac{\partial\psi^*(x)}{\partial x}\right] \tag{5.83}$$

Applying Eq. (5.83) to regions I and III the probability currents are

$$j_I(x) = \frac{\hbar k}{m}\left(|A|^2 - |B|^2\right) \tag{5.84}$$

and

$$j_{III}(x) = \frac{\hbar k}{m}|F|^2 \tag{5.85}$$

Because the velocities, $\hbar k/m$, are the same in regions I and III

$$\left(|A|^2 - |B|^2\right) = |F|^2 \tag{5.86}$$

The two probability terms in $j_I(x)$ represent the incident and reflected probability currents. It is then clear that the transmission coefficient T must be the ratio of the transmitted probability current to the incident current so that

$$T = \frac{|F|^2}{|A|^2}$$

Similarly, the reflection coefficient is

$$R = \frac{|B|^2}{|A|^2} \tag{5.87}$$

Because the wave must be either reflected or transmitted, we have

$$R + T = 1 \tag{5.88}$$

In this problem the potential, and therefore the wave number k, is the same in regions I and III. If the potential in region III were different from that in region I, the definitions of R and T would be the same, but their relative values would change.

Before evaluating the constants we can sketch the wave functions in each region. The nature of these wave functions is shown in Fig. 5.2.

Fig. 5.2
Problem 11—solution.
Sketches of the wave
functions that apply to
penetration of the rectangular
potential barrier

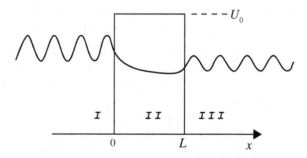

The de Broglie wavelengths in regions I and III are identical. The amplitude in region III is smaller because there will be some reflection at the barrier. The curvature of the wave function inside the classically forbidden barrier is away from the x-axis.

To evaluate the constants A, B, and F, we require the wave functions and their derivatives to be continuous at the boundaries, $x = 0$ and $x = L$. We have

$$A + B = C + D$$

$$ik\,(A - B) = \kappa\,(C - D)$$

$$Ce^{\kappa L} + De^{-\kappa L} = Fe^{ikL}$$

$$\kappa\left(Ce^{\kappa L} - De^{-\kappa L}\right) = ikFe^{ikL} \tag{5.89}$$

These are four simultaneous equations with five unknown coefficients. We are free to choose any one of them so we set $A = 1$ for convenience. Adding and subtracting the third and fourth relations in Eq. (5.89) allows us to express C and D in terms of F. Using these expressions in the first and second relations, with $A = 1$, allows evaluation of A and B, and thus the reflection and transmission amplitudes, R and T. After some algebra we arrive at

$$T_{E<U_0} = \left[1 + \frac{1}{4}\left(\frac{k^2 + \kappa^2}{k\kappa}\right)^2 \sinh^2(\kappa L)\right]^{-1} \tag{5.90}$$

After substituting the κ and k into Eq. (5.90) we obtain

$$T_{E<U_0} = \frac{1}{\left[1 + \dfrac{U_0^2}{4E\,(U_0 - E)}\sinh^2\left(\dfrac{L}{\hbar}\sqrt{2m\,(U_0 - E)}\right)\right]} \tag{5.91}$$

and, using Eq. (5.88), we have

$$R_{E<U_0} = \left[1 + \frac{4E\,(U_0 - E)}{U_0^2\sinh^2\left(\dfrac{L}{\hbar}\sqrt{2m\,(U_0 - E)}\right)}\right]^{-1} \tag{5.92}$$

Equation (5.91) shows that, as expected, there is transmission through the barrier even when $E < U_0$. It is instructive to take this system to the classical limit, $\hbar \to 0$. In this limit the hyperbolic sines become infinite and $T_{E<U_0} \to 0$ while $R_{E<U_0} \to 1$.

It is relatively easy to obtain $T_{E>U_0}$ and $R_{E>U_0}$ using the same equations as above with appropriate modifications. In essence, the hyperbolic functions become conventional circular trigonometric functions and variations in these coefficients occur as functions of the de Broglie wavelengths relation to the barrier width.

$$\langle \hat{x}\,(t)\rangle = \frac{4}{5\alpha}\cos\omega t$$

12. A beam of particles each of mass m and kinetic energy $E > 0$ is incident from the left on a potential *well*, the parameters of which are

$$U(x) = 0 \qquad x < 0$$
$$= -U_0 \quad 0 < x < L$$
$$= 0 \qquad x > L \tag{5.93}$$

where U_0 is a real positive number.

(a) Sketch the de Broglie wavelength of the particles in each region of the potential well.
(b) Find the transmission and reflection coefficients, $T_{E>0}$ and $R_{E>0}$. [Hint: Use the results of Problem 11.]
(c) Find the values of E for which $T_{E>0} \equiv 1$ and show that for these kinetic energies $R_{E>0} \equiv 0$.
(d) Show that the values of E for which $T_{E>0} \equiv 1$ are the same as the quantized energies of an L-box (infinitely deep)

Solution

(a) The de Broglie wavelength will be shorter inside the well, region II, than in region I because the kinetic energy of the particles in region II is greater than E. The wavelength in region III will be the same as that in region I, but the amplitude will be smaller because of reflection (see Fig. 5.3).

Fig. 5.3
Problem 12—solution

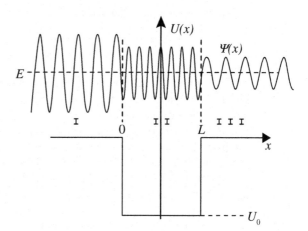

(b) The potential well of this problem can be constructed by inverting the barrier in Problem 11, i.e. $U_0 \rightarrow -U_0$. Thus, the transmission and reflection coefficients of Problem 11 are identical with those of this problem with the substitution $U_0 \rightarrow -U_0$ in Eqs. (5.91) and (5.92). Making this substitution we have

$$T_{E>0} = \frac{4E\,(E+U_0)}{4E\,(E+U_0) + U_0^2 \sin^2\left(\dfrac{L}{\hbar}\sqrt{2m\,(E+U_0)}\right)} \tag{5.94}$$

and

$$R_{E>0} = \left[1 + \frac{4E\,(E+U_0)}{U_0^2 \sin^2\left(\dfrac{L}{\hbar}\sqrt{2m\,(E+U_0)}\right)}\right]^{-1} \tag{5.95}$$

(c) For $T_{E>0}$ to be a maximum the sine in the denominator of Eq. (5.94) must vanish. To meet this condition

$$\frac{L}{\hbar}\sqrt{2m\,(E+U_0)} = n\pi \tag{5.96}$$

where n is an integer. Notice that when $T_{E>0} \equiv 1$ the reflection coefficient $R_{E>0} \equiv 0$, Eq. (5.95), as it must.

(d) The kinetic energy inside the box is $E + U_0 = p^2/2m$ where p is the momentum of the particles when over the box. Solving Eq. (5.96) for the kinetic energies that maximize the transmission we have

$$p^2/2m = \frac{n^2\pi^2\hbar^2}{2mL^2} \tag{5.97}$$

The de Broglie wavelength of the particles when over the well is

$$\lambda = \frac{2\pi\hbar}{p} = \frac{2L}{n} \tag{5.98}$$

Thus, the de Broglie wavelengths associated with $T_{E>0} = 1$ are $2/n$ times L, the width of the well. This is precisely the condition required to quantize the energies of an L-box (see Appendix M), which can be deduced by fitting half de Broglie waves in the box of width L (see Problem 1 of Chap. 6). This quantum mechanical transmission behavior is in contrast to that of an analogous classical system. Classically, the transmission across the potential well is unity for any $E > 0$, although the particle does speed up when it is over the well.

13. A monoenergetic beam of particles moving from $-x \to +x$, each of mass m, is incident on a potential barrier at $x = 0$. This potential energy $U(x)$ is

$$U(x) = V\delta(x) \quad \text{where} \quad U_0 > 0 \tag{5.99}$$

The constant V has units $(\mathrm{J \cdot m})$ because the units of $\delta(x)$ are $\mathrm{m^{-1}}$ (see Appendix J). Find the transmission and reflection coefficients, T_δ and R_δ. [Hint:

Integrate the TISE across the discontinuity in $U(x)$ at $x = 0$ from $-\epsilon \to +\epsilon$ and then take the limit as $\epsilon \to 0$.]

Solution

In general, the wave function is

$$\psi_I(x) = e^{ikx} + Be^{-ikx} \quad x < 0$$
$$\psi_{II}(x) = Ce^{ikx} + De^{-ikx} \quad x > 0 \tag{5.100}$$

where

$$k^2 = 2mE/\hbar^2 \tag{5.101}$$

The terms e^{ikx} represent plane waves travelling in the $+x$ direction and the terms e^{-ikx} are plane waves travelling in the $-x$ direction. While it is possible that $B \neq 0$ in region I (reflection), we must have $D = 0$ in region II because only transmitted waves, necessarily travelling in the $+x$ direction, are present there. Also, in Eq. (5.100) we have set the coefficient of e^{ikx} in region I to unity so the transmission coefficient T and the reflection coefficient R are

$$T_\delta = |C|^2$$
$$R_\delta = |B|^2 \tag{5.102}$$

Because the wave function must be continuous even if the derivative of the wave function is discontinuous (as it is in the case of the δ-function potential), we must have

$$\psi_I(0) = \psi_{II}(0) \implies 1 + B = C \tag{5.103}$$

The TISE for this potential is

$$\frac{d^2\psi(x)}{dx^2} - \lambda\delta(x)\psi(x) = k^2\psi(x) \quad \text{for all } x \tag{5.104}$$

where

$$\lambda = 2mV/\hbar^2 \tag{5.105}$$

We can find the magnitude of the discontinuity in the derivative of $\psi(x)$ at $x = 0$ by integrating the TISE from $-\epsilon$ to $+\epsilon$ (across the δ-function) and taking the limits as $\epsilon \to 0$ from the left and from the right. We have

$$\int_{-\epsilon}^{\epsilon} \frac{d^2\psi(x)}{dx^2} dx - \lambda \int_{-\epsilon}^{\epsilon} \delta(x)\,\psi(x)\,dx = k^2 \int_{-\epsilon}^{\epsilon} \psi(x)\,dx \qquad (5.106)$$

or

$$\lim_{\epsilon\to0} \frac{d\psi(x)}{dx}\bigg|_{x=\epsilon} - \lim_{\epsilon\to0} \frac{d\psi(x)}{dx}\bigg|_{x=-\epsilon} - \lambda\psi(0) = 0 \qquad (5.107)$$

where the right-hand side vanishes because $\lim_{x\to0-} \psi(x) = \lim_{x\to0+} \psi(x)$. Substituting the wave function and taking the limit, we have

$$ik(1 - B - C) - \lambda C = 0 \qquad (5.108)$$

Together with Eq. (5.103) we have two simultaneous equations and two unknowns. After eliminating C using Eq. (5.103) and replacing k and λ with the system parameters that constitute them we have

$$R_\delta = |B|^2 = 1/\left(1 + \frac{2E\hbar^2}{mV^2}\right)$$

$$= \frac{mV^2}{2\hbar^2E + mV^2} \qquad (5.109)$$

Eliminating B and solving for C we obtain

$$T_\delta = |C|^2 = 1/\left(1 + \frac{mV^2}{2\hbar^2E}\right)$$

$$= \frac{2\hbar^2E}{2\hbar^2E + mV^2} \qquad (5.110)$$

It is comforting to note that the sum $R + T \equiv 1$ as it should. Our comfort should not lead to complacence because the unusual units of V suggest that we check the units of T_δ and R_δ, both of which should be unitless. Since Eqs. (5.109) and (5.110) each contains the same factors we need only check the units of $mV^2/\hbar^2E$. We have

$$\frac{mV^2}{\hbar^2E} = \frac{kg\cdot(J\cdot m)^2}{(J\cdot s)^2\cdot J} = \frac{kg\cdot m^2}{s^2\cdot J} = \frac{kg\cdot m^2}{s^2\cdot \frac{kg\cdot m^2}{s^2}} = 1 \qquad (5.111)$$

This is not conclusive proof that our answer is correct, but at least we know that the units are correct, which is encouraging.

14. Show that the transmission coefficient $T_{E<U_0}$ for the rectangular barrier of Problem 11 of this chapter, Eq. (5.91), reduces to the transmission coefficient T_δ for the δ-function barrier of Problem 13 of this chapter in the limits $U_0 \to \infty$, $L \to 0$ with $U_0L = $ constant.

Solution

The quantity U_0L has the same units as V (Problem 13) so these quantities are comparable in this limit. We apply the limits given in the statement of the problem to the terms in Eq. (5.91) to effect the conversion. In the limit $U_0 \to \infty$ the argument of the hyperbolic sine is

$$\frac{L}{\hbar}\sqrt{2mU_0\left(1 - \frac{E}{U_0}\right)} \to \frac{L}{\hbar}\sqrt{2mU_0} \tag{5.112}$$

The leading term in the $\sinh x$ is x (see Appendix H.1) so

$$\sinh^2\left(\frac{L}{\hbar}\sqrt{2mU_0}\right) \simeq \frac{L^2}{\hbar^2}2mU_0 \tag{5.113}$$

In these limits Eq. (5.91) becomes

$$T_{E<U_0} = \frac{1}{\left[1 + \dfrac{U_0L^2}{4E\hbar^2}2mU_0\right]}$$

$$= \frac{2\hbar^2E}{2\hbar^2E + m\left(U_0L\right)^2} \tag{5.114}$$

T_δ is given in Eq. (5.110). It is

$$T_\delta = \frac{2\hbar^2E}{2\hbar^2E + mV^2} \tag{5.115}$$

which is identical to Eq. (5.114).

15. A semi-infinite potential barrier has a beam of particles each of mass m and kinetic energy E incident on it from the left as shown in Fig. 5.4 for $E < U_0$.

Fig. 5.4 Problem 15

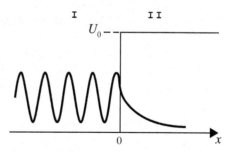

(a) Use the results of Problem 11 of this chapter to find the reflection and transmission coefficients, R and T for $E < U_0$.

(b) Find R and T for $E > U_0$ by matching the wave function and its derivatives at the step as was done using Eq. (5.82).

Solution

(a) This is a special case of the barrier of width L with $L \to \infty$. For this part of the problem, $E < U_0$, it is a simple matter to take the limit in Eqs. (5.91) and (5.92) as $L \to \infty$ because $\lim_{x \to \infty} \sinh^2 x = \infty$. Thus, from Eqs. (5.91) and (5.92) we see that the reflection coefficient is unity while the transmission coefficient vanishes. This is sensible because, no matter how far the particle penetrates the semi-infinite barrier it must eventually be squirted back into region I as indicated by the sketched wave function in Fig. 5.4.

(b) The case for $E > U_0$ presents a bit of a mathematical dilemma. We might consider converting Eqs. (5.91) and (5.92) which pertain to $E < U_0$ to the case for $E > U_0$. This would, however, make the factor $\sqrt{(U_0 - E)}$ imaginary thus converting the hyperbolic sines to circular sines because [see Eq. (K.14)]

$$\sinh(ix) = i \sin x \qquad (5.116)$$

This makes it difficult to take the limit as $L \to \infty$. In addition, we must consider the fact that the particle velocities in the two regions of space are different. Recall that in regions I and III in the case of the finite barrier the particle velocities are the same. Using the same notation as that in Problem 11 we write the wave function for this step as

$$\psi_I(x) = Ae^{ikx} + Be^{-ikx} \quad x < 0$$
$$\psi_{II}(x) = Fe^{ik'x} \qquad 0 < x \qquad (5.117)$$

Because the wave numbers are real in both regions of space we designate them by $k = \sqrt{2mE/\hbar^2}$ and $k' = \sqrt{2m(E - U_0)/\hbar^2}$ for regions I and II, respectively. The probability currents are then

$$j_I(x) = \frac{\hbar k}{m}\left(|A|^2 - |B|^2\right)$$

$$j_{II}(x) = \frac{\hbar k'}{m}|F|^2 \qquad (5.118)$$

and, as before,

$$R = \frac{|B|^2}{|A|^2} \qquad (5.119)$$

In this case, however, T must be the ratio of the intensity of the transmitted probability current to the incident probability current, which is given by

$$T = \frac{\dfrac{\hbar k'}{m}|F|^2}{\dfrac{\hbar k}{m}|A|^2} \tag{5.120}$$

Solving for the coefficients A, B, and F yields R and T and, including the results for $E < U_0$, we have

$$R = \frac{\left(1 - \sqrt{1 - U_0/E}\right)^2}{\left(1 + \sqrt{1 - U_0/E}\right)^2} \quad E > U_0$$

$$= 1 \qquad E < U_0 \tag{5.121}$$

and

$$T = \frac{4\sqrt{1 - U_0/E}}{\left(1 + \sqrt{1 - U_0/E}\right)^2} \quad E > U_0$$

$$= 0 \qquad E < U_0 \tag{5.122}$$

From Eqs. (5.121) and (5.122) it can be seen that there is complete reflection and zero transmission when $E = U_0$, but that the reflection decreases monotonically for higher values of the incident energy. As expected, the transmission approaches unity for high values of the incident energy. Figure 5.5 shows graphs of the reflection and transmission coefficients. The feature of this problem that is not present in Problem 11 is the necessity to account for the different particle momenta in the two regions of space.

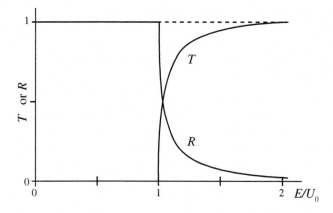

Fig. 5.5 Problem 15—solution

Chapter 6
Bound States in One Dimension

6.1 Degeneracy

Before beginning with the problems it is worthwhile to point out some features of one-dimensional bound states. First, there is no complication arising from degenerate states. Degeneracy exits when two or more states have the same energy. Although a very important consideration in two- and three-dimensional problems, there is no degeneracy in one-dimensional problems. We can prove this by assuming that the assertion is not true. We assume that $\psi_1(x)$ and $\psi_2(x)$ are linearly independent eigenfuntions that have the same eigenvalue E and write the TISE (time independent Schrödinger equation) in the form

$$\frac{1}{\psi(x)} \frac{d^2\psi(x)}{dx^2} = \frac{2m}{\hbar^2}[E - U(x)] \qquad (6.1)$$

Now, the right-hand side of Eq. (6.1) will be the same for both $\psi_1(x)$ and $\psi_2(x)$. Therefore

$$\frac{1}{\psi_1(x)} \frac{d^2\psi_1(x)}{dx^2} = \frac{1}{\psi_2(x)} \frac{d^2\psi_2(x)}{dx^2}$$

or

$$\frac{d}{dx}\left[\frac{d\psi_1(x)}{dx}\psi_2(x)\right] = \frac{d}{dx}\left[\frac{d\psi_2(x)}{dx}\psi_1(x)\right]$$

which, upon integration, gives

$$\left[\frac{d\psi_1(x)}{dx}\psi_2(x)\right] - \left[\frac{d\psi_2(x)}{dx}\psi_1(x)\right] = C$$

© Springer International Publishing AG 2017
J.D. Kelley, J.J. Leventhal, *Problems in Classical
and Quantum Mechanics*, DOI 10.1007/978-3-319-46664-4_6

where C is a constant. C may be evaluated by taking the limit of both sides as $x \to \infty$ because for bound states $\psi(\infty) = 0$. This indicates that $C = 0$ leaving

$$\frac{1}{\psi_1(x)} \frac{d\psi_1(x)}{dx} = \frac{1}{\psi_2(x)} \frac{d\psi_2(x)}{dx}$$

Another integration gives

$$\ln \psi_1(x) = \ln \psi_2(x) + \ln K$$

where we have written the constant of integration as $\ln K$ for convenience. Taking antilogs we have

$$\psi_1(x) = K\psi_2(x)$$

which shows that $\psi_1(x)$ and $\psi_2(x)$ are *not* linearly independent thus contradicting the original assumption of two different degenerate states.

6.2   Parity

If the potential has even parity so that $U(x) = U(-x)$, then the eigenfunctions must have either even or odd parity. The number of nodes in the eigenfunctions increases with the energy of the state; the ground state has no nodes and is therefore of even parity. To see this we write the TISE with the assumption that $U(x)$ is an even function. If we now let $x \to -x$ we obtain

$$\left[-\frac{\hbar^2}{2m} \frac{d^2}{dx^2} + U(x) \right] \psi_n(-x) = E\psi_n(-x) \tag{6.2}$$

where the subscripts emphasize that the ψs are eigenfunctions. It is obvious that $\psi_n(-x)$ and $\psi_n(x)$ are solutions of the same TISE, and that they have the same eigenvalue E_n. Because $\psi_n(-x)$ and $\psi_n(x)$ have the same eigenvalue, they can differ only by a constant. That is,

$$\psi_n(-x) = \beta\psi_n(x) \tag{6.3}$$

If we change the sign of x again, we have

$$\psi_n(x) = \beta\psi_n(-x)$$
$$= \beta\left[\beta\psi_n(x)\right]$$
$$= \beta^2\psi_n(x) \tag{6.4}$$

Thus $\beta = \pm 1$ so that if the potential energy is an even function, the eigenfunctions of the TISE have definite parity, i.e. $\psi(x) = \pm\psi(-x)$.

6.3 Characteristics of the Eigenfunctions

A graph of the potential energy function allows one to qualitatively sketch its eigenfunctions without detailed calculations. To do this it is necessary to have knowledge of the characteristics of bound state eigenfunctions. These characteristics include curvature of the eigenfunction, number of nodes, symmetry and limiting behavior. While we will not derive any of these characteristics we present a summary in Table 6.1. They are developed in most textbooks on the subject.

6.4 Superposition Principle

A quantum mechanical system need not be in a single eigenstate. In general it is not! Inasmuch as the eigenfunctions constitute a complete set, any allowable wave function may be expanded in terms of this complete set of eigenstates which we will designate by $\psi_i(x)$. From this point of view systems are in a superposition of (eigen)states. The state of the system, $\psi(x)$, is thus represented as

$$\psi(x) = \sum_i c_i \psi_i(x) \tag{6.5}$$

where the sum is over all eigenstates. For $\psi(x)$ to be normalized we must have

$$\int_{\text{all space}} \psi^*(x) \psi(x) \, dx \equiv 1 \tag{6.6}$$

Table 6.1 Properties of bound state one-dimensional eigenfunctions

Asymptotic behavior of $\psi_n(x)$	$\lim_{x \to \pm\infty} \psi_n(x) = 0$
Continuity of $\psi_n(x)$	All x
Continuity of $\frac{d\psi_n(x)}{dx}$	All x except at infinite discontinuity in $U(x)$
Curvature of $\psi_n(x)$ (classically allowed region)	Toward x-axis
Curvature of $\psi_n(x)$ (classically forbidden region)	Away from x-axis
Number of nodes	Increases with energy ($n = 0, 1, 2 \ldots$)
Symmetry	Even/odd only for even potentials

As a result of orthogonality of the $\psi_i(x)$ we have

$$\int_{\text{all space}} \psi^*(x)\, \psi(x)\, dx = \sum_i |c_i|^2 = 1 \qquad (6.7)$$

Moreover, the probability of measuring a particular eigenvalue is given by

$$P_i = |c_i|^2 \qquad (6.8)$$

Problems

1. A particle is in one of the bound stationary states of a one-dimensional potential well. An eigenfunction for this potential well is shown in the graph below (Fig. 6.1).

Fig. 6.1 Problem 1

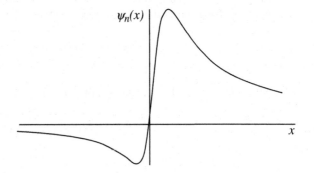

(a) If a measurement of the location of the particle in space were made, would it most likely be at a positive value of x or negative value of x? Explain.
(b) Sketch a possible potential on the same axes as that on which the eigenfunction is displayed. Point out salient features?
(c) To which of the allowed energies, i.e. ground state, first excited state, etc., does this wave function correspond? Explain.

Solution

(a) Positive x because

$$\int_{-\infty}^{0} \psi^*(x)\, \psi(x)\, dx < \int_{0}^{\infty} \psi^*(x)\, \psi(x)\, dx \qquad (6.9)$$

(b) $U(x)$ is skewed because the wave function $\psi(x)$ is skewed. This is only a sketch (Fig. 6.2).

Fig. 6.2
Problem 1b—solution

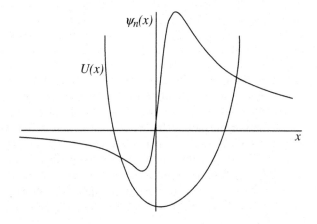

(c) The first excited state because there is one node.

2. This problem illustrates some of the characteristics of one-dimensional bound states and their eigenfunctions.

  (a) Obtain the expression for the energy eigenvalues of a particle of mass m trapped in an infinitely deep box of width L, an L-box (Appendix M), by fitting de Broglie waves in the box. Do not use Schrödinger's wave equation.

  (b) The graph shown in Fig. 6.3 is that of an energy eigenfunction $\psi_n(x)$ which is not an eigenfunction for the particle in-a-box in part (a) of this problem. The solution of part (a) should, however, help you solve this part of the problem. Note that $\psi_n(x) \equiv 0$ for $x \leq 0$. Which state of the (unknown) potential energy function does $\psi_n(x)$ represent? That is, ground state, first excited state, etc.?

Fig. 6.3 Problem 2

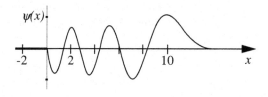

  (c) On the same graph as the energy eigenfunction of part (b) sketch a potential energy function $U(x)$ for which $\psi_n(x)$ could be an eigenfunction. Point out salient features of $\psi_n(x)$ that lead you to represent the potential energy function as you sketch it. Pay particular attention to the classically allowed and forbidden regions and the curvature of $\psi_n(x)$.

Solution

(a) The potential energy for an *L*-box is

$$U(x) = 0 \qquad 0 \le x \le L$$

$$= \infty \qquad -\infty < x < 0 \, ; L < x < \infty \qquad (6.10)$$

Because of the infinite walls at $x = 0$ and $x = L$ that stretch to $x = \pm\infty$ the eigenfunctions must vanish in the regions $-\infty < x < 0$ and $L < x < \infty$. Thus, the eigenfunctions $\psi_n(x)$ vanish outside the box and the probability of finding the particle in these regions is zero. (This would not be the case if the walls were of finite height.) These eigenfunctions and the de Broglie waves associated with them must vanish at both $x = 0$ and $x = L$. The only way the de Broglie waves can vanish at $x = 0$ *and* $x = L$ is if the wavelengths are integral or half-integral multiples of L so they "fit" into the box. This is illustrated in Fig. 6.4 which shows the *L*-box and the first two eigenfunctions $\psi_1(x)$ and $\psi_2(x)$.

Fig. 6.4
Problem 2a—solution

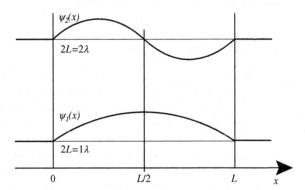

The situation is identical with standing waves or "resonances" set up by a vibrating string of length *L* with both ends fixed. To comply with the criterion that the wavelength must be integral or half-integral multiples of *L* we must have

$$\lambda_n = 2\frac{L}{n} = \frac{h}{p_n} \, ; n = 1, 2, 3 \cdots \Longrightarrow p_n = \frac{hn}{2L} \qquad (6.11)$$

Because the potential energy inside the box is zero the total energy is

$$E_n = \frac{p_n^2}{2m} = \frac{1}{2m}\left(\frac{hn}{2L}\right)^2$$

$$= \frac{\hbar^2 \pi^2 n^2}{2mL^2} \qquad (6.12)$$

which is indeed the *L*-box energy eigenvalue for state *n*. Notice the similarity between this problem and Problem 3 of Chap. 5 in which it was shown that de Broglie waves of fixed wavelength fit into the quantized Bohr orbits.

(b) Because $\psi_n(x) = 0$ for $x \leq 0$ we must have $U(x) = \infty$ for $x \leq 0$. There are five other nodes in the classically allowed region. Therefore, this eigenfunction represents the fifth excited state.

(c) The graph below points out the salient features of a possible potential energy curve based on the given wave function (Fig. 6.5).

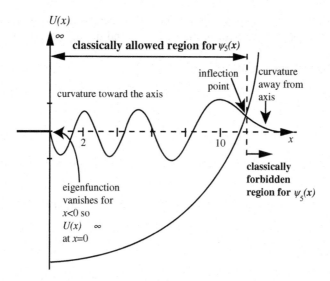

Fig. 6.5 Problem 2c—solution

This problem illustrates some important characteristics of one-dimensional bound state eigenfunctions. We may generalize from the simple standing wave picture of the particle-in-box of part (a) where the eigenfunctions have a fixed de Broglie wavelength throughout the box. If we imagine deforming the side walls of the box to some arbitrary shape, it is clear that the finite potential energy of the particle in the vicinity of these non-vertical walls requires the kinetic energy T of the particle to vary as the potential varies. A thorough understanding of the concepts introduced in this problem will greatly facilitate the student's understanding of eigenfunctions and the significance of fitting de Broglie wave into potential energy curves. If $T = T(x)$, then the momentum $p = p(x)$ and we may regard the de Broglie wavelength λ as also being a function of position when the potential energy is not constant. Thus, fitting de Broglie waves in a potential well of arbitrary shape is obviously not as simple a task as fitting them in an infinite square well, but the principle is the same. Additional features appear because "soft" walls make it possible for the particle to penetrate into the classically forbidden region. As shown in most elementary textbooks, this penetration is marked by a change in the second derivative of the eigenfunction at the boundary between the classically allowed and forbidden regions.

3. A one-dimensional attractive δ-function potential is given by

$$U(x) = -U_0\delta(x) \tag{6.13}$$

where U_0 is a positive constant.

(a) Find the bound state energy E. Show that there is only one eigenvalue no matter what the strength of the well U_0. Find the unnormalized eigenfunction.
(b) Normalize the eigenfunction.

Solution

(a) To find the bound state energy (or energies) we first examine the TISE.

$$-\frac{\hbar^2}{2m}\frac{d^2\psi(x)}{dx^2} - U_0\delta(x)\,\psi(x) = E\psi(x) \quad \text{for all } x \tag{6.14}$$

Integrating Eq. (6.14) across $x = 0$ (from $-\epsilon$ to ϵ) we have

$$\int_{-\epsilon}^{\epsilon}\frac{d^2\psi(x)}{dx^2}\,dx + \lambda\int_{-\epsilon}^{\epsilon}\delta(x)\,\psi(x)\,dx = \kappa^2\int_{-\epsilon}^{\epsilon}\psi(x)\,dx \tag{6.15}$$

where

$$\lambda = \frac{2mU_0}{\hbar^2} \tag{6.16}$$

Taking the limit of all factors in Eq. (6.15) as $\epsilon \to 0$ we have

$$\lim_{\epsilon\to 0}\frac{d\psi(x)}{dx}\bigg|_{x=\epsilon} - \lim_{\epsilon\to 0}\frac{d\psi(x)}{dx}\bigg|_{x=-\epsilon} + \lambda\psi(0) = 0 \tag{6.17}$$

where the right-hand side vanishes because the wave function must be a continuous function. That is, $\lim_{x\to 0-}\psi(x) = \lim_{x\to 0+}\psi(x)$.

From Eq. (6.14) it is seen that in the region for which $x \neq 0$ the term containing the delta function vanishes. The solution of the resulting equation for negative energies is

$$\psi(x) = Ae^{\kappa x} + Be^{-\kappa x} \tag{6.18}$$

where

$$\kappa^2 = \frac{2m\,|E|}{\hbar^2} \tag{6.19}$$

Because the eigenfunction must be exponentially decaying on both sides of $x = 0$, the wave function of the bound state must be of the form

$$\psi(x) = Ae^{-\kappa x} \quad x > 0$$
$$= Ae^{\kappa x} \quad x < 0 \tag{6.20}$$

where $A = B$ because of continuity of the wave function.

Inserting the eigenfunction given in Eq. (6.20) into Eq. (6.17), we have

$$\lim_{\epsilon \to 0} (-\kappa A e^{-\kappa \epsilon}) - \lim_{\epsilon \to 0} (\kappa A e^{\kappa \epsilon}) + \lambda A = 0 \tag{6.21}$$

or

$$\kappa = \frac{\lambda}{2} \tag{6.22}$$

so that

$$\kappa^2 = \frac{2m\,|E|}{\hbar^2} = \frac{m^2 U_0^2}{\hbar^4} \tag{6.23}$$

from which we obtain the eigenvalue

$$E = -\frac{mU_0^2}{2\hbar^2} \tag{6.24}$$

We see that there are no other eigenvalues because there are no other solutions for E. That is, there are no quantum numbers that arise naturally from the solution as appeared in the cases of, for example, the square well or the harmonic oscillator (SHO). Thus, there is one and only one bound state for a δ-function potential well no matter what the strength of the well U_0.

The key to working this problem was understanding and using the characteristics of the one-dimensional bound state eigenfunctions as summarized in Table 6.1. In particular, knowledge that the eigenfunction must be continuous, even at an infinite discontinuity in the potential energy (the δ-function at $x = 0$), was crucial [see Eq. (6.17)].

(b) To normalize the eigenfunction, we find A as follows.

$$\int_{-\infty}^{0} |A|^2 e^{2\kappa x} dx + \int_{0}^{\infty} |A|^2 e^{-2\kappa x} dx = 1 \tag{6.25}$$

$$|A|^2 \left\{ \frac{1}{2\kappa} e^{2\kappa x} \Big|_{-\infty}^{0} + \left(\frac{1}{-2\kappa} \right) e^{-2\kappa x} \Big|_{0}^{\infty} \right\} = 1$$

$$|A|^2 \left\{ \frac{1}{2\kappa} + \left(\frac{1}{2\kappa} \right) \right\} = 1$$

Therefore

$$|A|^2 = \kappa \tag{6.26}$$

$$= \sqrt{\frac{2m}{\hbar^2} \cdot \frac{mU_0^2}{2\hbar^2}}$$

$$= \frac{mU_0}{\hbar^2}$$

and the *normalized* eigenfunction is

$$\psi(x) = \sqrt{\frac{mU_0}{\hbar^2}} e^{-\kappa x} \quad x > 0 \tag{6.27}$$

$$= \sqrt{\frac{mU_0}{\hbar^2}} e^{\kappa x} \quad x < 0$$

4. Shown in Fig. 6.6 are two potential wells $U_1(x) [= U_1(-x)]$ and $U_2(x)$ that are drawn on the same horizontal and vertical scales. $U_2(x)$ is the same as $U_1(x)$ for $x > 0$. The left-hand wall of $U_2(x)$ is, however, impenetrable. Sketch the ground state and first excited state eigenfunctions on the well on the left at their approximate energies. Sketch the ground state eigenfunction at the appropriate energy on the well on the right. Keep the same energy scale for both wells. Comment on the following features of all three eigenfunctions: parity, number of nodes, curvature in different regions of x.

Fig. 6.6 Problem 4

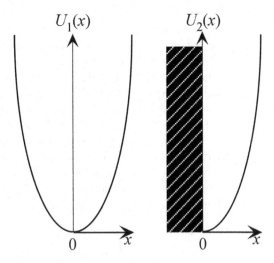

Solution

For $U_1(x)$: The ground state is an even function because $U_1(x)$ is evidently an even function. The eigenfunction has no nodes in the classically allowed region and it has inflection points at the classical turning points as shown in Fig. 6.7. The first excited state is an odd function with a single node at $x = 0$. It too has inflection points at the classical turning points. All eigenfunctions curve toward the x-axis in the classically allowed region and away from it in the classically forbidden region.

Fig. 6.7
Problem 4—solution

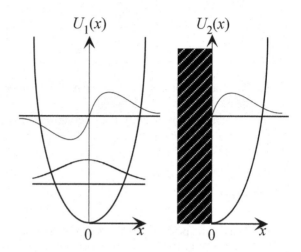

$$U_1(x) \qquad\qquad U_2(x)$$

For $U_2(x)$: For $x \geq 0$ the ground state eigenfunction is the right-hand side of the first excited state eigenfunction of the well on the left. For $x < 0$ the eigenfunction is zero because the potential energy is infinite. The energy is that of the first excited state of the well on the left. This ground state has no definite parity, no nodes in the classically allowed region, one inflection point at the right hand classical turning point and a zero at $x = 0$ at which the derivative of the eigenfunction is discontinuous. The eigenfunctions are the odd eigenfunctions of $U_1(x)$ for $x > 0$. For $x < 0$ they vanish. The curvature of the eigenfunctions is the same as described in the previous paragraph.

5. A particle of mass m is in the ground state of an a-box, i.e. a potential well of width a for which the potential energy is given by

$$U(x) = 0 \quad -a/2 \leq x \leq a/2$$
$$= \infty \quad x > a/2 \,;\, x < -a/2 \qquad\qquad (6.28)$$

The width of the box is suddenly doubled symmetrically so that the walls of the box are located at $x = -a$ and $x = +a$ as illustrated in Fig. 6.8.

Fig. 6.8 Problem 5

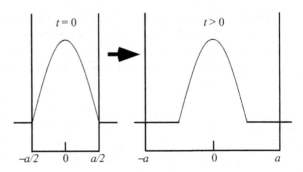

(a) Calculate the probability that the particle will be found in the ground state of the new box.

(b) Calculate the probability that the particle will be found in the first excited state of the new box.

Solution

This is a classic problem the solution to which should be understood thoroughly. It is fundamental to an understanding of eigenstates (stationary states), non-stationary states, and the superposition theorem.

(a) First we must interpret the word "suddenly" in this context. It means that the transition from the original box to the new box occurs in such a short time that the wave function cannot adjust to this change. Thus, the wave function in the new box is the same as that in the original box, the "sudden approximation." We are told, however, that the wave function in the original a-box represents the ground state in that box. While the wave function was an eigenfunction of the old box, it is not an eigenfunction of the new box. The transition between boxes is illustrated in Fig. 6.8. In keeping with the superposition theorem we can expand this wave function in terms of the complete set of eigenfunctions of the new box. The probability of finding the particle in *any* state of the new box will be the square of the expansion coefficient of the corresponding eigenfunction.

The eigenfunctions for the original a-box with the potential energy is given by Eq. (6.28) are

$$\psi_n(x) = \sqrt{\frac{2}{a}} \cos\left(\frac{n\pi x}{a}\right) \qquad -\frac{a}{2} \leq x \leq \frac{a}{2} \quad n \text{ odd (even parity)}$$

$$\psi_n(x) = \sqrt{\frac{2}{a}} \sin\left(\frac{n\pi x}{a}\right) \qquad -\frac{a}{2} \leq x \leq \frac{a}{2} \quad n \text{ even (odd parity)}$$

$$= 0 \qquad x < -\frac{a}{2} ; x > \frac{a}{2} \text{ for all } n \qquad\qquad (6.29)$$

so the wave function (not eigenfunction) of the new box (which is the ground state eigenfunction of the old box) is

$$\psi^{new}(x) = \sqrt{\frac{2}{a}}\cos\left(\frac{\pi x}{a}\right) \quad -\frac{a}{2} \leq x \leq \frac{a}{2}$$

$$\psi^{new}(x) = 0 \quad x \leq -\frac{a}{2} \text{ and } x \geq \frac{a}{2} \tag{6.30}$$

Noting that this wave function, Eq. (6.30), is an even function, its expansion in terms of the eigenfunctions of the new box cannot contain the sine eigenfunctions. Using Eq. (6.30), this expansion is

$$\sqrt{\frac{2}{a}}\cos\left(\frac{\pi x}{a}\right) = \sqrt{\frac{1}{a}}\sum_{j\,odd}^{\infty} c_j \cos\left(\frac{j\pi x}{2a}\right) ; \quad -\frac{a}{2} \leq x \leq \frac{a}{2} \tag{6.31}$$

The probability P_j of finding the particle in the jth eigenstate of the new box is given by the square of the jth expansion coefficient in Eq. (6.31),

$$P_j = |c_j|^2 \tag{6.32}$$

We seek

$$P_1 = |c_1|^2 \tag{6.33}$$

To find c_1 we multiply both sides of Eq. (6.31) for $\psi^{new}(x)$ by the ground state *eigenfunction* of the new box, which is

$$\psi_1^{new}(x) = \sqrt{\frac{1}{a}}\cos\left(\frac{\pi x}{2a}\right) \tag{6.34}$$

Integrating over all space leads to c_1.

$$\int_{-\infty}^{-a/2} 0\cdot dx + \frac{\sqrt{2}}{a}\int_{-a/2}^{a/2}\cos\left(\frac{\pi x}{a}\right)\cos\left(\frac{\pi x}{2a}\right)dx + \int_{a/2}^{\infty} 0\cdot dx = c_1 \tag{6.35}$$

To evaluate the integral on the lhs we let

$$y = \frac{\pi x}{2a} \tag{6.36}$$

so the non-zero integral becomes

$$I = \int_{-a/2}^{a/2}\cos\left(\frac{\pi x}{a}\right)\cos\left(\frac{\pi x}{2a}\right)dx$$

$$= \frac{4a}{\pi}\int_0^{\pi/4}\cos(2y)\cos y\,dy \tag{6.37}$$

where we have used the fact that the integrand is even and integrated over symmetric limits.

To perform the last integral we use the trigonometric identity in Eq. (E.6)

$$\cos A \cos B = \frac{1}{2} [\cos (A - B) + \cos (A + B)] \tag{6.38}$$

The integral is

$$
\begin{aligned}
I &= \frac{4a}{\pi} \int_0^{\pi/4} \frac{1}{2} [\cos y + \cos 3y]\, dy \\
&= \frac{2a}{\pi} \left[\sin y + \frac{1}{3} \sin 3y \right]_0^{\pi/4} \\
&= \frac{2a}{\pi} \left[\frac{1}{\sqrt{2}} + \frac{1}{3} \frac{1}{\sqrt{2}} \right] = \frac{8a}{3\sqrt{2\pi}} \tag{6.39}
\end{aligned}
$$

Returning to Eq. (6.35) we have

$$c_1 = \frac{\sqrt{2}}{a} \frac{8a}{3\sqrt{2\pi}} = \frac{8}{3\pi} \tag{6.40}$$

so the probability of finding m in the ground state is

$$
\begin{aligned}
P_1 &= |c_1|^2 = \left(\frac{8}{3\pi} \right)^2 \\
&= \frac{64}{9\pi^2} \approx 0.72 \tag{6.41}
\end{aligned}
$$

(b) As noted above, the eigenfunction for the old box is even while that for the first excited state of the new box is odd. Therefore, the overlap integral vanishes and the probability of finding the particle in *any* of the odd states of the new box is zero.

6. Let us now modify Problem 5. We start with the particle in the ground state of an infinitely deep L-box (see Appendix M) for which the potential energy function is

$$
\begin{aligned}
U(x) &= 0 \qquad 0 \le x \le L \\
&= \infty \qquad -\infty < x < 0; L < x < \infty \tag{6.42}
\end{aligned}
$$

The width of the box is doubled by suddenly moving only the right-hand wall. Figure 6.9 shows the wave functions before and after the box is suddenly expanded. Use the sudden approximation to find the probability of observing the particle in the ground state of the new box.

Fig. 6.9 Problem 6

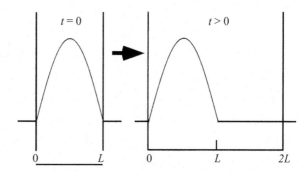

Solution

In contrast to the situation in Problem 5, the spatial symmetry about $x = L/2$ that existed originally was destroyed by the asymmetric way in which the box was expanded. Nonetheless, the calculation is relatively straightforward. Again we expand the old wave function, which is not an eigenfunction of the new box, in a series of eigenfunctions of the new box. Thus, omitting the regions for which the functions are zero

$$\psi^{\text{new}}(x) = \sqrt{\frac{2}{L}} \sin\left(\frac{\pi x}{L}\right)$$

$$\sqrt{\frac{2}{L}} \sin\left(\frac{\pi x}{L}\right) = \sqrt{\frac{1}{L}} \sum_j^{\infty} c_j \left(\frac{j\pi x}{2L}\right) ; 0 \le x \le 2L \qquad (6.43)$$

We seek $P_1 = |c_1|^2$ so we multiply both sides of Eq. (6.43) by the ground state eigenfunction of the new box and obtain

$$c_1 = \int_0^L \left[\sqrt{\frac{2}{L}} \sin\left(\frac{\pi x}{L}\right)\right] \cdot \left[\sqrt{\frac{1}{L}} \sin\left(\frac{\pi x}{2L}\right)\right] dx \qquad (6.44)$$

Note that the upper limit of integration is L because $\psi^{\text{old}}(x) = 0$ for $x > L$. To perform the last integral integral we use the trigonometric identity in Eq. (E.5)

$$\sin A \sin B = \frac{1}{2} [\cos(A - B) - \cos(A + B)] \qquad (6.45)$$

so that

$$c_1 = \frac{1}{\sqrt{2L}} \int_0^L \left[\cos\left(\frac{\pi x}{2L}\right) - \cos\left(\frac{3\pi x}{2L}\right) \right] dx$$

$$= \frac{1}{\sqrt{2L}} \left[\frac{2L}{\pi} \sin\left(\frac{\pi x}{2L}\right) - \frac{2L}{3\pi} \sin\left(\frac{3\pi x}{2L}\right) \right]_0^L$$

$$= \frac{1}{\sqrt{2L}} \left[\frac{2L}{\pi} - \frac{2L}{3\pi}(-1) \right] = \frac{4\sqrt{2}}{3\pi} \qquad (6.46)$$

and

$$P_1 = |c_1|^2 = \left(\frac{4\sqrt{2}}{3\pi} \right)^2 \approx 0.36 \qquad (6.47)$$

Notice that this result is different from that of part (a) of Problem 5. This is because in this problem there are both even and odd terms (with respect to $x = 0$) in the wave function expansion while the symmetry of Problem 5 demands that only the even terms are present.

7. A particle is trapped in an a-box potential

$$U(x) = 0 \qquad -a/2 \le x \le a/2$$

$$= \infty \qquad -\infty < x < -a/2 \,;\, a/2 < x < \infty \qquad (6.48)$$

The state of the particle is described by the wave function

$$\psi(x) = A \left\{ \left[\sqrt{\frac{2}{a}} \cos\left(\frac{\pi x}{a}\right) \right] - \frac{1}{m} \left[\sqrt{\frac{2}{a}} \cos\left(\frac{3\pi x}{a}\right) \right] \right\} \;;\; -a/2 \le x \le a/2$$

$$= 0 \qquad -\infty < x < -a/2 \,;\, a/2 < x < \infty \qquad (6.49)$$

where m is a positive constant. A graph of Eq. (6.49) with $m = 3.5$ is shown in Fig. 6.10 together with graphs of each of the constituent cosine terms.

Fig. 6.10 Problem 7

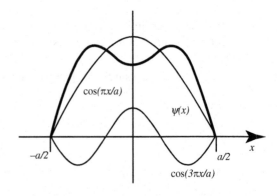

(a) Find the value of A that normalizes $\psi(x)$ in Eq. (6.49) for arbitrary m.
(b) Find the probabilities that a measurement of the energy will yield E_1 and E_3, the energy eigenvalues of the ground and first excited states of the a-box. Note that there are no states with even quantum numbers because $\psi(x)$ is an even function of x (see Fig. 6.10).

Solution

(a) To normalize $\psi(x)$ we must force the integral of $\psi^*(x)\,\psi(x)$ over all space to equal unity.

$$\int_{-a/2}^{a/2} \psi^*(x)\,\psi(x)\,dx = 1 \tag{6.50}$$

We note that the two terms in square brackets in Eq. (6.49) are normalized eigenfunctions of the a-box potential, $\psi_1(x)$ and $\psi_3(x)$. Therefore, $\psi^*(x)\,\psi(x)$ is the sum of three terms two of which are the squared eigenfunctions. The integral of the cross term is manifestly zero because of orthogonality so we have

$$\int_{-a/2}^{a/2} \psi^*(x)\,\psi(x)\,dx = A^2 \left(1 + \frac{1}{m^2}\right) \tag{6.51}$$

so

$$A^2 = \left(\frac{m^2}{1+m^2}\right) \tag{6.52}$$

Thus, the wave function on the interval $-a/2 \le x \le a/2$ is

$$\psi(x) = \left(\frac{m}{\sqrt{1+m^2}}\right) \sqrt{\frac{2}{a}} \left[\cos\left(\frac{\pi x}{a}\right) - \frac{1}{m}\cos\left(\frac{3\pi x}{a}\right)\right] \tag{6.53}$$

(b) Because $\psi(x)$ is a normalized linear combination of the two even eigenstates of the a-box the probabilities of measuring their eigenvalues, E_1 and E_3, are simply the squares of their expansion coefficients in Eq. (6.53). The individual probabilities P_i are therefore

$$P_1 = \left(\frac{m^2}{1+m^2}\right) \quad \text{and} \quad P_3 = \left(\frac{1}{1+m^2}\right) \tag{6.54}$$

It is clear that as m increases, the probability of measuring E_1 increases and that of measuring E_3 decreases as illustrated in Fig. 6.11.

Fig. 6.11
Problem 7—solution

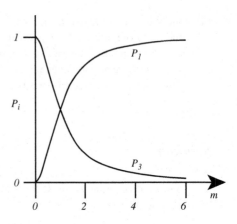

For this particular wave function the only two eigenstates represented in $\psi(x)$ are those with eigenvalues E_1 and E_3 so that these are the only measurable values of a measurement of energy.

8. A particle is subjected to an a-box potential given by

$$U(x) = 0 \qquad -a/2 \le x \le a/2$$
$$= \infty \qquad x < -\frac{a}{2} ; x > \frac{a}{2} \tag{6.55}$$

The particle is in a state such that the wave function is

$$\psi(x) = A \cos^2\left(\frac{\pi x}{a}\right) \qquad -a/2 \le x \le a/2$$
$$= 0 \qquad x < -\frac{a}{2} ; x > \frac{a}{2} \tag{6.56}$$

A graph of Eq. (6.56) is shown in Fig. 6.12.

Fig. 6.12 Problem 8

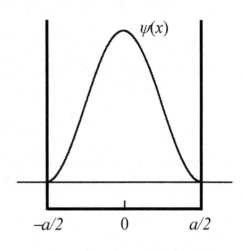

(a) Find the value of A that normalizes $\psi(x)$ in Eq. (6.56).
(b) Find the probability that a measurement of the energy will yield E_1, the eigenvalue of the first eigenstate.

Solution

(a) To normalize $\psi(x)$ we must force the integral of $\psi^*(x)\,\psi(x)$ over all space to equal unity.

$$\int_{-a/2}^{a/2} \psi^*(x)\,\psi(x)\,dx = 1 \tag{6.57}$$

so that, using Eq. (G.12) we have

$$A^2 \int_{-a/2}^{a/2} \cos^4\left(\frac{\pi x}{a}\right) dx = 1$$

$$A^2 \cdot \frac{3a}{8} = 1 \tag{6.58}$$

so

$$A^2 = \frac{8}{3a} \tag{6.59}$$

and therefore

$$\psi(x) = \sqrt{\frac{8}{3a}} \cos^2\left(\frac{\pi x}{a}\right) \quad -a/2 \le x \le a/2$$

$$= 0 \quad x < -\frac{a}{2}; x > \frac{a}{2} \tag{6.60}$$

(b) Now for the physics. To exploit of the superposition theorem we must expand the wave function given by Eq. (6.60) in terms of the complete set of a-box eigenfunctions $\psi_n(x)$.

$$\psi(x) = \sum_{n=1}^{\infty} C_n \psi_n(x) \quad -a/2 \le x \le a/2$$

$$= 0 \quad x < -\frac{a}{2}; x > \frac{a}{2} \tag{6.61}$$

The square of the nth expansion coefficient C_n is then the probability of measuring the energy eigenvalue of the nth state. The eigenfunctions are

$$\psi_n(x) = \sqrt{\frac{2}{a}}\cos\left(\frac{n\pi x}{a}\right) \; ; \; -a/2 \le x \le a/2 \; ; \; n \text{ odd (even)}$$

$$\psi_n(x) = \sqrt{\frac{2}{a}}\sin\left(\frac{n\pi x}{a}\right) \; ; \; -a/2 \le x \le a/2 \; ; \; n \text{ even (odd)}$$

$$= 0 \qquad x < -a/2 \text{ and } x > a/2 \quad \text{all } n \tag{6.62}$$

The wave function, Eq. (6.60), is an even function so there can be no sines in the expansion. Thus, the solution will contain only odd values of n so the series representation of $\psi(x)$ is

$$\sqrt{\frac{8}{3a}}\cos^2\left(\frac{\pi x}{a}\right) = \sum_{\text{odd } n}^{\infty} C_n\left[\sqrt{\frac{2}{a}}\cos\left(\frac{n\pi x}{a}\right)\right] \; ; \; -a/2 \le x \le a/2$$

$$\tag{6.63}$$

Multiplying both sides of Eq. (6.63) by $\psi_1^*(x) = \sqrt{\frac{2}{a}}\cos\left(\frac{\pi x}{a}\right)$ and integrating over the interval $-a/2 \le x \le a/2$ we have, using Eq. (G.11)

$$C_1 = \sqrt{\frac{16}{3a^2}} \int_{-a/2}^{a/2} \cos^3\left(\frac{\pi x}{a}\right) dx$$

$$= 2\sqrt{\frac{16}{3a^2}}\left[\frac{a\sin(\pi x/a)}{\pi} - \frac{a\sin^3(\pi x/a)}{3\pi}\right]_0^{a/2} = \frac{2}{\pi}\sqrt{\frac{16}{3a^2}}\left[\frac{2}{3}\right]$$

$$= \frac{16}{3\sqrt{3}\pi} \tag{6.64}$$

The probability of finding the system in the ground state and therefore measuring the energy to be E_1 is

$$|C_1|^2 = \frac{256}{27\pi^2}$$

$$\approx 0.961 \tag{6.65}$$

The major component of the given wave function $\psi(x)$ is the ground state of the a-box. This is illustrated in Fig. 6.13 which shows that the original wave function and the ground state eigenfunction $\psi_1(x)$ are very similar.

Fig. 6.13
Problem 8—solution. The
heavy line is the given wave
function and the *lighter line* is
the ground state
eigenfunction

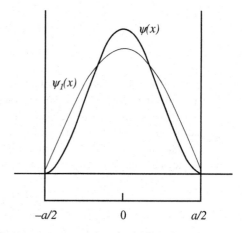

$-a/2$ 0 $a/2$

9. A particle of mass m is initially in the ground state of a parabolic potential given by $U_1(x) = (2k)x^2$. The force constant is suddenly halved so that the new potential becomes $U_2(x) = kx^2$.

(a) Use the sudden approximation to find the probability that the particle is in the ground state of the *new* potential $U_2(x)$.
(b) What is the probability that the particle will be in the first excited state of the new potential?

Solution

(a) The normalized ground state eigenfunction for the parabolic potential (SHO, see Appendix O) has the form

$$\psi_0(x) = \frac{\sqrt{\alpha}}{\pi^{1/4}} e^{-\alpha^2 x^2/2} \qquad (6.66)$$

where for $U_1(x)$

$$\alpha = \sqrt{\frac{m\omega}{\hbar}} = \sqrt{\frac{m}{\hbar}\left[\frac{(2k)^{1/2}}{m^{1/2}}\right]} = m^{1/4}\,(2k)^{1/4}\,\sqrt{\frac{1}{\hbar}} \qquad (6.67)$$

The ground state eigenfunction for $U_2(x)$ is

$$\psi_0'(x) = \frac{\sqrt{\beta}}{\pi^{1/4}} e^{-\beta^2 x^2/2} \qquad (6.68)$$

where

$$\beta = m^{1/4} (k)^{1/4} \sqrt{\frac{1}{\hbar}} = 2^{-1/4}\alpha \qquad (6.69)$$

After the potential suddenly changes from $U_1(x)$ to $U_2(x)$ the wave function remains that in Eq. (6.66), but it is *not* an eigenfunction for the new potential. As in Problem 5 we can expand the original wave function in Eq. (6.66) in terms of the eigenfunctions for the new potential. We use the bra/ket notation commonly used for the harmonic oscillator to make our equations more concise. In terms of kets we may write

$$|0\rangle = \sum_i c_i |n'\rangle \qquad (6.70)$$

where $|0\rangle$ corresponds to the eigenfunction in Eq. (6.66) and the $|n'\rangle$ are the eigenkets of the harmonic oscillator in the potential $U_2(x)$.

We require only c_0 because $|c_0|^2$ is the desired probability so we take the inner product of both sides with $\langle 0'|$. Because the $|n'\rangle$ are orthonormal we have

$$\langle 0' |0\rangle = c_0$$

$$= \int_{-\infty}^{\infty} \frac{\sqrt{\beta}}{\pi^{1/4}} e^{-\beta^2 x^2/2} \frac{\sqrt{\alpha}}{\pi^{1/4}} e^{-\alpha^2 x^2/2} dx$$

$$= \sqrt{\frac{\alpha\beta}{\pi}} \int_{-\infty}^{\infty} e^{-(\alpha^2+\beta^2)x^2/2} dx \qquad (6.71)$$

This integral is given in Eq. (G.3).

$$\int_{-\infty}^{\infty} e^{-\gamma^2 y^2} dy = \sqrt{\frac{\pi}{\gamma^2}} \qquad (6.72)$$

In this case $\gamma^2 = \left(\alpha^2 + \beta^2\right)/2$ so

$$c_0 = \sqrt{\frac{\alpha\beta}{\pi}} \frac{\sqrt{\pi}}{\sqrt{\frac{\left(\alpha^2+\beta^2\right)}{2}}} = \sqrt{\frac{2\alpha\beta}{\alpha^2+\beta^2}}$$

$$= \sqrt{\frac{2\alpha\left(2^{-1/4}\alpha\right)}{\alpha^2+\alpha^2/\sqrt{2}}} = \sqrt{\frac{2^{3/4}}{1+1/\sqrt{2}}} \qquad (6.73)$$

Squaring, we have the probability P that the system will be found in the ground state on the new oscillator.

$$P = |c_0|^2 = \frac{2^{3/4}}{1 + 1/\sqrt{2}} \approx 0.98 \qquad (6.74)$$

Thus, the overlap between the ground-state eigenfunctions of the two parabolic potential wells is nearly unity as may be seen in Fig. 6.14.

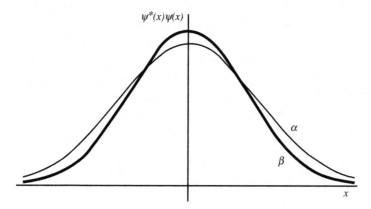

Fig. 6.14 Problem 9—solution

(b) As in Problem 7 parity considerations make the answer to this part very simple. It is impossible for there to be any odd contribution to an even wave function so the answer is zero.

Chapter 7
Ladder Operators for the Harmonic Oscillator

The ladder operator method of solving the harmonic oscillator problem is not only elegant, but extremely useful. In fact, the general method transcends the harmonic oscillator inasmuch as there are other systems for which ladder operators exist, most notably angular momentum. We devote an entire section to these operators because of their importance for solving problems and for understanding how to use ladder operators (also known as raising and lowering operators).

Because these operators simplify many problems it is usually advisable to employ them whenever possible. We state this as a general rule:

- **Always attempt to solve harmonic oscillator problems using ladder operators before embarking on arduous calculations.**

Although the definition of harmonic oscillator ladder operators is not universal the most commonly used definition is

$$\hat{a} = \sqrt{\frac{m\omega}{2\hbar}} \left(\hat{x} + \frac{i}{m\omega} \hat{p} \right) = \frac{1}{\sqrt{2}} \left(\alpha \hat{x} + i\frac{1}{\alpha\hbar} \hat{p} \right) \tag{7.1}$$

and its complex conjugate

$$\hat{a}^\dagger = \sqrt{\frac{m\omega}{2\hbar}} \left(\hat{x} - \frac{i}{m\omega} \hat{p} \right) = \frac{1}{\sqrt{2}} \left(\alpha \hat{x} - i\frac{1}{\alpha\hbar} \hat{p} \right) \tag{7.2}$$

where

$$\alpha = \sqrt{\frac{m\omega}{\hbar}} \tag{7.3}$$

© Springer International Publishing AG 2017
J.D. Kelley, J.J. Leventhal, *Problems in Classical and Quantum Mechanics*, DOI 10.1007/978-3-319-46664-4_7

As will be shown in Problem 2

$$\hat{a} = \text{lowering operator} \tag{7.4}$$

$$\hat{a}^\dagger = \text{raising operator} \tag{7.5}$$

A number of relations involving the harmonic oscillator oscillator ladder operators are summarized in Table O.2.

Problems

1. In preparation for Problem 2 of this chapter you are asked to verify a fundamental commutation relation, that between the position operator $\hat{x}$ and the momentum operator $\hat{p}$. Commutation relations are discussed in Appendix N. In one dimension the $\hat{x}, \hat{p}$ commutation relation is

$$[\hat{x}, \hat{p}] = \hat{x}\hat{p} - \hat{p}\hat{x} = i\hbar \tag{7.6}$$

To verify this relation operate on an arbitrary wave function $\psi(x)$ with this commutator. It will be helpful to recall that the momentum operator equivalent of $\hat{p}$ is

$$\hat{p} = \frac{\hbar}{i}\frac{d}{dx} \tag{7.7}$$

Also, verify Eq. (7.6) for an arbitrary wave function $\phi(p)$ using operator equivalent of $\hat{x}$ in momentum space

$$\hat{x} = -\frac{\hbar}{i}\frac{d}{dp} \tag{7.8}$$

Solution

Using the wave function $\psi(x)$ demands that we replace the momentum operator $\hat{p}$ with it position space differential operator so we write

$$[\hat{x}, \hat{p}]\psi(x) = (\hat{x}\hat{p} - \hat{p}\hat{x})\psi(x)$$
$$= x\frac{\hbar}{i}\frac{d\psi(x)}{dx} - \frac{\hbar}{i}\frac{d[x\psi(x)]}{dx}$$
$$= x\frac{\hbar}{i}\frac{d\psi(x)}{dx} - x\frac{\hbar}{i}\frac{d\psi(x)}{dx} - \frac{\hbar}{i}\psi(x)$$
$$= i\hbar\psi(x) \tag{7.9}$$

Analogously we write

$$[\hat{x}, \hat{p}] \, \phi \, (p) = (\hat{x}\hat{p} - \hat{p}\hat{x}) \, \phi \, (p)$$

$$= -\frac{\hbar}{i} \frac{d \, [p\phi \, (p)]}{dp} + p\frac{\hbar}{i} \frac{d\phi \, (p)}{dp}$$

$$= -\frac{\hbar}{i} p \frac{d\phi \, (p)}{dp} - \frac{\hbar}{i} \phi \, (p) + p\frac{\hbar}{i} \frac{d\phi \, (p)}{dp}$$

$$= i\hbar \phi \, (p) \tag{7.10}$$

2. This problem demonstrates some of the essential features of the ladder operators.

(a) Show that $\left[\hat{a}, \hat{a}^{\dagger}\right] = 1$.

(b) Show that $\hat{H} = \hbar\omega \left(\hat{a}^{\dagger}\hat{a} + \frac{1}{2}\right)$.

(c) Define $\hat{N} = \hat{a}^{\dagger}\hat{a}$ and show that $\left[\hat{N}, \hat{H}\right] = 0$ and therefore $\hat{N}$ and $\hat{H}$ have simultaneous eigenkets.

(d) Show that $\left[\hat{N}, \hat{a}\right] = -\hat{a}$ and $\left[\hat{N}, \hat{a}^{\dagger}\right] = \hat{a}^{\dagger}$.

(e) Show that, if $|n\rangle$ is an eigenstate of $\hat{N}$ with eigenvalue n, i.e. $\hat{a}^{\dagger}\hat{a} \, |n\rangle = n \, |n\rangle$ then $\hat{a}^{\dagger} \, |n\rangle$ and $\hat{a} \, |n\rangle$ are also eigenstates of $\hat{N}$ with eigenvalues $\sqrt{(n+1)}$ and $\sqrt{(n-1)}$, respectively. Moreover,

$$\hat{a} \, |n\rangle = \sqrt{n} \, |n-1\rangle \tag{7.11}$$

and

$$\hat{a}^{\dagger} \, |n\rangle = \sqrt{n+1} \, |n+1\rangle \tag{7.12}$$

justifying the designations of $\hat{a}^{\dagger}$ and $\hat{a}$ as raising and lowering operators.

(f) Show that

$$\hat{H} \, |n\rangle = \hbar\omega \left(n + \frac{1}{2}\right) |n\rangle \tag{7.13}$$

so that the energy eigenvalues are given by $E_n = \left(n + \frac{1}{2}\right) \hbar\omega$.

(g) Construct the ground state eigenfunction in coordinate space.

Solution

(a) Using the definitions of the ladder operators, Eqs. (7.1) and (7.2)

$$[\hat{a}, \hat{a}^\dagger] = \frac{1}{2} \left[\left(\alpha \hat{x} + i \frac{1}{\alpha \hbar} \hat{p} \right), \left(\alpha \hat{x} - i \frac{1}{\alpha \hbar} \hat{p} \right) \right]$$

$$= \frac{1}{2} \left\{ \left(-\frac{i}{\hbar} \right) [\hat{x}, \hat{p}] + \left(\frac{i}{\hbar} \right) [\hat{p}, \hat{x}] \right\}$$

$$= \frac{-i}{2\hbar} \{ 2 [\hat{x}, \hat{p}] \}$$

$$= 1 \qquad\qquad\qquad (7.14)$$

(b) Solving Eqs. (7.1) and (7.2) for $\hat{x}$ and $\hat{p}$ we have

$$\hat{x} = \sqrt{\frac{\hbar}{2m\omega}} \, (\hat{a} + \hat{a}^\dagger) = \frac{1}{\sqrt{2}\alpha} (\hat{a} + \hat{a}^\dagger) \qquad\qquad (7.15)$$

and

$$\hat{p} = -i\sqrt{\frac{m\omega\hbar}{2}} \, (\hat{a} - \hat{a}^\dagger) = -i \frac{\alpha\hbar}{\sqrt{2}} (\hat{a} - \hat{a}^\dagger) \qquad\qquad (7.16)$$

Inserting these in the Hamiltonian we have

$$\hat{H} = \frac{\hat{p}^2}{2m} + \frac{1}{2} m\omega^2 \hat{x}^2$$

$$= -\frac{\hbar\omega}{4} (\hat{a} - \hat{a}^\dagger)^2 + \frac{\hbar\omega}{4} (\hat{a} + \hat{a}^\dagger)^2$$

$$= \frac{\hbar\omega}{4} \left\{ -(\hat{a}\hat{a} - \hat{a}\hat{a}^\dagger - \hat{a}^\dagger\hat{a} + \hat{a}^\dagger\hat{a}^\dagger) + (\hat{a}\hat{a} + \hat{a}\hat{a}^\dagger + \hat{a}^\dagger\hat{a} + \hat{a}^\dagger\hat{a}^\dagger) \right\}$$

$$= \frac{\hbar\omega}{2} (\hat{a}^\dagger\hat{a} + \hat{a}\hat{a}^\dagger) \qquad\qquad (7.17)$$

From Eq. (7.14) we have

$$\hat{a}\hat{a}^\dagger = 1 + \hat{a}^\dagger\hat{a} \qquad\qquad (7.18)$$

so

$$\hat{H} = \frac{\hbar\omega}{2} (\hat{a}^\dagger\hat{a} + 1 + \hat{a}^\dagger\hat{a})$$

$$= \hbar\omega \left(\hat{a}^\dagger\hat{a} + \frac{1}{2} \right) \qquad\qquad (7.19)$$

(c) Using part (b) we have

$$\left[\hat{N}, \hat{H}\right] = \left[\hat{a}^\dagger \hat{a}, \hat{H}\right] = \hbar\omega \left[\hat{a}^\dagger \hat{a}, \left(\hat{a}^\dagger \hat{a} + \frac{1}{2}\right)\right] \equiv 0 \qquad (7.20)$$

(d) Using Eq. (N.3) we have

$$\left[\hat{N}, \hat{a}\right] = \left[\hat{a}^\dagger \hat{a}, \hat{a}\right] = \left[\hat{a}^\dagger, \hat{a}\right]\hat{a} + \hat{a}^\dagger \left[\hat{a}, \hat{a}\right] = -\hat{a} \qquad (7.21)$$

and

$$\left[\hat{N}, \hat{a}^\dagger\right] = \left[\hat{a}^\dagger \hat{a}, \hat{a}^\dagger\right] = \left[\hat{a}^\dagger, \hat{a}^\dagger\right]\hat{a} + \hat{a}^\dagger \left[\hat{a}, \hat{a}^\dagger\right] = \hat{a}^\dagger \qquad (7.22)$$

(e) Operate on the vector $\hat{a}\,|n\rangle$ with $\hat{N}$ and use the commutation relation Eq. (7.21) to obtain

$$\hat{N}\hat{a}\,|n\rangle = \left(-\hat{a} + \hat{a}\hat{N}\right)|n\rangle$$
$$\hat{N}\{\hat{a}\,|n\rangle\} = (-\hat{a} + \hat{a}n)\,|n\rangle$$
$$= (n - 1)\{\hat{a}\,|n\rangle\} \qquad (7.23)$$

Similarly, operate on $\hat{a}^\dagger\,|n\rangle$ with $\hat{N}$ and use the commutation relation Eq. (7.22) to obtain

$$\hat{N}\hat{a}^\dagger\,|n\rangle = \left(\hat{a}^\dagger + \hat{a}^\dagger \hat{N}\right)|n\rangle$$
$$\hat{N}\{\hat{a}^\dagger\,|n\rangle\} = (\hat{a}^\dagger + \hat{a}^\dagger n)\,|n\rangle$$
$$= (n + 1)\{\hat{a}^\dagger\,|n\rangle\} \qquad (7.24)$$

Equations (7.23) and (7.24) show that

$$\hat{a}\,|n\rangle = c_1\,|n - 1\rangle \qquad (7.25)$$

and

$$\hat{a}^\dagger\,|n\rangle = c_2\,|n + 1\rangle \qquad (7.26)$$

Now we must find c_1 and c_2. Because the eigenvectors must be normalized we must have

$$1 = \langle n - 1\,|n - 1\rangle$$
$$= \frac{1}{|c_1|^2}\,\langle n|\,\hat{a}^\dagger \hat{a}\,|n\rangle$$

$$= \frac{1}{|c_1|^2} \langle n| \hat{N} |n \rangle$$

$$= \frac{n}{|c_1|^2} \tag{7.27}$$

Choose c_1 to be real so $c_1 = \sqrt{n}$ and

$$\hat{a} |n\rangle = \sqrt{n} |n - 1\rangle \tag{7.28}$$

Similarly

$$\hat{a}^\dagger |n\rangle = \sqrt{n + 1} |n + 1\rangle \tag{7.29}$$

(f) From parts 2(b) and 2(e) of this problem we know that

$$\hat{H} = \hbar \omega \left(\hat{a}^\dagger \hat{a} + \frac{1}{2} \right) \quad \text{and} \quad \hat{N} = \hat{a}^\dagger \hat{a} \tag{7.30}$$

so

$$\hat{H} = \hbar \omega \left(\hat{N} + \frac{1}{2} \right) \tag{7.31}$$

Also

$$\hat{a}^\dagger \hat{a} |n\rangle = \hat{N} |n\rangle = n |n\rangle \tag{7.32}$$

Therefore

$$\hat{H} |n\rangle = \hbar \omega \left(\hat{N} + \frac{1}{2} \right) |n\rangle$$

$$\hbar \omega \left(n + \frac{1}{2} \right) |n\rangle \tag{7.33}$$

so, clearly

$$E_n = \left(n + \frac{1}{2} \right) \hbar \omega \tag{7.34}$$

(g) Use the fact that the lowering operator lowers the ground state eigenket to zero. Therefore

$$\hat{a} |\psi_0(x)\rangle = 0 \tag{7.35}$$

In coordinate space this becomes

$$\sqrt{\frac{m\omega}{2\hbar}} \left(x + \frac{\hbar}{m\omega} \frac{d}{dx} \right) \psi_0(x) = 0 \tag{7.36}$$

The last equation is a separable first-order differential equation which is easily solvable. The solution is

$$\psi_0(x) = Ke^{-\alpha^2 x^2/2} \tag{7.37}$$

where K is a constant of integration, which we obtain by normalization. Assuming K is real

$$\int_{-\infty}^{\infty} |\psi_0(x)|^2 \, dx = 1$$

$$K^2 \int_{-\infty}^{\infty} e^{-\alpha^2 x^2} = 1 \tag{7.38}$$

Using the definite integral given in Eq. (G.3) we have

$$K^2 \frac{\sqrt{\pi}}{\alpha} = 1 \Longrightarrow K = \left(\frac{\alpha^2}{\pi}\right)^{1/4} \tag{7.39}$$

so

$$\psi_0(x) = \left(\frac{\alpha^2}{\pi}\right)^{1/4} e^{-\alpha^2 x^2/2} \tag{7.40}$$

3. Aspects of this problem, particularly part (c), are identical with Problem 9 of Chap. 5. The method of solution is, however, different.

The state vector at $t = 0$ for a one-dimensional harmonic oscillator is given by

$$|\Psi(0)\rangle = \frac{1}{\sqrt{5}}|1\rangle + \frac{2}{\sqrt{5}}|2\rangle \tag{7.41}$$

where $|n\rangle$ is an eigenket of the time independent harmonic oscillator Hamiltonian. Use the ladder operators

$$\hat{a} = \frac{1}{\sqrt{2}}\left(\alpha\hat{x} + i\frac{1}{\alpha\hbar}\hat{p}\right)$$

$$\hat{a}^\dagger = \frac{1}{\sqrt{2}}\left(\alpha\hat{x} - i\frac{1}{\alpha\hbar}\hat{p}\right) \tag{7.42}$$

and their properties as given in Problem 2 of this chapter for the computations in part (c).

(a) Find $|\Psi(t)\rangle$, the state vector for $t > 0$.
(b) Find the expectation value of the energy for $t > 0$ using Dirac notation.
(c) Find $\langle \hat{x}(t) \rangle$, the expectation value of the position for $t > 0$ using ladder operators.

Solution

(a) Simply multiply each term in the expansion by the time dependent part of the eigenfunction. Therefore

$$|\Psi(t)\rangle = \frac{1}{\sqrt{5}} |1\rangle\, e^{-3i\omega t/2} + \frac{2}{\sqrt{5}} |2\rangle\, e^{-5i\omega t/2} \qquad (7.43)$$

(b)

$$\langle E \rangle = \langle \Psi(t)| \hat{H} |\Psi(t)\rangle$$

$$= \frac{1}{5} E_1 \langle 1 |1\rangle + \frac{4}{5} E_2 \langle 2 |2\rangle$$

$$= \frac{1}{5}\frac{3}{2}\hbar\omega + \frac{4}{5}\frac{5}{2}\hbar\omega = \frac{23}{10}\hbar\omega \qquad (7.44)$$

(c)

$$\langle \hat{x}(t)\rangle = \langle \Psi(t)| \hat{x} |\Psi(t)\rangle \qquad (7.45)$$

Solving Eq. (7.42) for $\hat{x}$ we have

$$\hat{x} = \frac{1}{\sqrt{2}\alpha}\left(\hat{a} + \hat{a}^\dagger\right) \qquad (7.46)$$

Inserting $\hat{x}$ from Eq. (7.46) into Eq. (7.45) we have

$$\sqrt{2}\alpha\,\langle \hat{x}(t)\rangle = \frac{1}{5} \langle 1|\left(\hat{a}+\hat{a}^\dagger\right)|1\rangle + \frac{2}{5} \langle 1|\left(\hat{a}+\hat{a}^\dagger\right)|2\rangle\, e^{i\omega t}$$

$$+ \frac{2}{5} \langle 2|\left(\hat{a}+\hat{a}^\dagger\right)|1\rangle\, e^{-i\omega t} + \frac{4}{5} \langle 2|\left(\hat{a}+\hat{a}^\dagger\right)|2\rangle \quad (7.47)$$

The first and last terms of Eq. (7.47) clearly vanish so, using Eqs. (7.28) and (7.29) on the remaining terms we have

$$\sqrt{2}\alpha\,\langle \hat{x}(t)\rangle = \frac{2}{5} \langle 1|\left(\hat{a}+\hat{a}^\dagger\right)|2\rangle\, e^{i\omega t} + \frac{2}{5} \langle 2|\left(\hat{a}+\hat{a}^\dagger\right)|1\rangle\, e^{-i\omega t}$$

$$= \frac{2}{5}\left[\langle 1|\hat{a}|2\rangle + \langle 1|\hat{a}^\dagger|2\rangle\right] e^{i\omega t}$$

$$+ \frac{2}{5}\left[\langle 2|\hat{a}|1\rangle + \langle 2|\hat{a}^\dagger|1\rangle\right] e^{-i\omega t}$$

$$= \frac{2}{5}\sqrt{2}\,\langle 1 |1\rangle\, e^{i\omega t} + \frac{2}{5}\sqrt{2}\,\langle 2 |2\rangle\, e^{-i\omega t}$$

$$= \frac{2}{5}\sqrt{2}e^{i\omega t}\left\{e^{i\omega t} + e^{-i\omega t}\right\}$$

$$= \frac{4}{5}\sqrt{2}\cos\omega t \qquad (7.48)$$

So

$$\langle\hat{x}(t)\rangle = \frac{4}{5\alpha}\cos\omega t = \frac{4}{5}\sqrt{\frac{m\omega}{\hbar}}\cos\omega t \qquad (7.49)$$

which is the result that was obtained in Problem 9 of Chap. 5, but here the matrix element $\langle 1|\hat{x}|2\rangle$ was evaluated using ladder operators. This was not only more expeditious, but it also reduced the possibility of error. The comparison of the solution to this problem with that to Problem 9 of Chap. 5 illustrates the utility of the "rule" stated in the introduction to this chapter: **Always attempt to solve harmonic oscillator problems using ladder operators before embarking on arduous calculations.**

4. The normalized state vector at $t = 0$ for a particle subject to a one-dimensional harmonic oscillator potential is

$$|\Psi(x,0)\rangle = \frac{1}{\sqrt{5}}|2\rangle + \frac{2}{\sqrt{5}}|4\rangle \qquad (7.50)$$

where the $|n\rangle$ are eigenvectors of the Hamiltonian and the number operator N. Let $\omega = \sqrt{k/m}$.

(a) Find the state vector as a function of time $|\Psi(x,t)\rangle$.
(b) Find the expectation value of the energy as a function of time. Does the answer agree with that predicted by the Ehrenfest theorem [1]?
(c) Find the expectation value of the position as a function of time using ladder operators.

Solution

(a) The time dependent state vector is obtained by multiplying each "component" of $|\Psi(x,0)\rangle$ by

$$\exp(-iE_n t/\hbar) = \exp\left[-i\left(n + \frac{1}{2}\right)t/\hbar\right] \qquad (7.51)$$

with the appropriate value of n. This gives

$$|\Psi(x,t)\rangle = \frac{1}{\sqrt{5}}|2\rangle\, e^{-(5/2)i\omega t} + \frac{2}{\sqrt{5}}|4\rangle\, e^{-(9/2)i\omega t} \qquad (7.52)$$

(b)

$$\langle E \rangle = \langle \Psi(x,t)| \hat{H} |\Psi(x,t)\rangle$$

$$= \frac{1}{5}\left(\frac{5}{2}\hbar\omega\right) + \frac{4}{5}\left(\frac{9}{2}\hbar\omega\right)$$

$$= \frac{1}{5}\left(\frac{5+36}{2}\right)\hbar\omega = \frac{41}{10}\hbar\omega \qquad (7.53)$$

The important point here is that $\langle E \rangle$ is independent of time. This is expected based on the Ehrenfest theorem which, for an operator $\hat{A}$ that does not contain the time explicitly, is [1]

$$\frac{d\langle \hat{A} \rangle}{dt} = \frac{i}{\hbar}\langle \Psi| \left[\hat{H}, \hat{A}\right]|\Psi\rangle \qquad (7.54)$$

Inasmuch as $\hat{H}$ commutes with itself, $\langle E \rangle = \langle \hat{H} \rangle$ is independent of time.

(c) From Eq. (7.15)

$$\hat{x} = \sqrt{\frac{\hbar}{2m\omega}}\left(\hat{a} + \hat{a}^{\dagger}\right) \qquad (7.55)$$

the expectation value $\langle \hat{x}(t) \rangle$ is

$$\langle \hat{x}(t) \rangle = \sqrt{\frac{\hbar}{2m\omega}}\langle \Psi(x,t)| \left(\hat{a} + \hat{a}^{\dagger}\right)|\Psi(x,t)\rangle \qquad (7.56)$$

We need not, however, write out these inner products. First we note that the exponential time factors cancel. Second, the only inner products that will occur are of the form $\langle 2| \hat{a}^{\dagger} |4\rangle$, $\langle 2| \hat{a}^{\dagger} |2\rangle$ and analogous inner products with the operator $\hat{a}$. All these inner products vanish because the actions of both $\hat{a}$ and $\hat{a}^{\dagger}$ only change the bra or ket by one [see Eq. (7.55)]. Therefore, $\langle \hat{x}(t) \rangle = 0$.

5. The state vector at $t = 0$ for a particle in a harmonic oscillator potential is given by

$$|\Psi(x,0)\rangle = \frac{1}{\sqrt{3}}|1\rangle + \sqrt{\frac{2}{3}}|2\rangle \qquad (7.57)$$

where the $|n\rangle$ are eigenvectors of the Hamiltonian and the number operator.

(a) Find the state vector as a function of time $|\Psi(x,t)\rangle$.
(b) Find the expectation value of the energy as a function of time. Does the answer agree with that predicted by the Ehrenfest theorem?
(c) Find the expectation value of the position as a function of time using ladder operators. Calculate this expectation value using the Ehrenfest theorem and compare the results.

Solution

(a) With $\omega = \sqrt{k/m}$ the time dependent state vector is obtained by multiplying each "component" of the state vector by the exponential factor containing the energy, that is

$$\exp\left(-iE_n t/\hbar\right) = \exp\left[-i\left(n + \frac{1}{2}\right)t/\hbar\right] \tag{7.58}$$

so

$$|\Psi\left(x,t\right)\rangle = \frac{1}{\sqrt{3}}|1\rangle\, e^{-(3/2)i\omega t} + \sqrt{\frac{2}{3}}\,|2\rangle\, e^{-(5/2)i\omega t} \tag{7.59}$$

(b)

$$\langle E \rangle = \langle \Psi\left(x,t\right)|\,\hat{H}\,|\Psi\left(x,t\right)\rangle$$
$$= \frac{1}{3}\left(\frac{3}{2}\hbar\omega\right) + \frac{2}{3}\left(\frac{5}{2}\hbar\omega\right)$$
$$= \left(\frac{13}{6}\hbar\omega\right) \tag{7.60}$$

which is independent of time and agrees with the Ehrenfest theorem as in Problem 4 of this chapter.

(c) In this problem we are not as fortunate as we were in Problem 4 of this chapter where the basis kets that constituted $|\Psi\left(x,t\right)\rangle$ were both even so that all cross (inner) products vanished. Here we must actually do the computations. We can, however, ignore those cross products with a ladder operator sandwiched between a bra and ket of equal number because they vanish. Using Eq. (7.15)

$$\hat{x} = \sqrt{\frac{\hbar}{2m\omega}}\left(\hat{a} + \hat{a}^\dagger\right) \tag{7.61}$$

we have

$$\langle \hat{x} \rangle = \sqrt{\frac{\hbar}{2m\omega}}\,\langle \Psi\left(x,t\right)|\left(\hat{a} + \hat{a}^\dagger\right)|\Psi\left(x,t\right)\rangle$$
$$= \sqrt{\frac{\hbar}{2m\omega}}\,\frac{\sqrt{2}}{3}\left\{\langle 2|\, e^{(5/2)i\omega t}\left(\hat{a} + \hat{a}^\dagger\right)|1\rangle\, e^{-(3/2)i\omega t}\right.$$
$$\left. + \langle 1|\, e^{(3/2)i\omega t}\left(\hat{a} + \hat{a}^\dagger\right)|2\rangle\, e^{-(5/2)i\omega t}\right\} \tag{7.62}$$

Again leaving out the terms that vanish due to orthogonality and using

$$\hat{a}\,|n\rangle = \sqrt{n}\,|n-1\rangle \quad \text{and} \quad \hat{a}^\dagger\,|n\rangle = \sqrt{n+1}\,|n+1\rangle \tag{7.63}$$

we have

$$
\begin{aligned}
\langle \hat{x} \rangle &= \sqrt{\frac{\hbar}{2m\omega}}\,\frac{\sqrt{2}}{3}\left[e^{i\omega t}\,\langle 2|\,\hat{a}^\dagger\,|1\rangle + e^{-i\omega t}\,\langle 1|\,\hat{a}\,|2\rangle\right] \\
&= \sqrt{\frac{\hbar}{2m\omega}}\,\frac{\sqrt{2}}{3}\left(e^{i\omega t}\sqrt{2} + e^{-i\omega t}\sqrt{2}\right) \\
&= \sqrt{\frac{\hbar}{m\omega}}\,\frac{\sqrt{2}}{3}2\cos\omega t
\end{aligned}
\tag{7.64}
$$

Now, let us do the computation using Ehrenfest's theorem, Eq. (7.54). We require the commutator $\left[\hat{H},\hat{x}\right] = \frac{1}{2m}\left[\hat{p}^2,\hat{x}\right]\left[\hat{p}^2,\hat{x}\right]$ which is easily calculated using the identity $\left[\hat{x},\hat{p}_x^n\right] = i\hbar n\hat{p}_x^{n-1}$ (see, e.g., [1]).

$$\left[\hat{H},\hat{x}\right] = -\frac{1}{2m}\left[\hat{x},\hat{p}^2\right] = -\frac{i\hbar}{m}\hat{p} \tag{7.65}$$

Therefore, according to the Ehrenfest theorem

$$
\begin{aligned}
\frac{d\langle\hat{x}\rangle}{dt} &= \frac{i}{\hbar}\,\langle\Psi|\left[\hat{H},\hat{x}\right]|\Psi\rangle \\
&= \left(\frac{i}{\hbar}\right)\left(-\frac{i\hbar}{m}\right)\langle\Psi|\hat{p}\,|\Psi\rangle \\
&= \frac{1}{m}\langle\Psi|\hat{p}\,|\Psi\rangle = \frac{\langle\hat{p}\rangle}{m}
\end{aligned}
\tag{7.66}
$$

The relationship between $d\langle\hat{x}\rangle/dt$ and $\langle\hat{p}\rangle$ is identical to that for a classical oscillator, i.e. $dx/dt = p/m$ [Eq. (2.33)]. This classical-quantum correspondence is unique to systems quadratic in p and x.
To evaluate this inner product we must use Eq. (7.16)

$$\hat{p} = -i\sqrt{\frac{m\omega\hbar}{2}}\,(\hat{a}-\hat{a}^\dagger) \tag{7.67}$$

The resulting equation will be identical with Eq. (7.64) except that the constant $\sqrt{\dfrac{\hbar}{2m\omega}} \rightarrow -i\sqrt{\dfrac{m\omega\hbar}{2}}$ and $\hat{a}^\dagger \rightarrow -\hat{a}^\dagger$. We have

$$\frac{d \langle \hat{x} \rangle}{dt} = -i \sqrt{\frac{m\omega\hbar}{2}} \frac{\sqrt{2}}{3} \left(\frac{1}{m}\right) \left[e^{i\omega t} \langle 2| \hat{a}^\dagger |1\rangle + e^{-i\omega t} \langle 1| \hat{a} |2\rangle\right]$$

$$= -i \sqrt{\frac{\hbar\omega}{m}} \frac{\sqrt{2}}{3} \left(-e^{i\omega t} + e^{-i\omega t}\right)$$

$$= -\sqrt{\frac{\hbar\omega}{m}} \frac{2\sqrt{2}}{3} \left(\frac{e^{i\omega t} - e^{-i\omega t}}{2i}\right)$$

$$= -\sqrt{\frac{\hbar\omega}{m}} \frac{\sqrt{2}}{3} 2\sin\omega t \tag{7.68}$$

Integrating, we have

$$\langle \hat{x} \rangle = \sqrt{\frac{\hbar\omega}{m}} \frac{\sqrt{2}}{3} 2 \left(\frac{1}{\omega} \cos\omega t\right)$$

$$= \sqrt{\frac{\hbar}{m\omega}} \frac{\sqrt{2}}{3} 2 \cos\omega t \tag{7.69}$$

where we have set the constant of integration equal to zero. This result is in agreement with that obtained using ladder operators, Eq. (7.64).

6. Computation of harmonic oscillator matrix elements using the eigenfunctions involving Hermite polynomials can be carried out, but it is tedious. Find the following matrix elements using any method you choose. Ladder operators are recommended.

(a) Use the $\langle 1| \hat{x} |2\rangle$ matrix element evaluated in Problem 3 of this chapter to find the general expression for $\langle m| \hat{x} |n\rangle$.
(b) Find the general expression for $\langle m| \hat{p} |n\rangle$.

Solution

(a) As shown in Eq. (7.15), the position operator may be written as

$$\hat{x} = \sqrt{\frac{\hbar}{2m\omega}} (\hat{a} + \hat{a}^\dagger) = \frac{1}{\sqrt{2}\alpha} (\hat{a} + \hat{a}^\dagger) \tag{7.70}$$

so the required matrix element is

$$\langle m| \hat{x} |n\rangle = \frac{1}{\sqrt{2}\alpha} \left(\langle m| \hat{a} |n\rangle + \langle m| \hat{a}^\dagger |n\rangle\right) \tag{7.71}$$

We already know the actions of $\hat{a}$ and $\hat{a}^\dagger$ on the eigenkets $|n\rangle$ of the harmonic oscillator. They are given by Eqs. (7.28) and (7.29) which are

$$\hat{a}\,|n\rangle = \sqrt{n}\,|n-1\rangle$$
$$\hat{a}^\dagger\,|n\rangle = \sqrt{n+1}\,|n+1\rangle \tag{7.72}$$

Therefore, the first term on the right-hand side of Eq. (7.71) is non-vanishing only when $m = n-1$ while the second terms vanishes unless $m = n+1$. This is written compactly in terms of Kronecker deltas as

$$\langle m|\,\hat{x}\,|n\rangle = \frac{1}{\sqrt{2}\alpha}\left(\sqrt{n}\,\delta_{m,n-1} + \sqrt{n+1}\,\delta_{m,n+1}\right) \tag{7.73}$$

Note the if $m = n$ then Eq. (7.73) gives the expectation value of x for any oscillator state $|n\rangle$, $\langle x\rangle_n \equiv 0$. This is the obvious answer because the oscillator potential is symmetric about the ordinate (even). Mathematically this is equivalent to saying that $\langle n|\,\hat{x}\,|n\rangle$ represents an odd integral over symmetric limits.

(b) Evaluation of $\langle m|\,\hat{p}\,|n\rangle$ proceeds in exactly the same manner as that for $\langle m|\,\hat{x}\,|n\rangle$. Using Eqs. (7.16) and (7.72) we have and

$$\langle m|\,\hat{p}\,|n\rangle = -i\frac{\alpha\hbar}{\sqrt{2}}\left(\langle m|\,\hat{a}\,|n\rangle - \langle m|\,\hat{a}^\dagger\,|n\rangle\right)$$

$$= -i\frac{\alpha\hbar}{\sqrt{2}}\left(\sqrt{n}\,\delta_{m,n-1} - \sqrt{n+1}\,\delta_{m,n+1}\right) \tag{7.74}$$

If $m = n$, then Eq. (7.74) gives the expectation value of the momentum, which is zero for the same reason $\langle m|\,\hat{x}\,|n\rangle$ is zero.

7. Find the expectation value of $\hat{p}^2$ for the nth eigenstate of the harmonic oscillator. Find the average value of the kinetic energy of a particle $\left(\hat{T}\right)$ in the nth eigenstate and relate it to the total energy of the nth state. Show that the result is consistent with the virial theorem, which, for any one-dimensional potential, is

$$2\left\langle \hat{T}\right\rangle = \left\langle x\frac{dU}{dx}\right\rangle \tag{7.75}$$

Solution

Again using Eqs. (7.16) and (7.72) we have

$$\langle n|\,\hat{p}^2\,|n\rangle = -\frac{\alpha^2\hbar^2}{2}\,\langle n|\,(\hat{a}-\hat{a}^\dagger)(\hat{a}-\hat{a}^\dagger)\,|n\rangle$$

$$= \left(\frac{m\omega}{\hbar}\right)\frac{\hbar^2}{2}\left(\langle n|\,\hat{a}\hat{a}^\dagger + \hat{a}^\dagger\hat{a}\,|n\rangle\right)$$

$$= \frac{m\omega\hbar}{2} \left(\langle n| \, \hat{a} \, \sqrt{n+1} \, |n+1\rangle + \langle n| \, \hat{a}^\dagger \, \sqrt{n} \, |n-1\rangle \right)$$

$$= \frac{m\omega\hbar}{2} (n+1+n)$$

$$= m \left[\left(n + \frac{1}{2} \right) \hbar\omega \right] \tag{7.76}$$

The average value of the kinetic energy is

$$\langle n| \, \hat{T} \, |n\rangle = \langle n| \, \frac{\hat{p}^2}{2m} \, |n\rangle$$

$$= \frac{1}{2} E_n \tag{7.77}$$

The one-dimensional virial theorem states

$$2\langle \hat{T} \rangle = \left\langle x \frac{dU}{dx} \right\rangle \tag{7.78}$$

so, for the harmonic oscillator

$$\langle \hat{T} \rangle = \frac{1}{2} \left\langle x \frac{d}{dx} \frac{1}{2} m\omega^2 x^2 \right\rangle$$

$$= \frac{1}{2} m\omega^2 \langle x^2 \rangle \tag{7.79}$$

Therefore

$$\langle x^2 \rangle = \left[\left(n + \frac{1}{2} \right) \frac{\hbar}{m\omega} \right] \tag{7.80}$$

Moreover, the expectation value of the potential energy is the other half of the total energy, that is,

$$\langle U \rangle = \frac{1}{2} k \langle x^2 \rangle = \frac{1}{2} m\omega^2 \langle x^2 \rangle = \langle \hat{T} \rangle \tag{7.81}$$

which is consistent with the virial theorem.

8. Use the matrix element for $\langle m| \, \hat{x}^2 \, |n\rangle$, Eq. (7.82) below, to find the expectation value of $\hat{x}^4$ for arbitrary state $|n\rangle$ of the harmonic oscillator.

$$\langle m| \, \hat{x}^2 \, |n\rangle = \frac{1}{2\alpha^2} \left[\sqrt{n(n-1)} \delta_{m,n-2} \right.$$

$$\left. + (2n+1) \, \delta_{m,n} + \sqrt{(n+1)(n+2)} \delta_{m,n+2} \right] \tag{7.82}$$

[Hint: Use the identity operator $\left(\sum_{k=0}^{\infty} |k\rangle \langle k| \right) = \hat{I}$ (See [1]).]

Solution

$$\langle n|\, \hat{x}^4 \, |n\rangle = \sum_{k=0}^{\infty} \langle n|\, x^2 \, |k\rangle \, \langle k|\, x^2 \, |n\rangle$$

$$= \frac{1}{4\alpha^4} \sum_{k=0}^{\infty} \left[(\sqrt{n\,(n-1)}\delta_{k,n-2} + (2n+1)\,\delta_{k,n} \right.$$

$$+ \left[\sqrt{(n+1)\,(n+2)}\delta_{k,n+2)} \right]$$

$$\times (\sqrt{n\,(n-1)}\delta_{k,n-2} + (2n+1)\,\delta_{k,n}$$

$$\left. + \sqrt{(n+1)\,(n+2)}\delta_{k,n+2)} \right] \tag{7.83}$$

When these six terms are multiplied the only survivors will be the squares of each of the three terms that make up $\langle k|\, x^2 \, |n\rangle$. Therefore

$$\langle n|\, \hat{x}^4 \, |n\rangle = \frac{1}{4\alpha^4} \left[n\,(n-1) + (2n+1)^2 + (n+1)\,(n+2) \right]$$

$$= \frac{1}{4\alpha^4} \left[n^2 - n + 4n^2 + 4n + 1 + n^2 + 3n + 2 \right]$$

$$= \frac{3}{4\alpha^4} \left(2n^2 + 2n + 1 \right) \tag{7.84}$$

Notice that the units of $\langle n|\, \hat{x}^4 \, |n\rangle$ in Eq. (7.84) are m^4, as they must be, because α has units m^{-1} (see the eigenfunction in Appendix O).

Chapter 8
Angular Momentum

In quantum mechanics angular momentum includes the usual angular momentum that we learn about in classical mechanics. This angular momentum is usually designated by L and defined as

$$L = r \times p \tag{8.1}$$

In quantum mechanics the term "angular momentum" has a much more general meaning. It is a "generalized angular momentum." A vector operator $\hat{J}$ is defined to be *an* angular momentum if its components obey the commutation rules

$$\left[\hat{J}_i, \hat{J}_j\right] = i\hbar \hat{J}_k \epsilon_{ijk} \tag{8.2}$$

where any of the i, j, and k represent Cartesian coordinates x, y, and z. The quantity ϵ_{ijk} is known as the Levi-Cevita symbol. If the indexes i, j, and k are in cyclic order (e.g. *jki*), $\epsilon_{ijk} = +1$. If they are out of order (such as *kji*), then $\epsilon_{ijk} = -1$. If any two indexes are the same, $\epsilon_{ijk} = 0$.

It is shown in introductory quantum mechanics textbooks [1] that

$$\left[\hat{J}^2, \hat{J}_i\right] = 0 \quad \text{where} \quad i = x, \ y \text{ or } z \tag{8.3}$$

Thus, the *magnitude* of the angular momentum $\left|\hat{J}\right|$ can be specified along with any one of the components of $\hat{J}$, customarily chosen to be $\hat{J}_z$. Additionally, $\hat{J}^2$ and $\hat{J}_z$ have simultaneous eigenkets which we designate by $|jm\rangle$. The eigenvalue equations are

$$\hat{J}^2 |jm\rangle = j(j+1)\hbar^2 |jm\rangle \text{ and } \hat{J}_z |jm\rangle = m\hbar |jm\rangle \tag{8.4}$$

where j and m are the quantum numbers associated with the operators $\hat{J}^2$ and $\hat{J}_z$.

© Springer International Publishing AG 2017
J.D. Kelley, J.J. Leventhal, *Problems in Classical and Quantum Mechanics*, DOI 10.1007/978-3-319-46664-4_8

It is useful to define angular momentum ladder operators $\hat{J}_{\pm}$ analogous to the harmonic oscillator ladder operators

$$\hat{J}_{\pm} = \hat{J}_x \pm i\hat{J}_y \tag{8.5}$$

These definitions together with the commutation relations given in Eq. (8.2) lead to the raising and lowering properties of the ladder operators :

$$\hat{J}_+ |jm\rangle = \hbar \sqrt{j(j+1) - m(m+1)} \, |j(m+1)\rangle$$
$$= \hbar \sqrt{(j-m)(j+m+1)} \, |j(m+1)\rangle \tag{8.6}$$

and

$$\hat{J}_- |jm\rangle = \hbar \sqrt{j(j+1) - m(m-1)} \, |j(m-1)\rangle$$
$$= \hbar \sqrt{(j+m)(j-m+1)} \, |j(m-1)\rangle \tag{8.7}$$

so that $\hat{J}_+$ and $\hat{J}_-$ are "raising" and "lowering" operators. They raise and lower the m-values of the eigenkets by unity.

For orbital angular momentum, the angular momentum that we learned about in elementary classical mechanics, $\hat{J}^2 \rightarrow \hat{L}^2$ and $\hat{J}_z \rightarrow \hat{L}_z$ and the orbital angular momentum *operators* can be written in coordinate representation (see Appendix Q). The eigenfunctions of these operators are the spherical harmonics $Y_{\ell m}(\theta, \phi)$ (see Appendix R) where $\ell(\ell+1)\hbar^2$ and $m\hbar$ are the eigenvalues of $\hat{L}^2$ and $\hat{L}_z$, respectively [see Eq. (8.4)].

Another important angular momentum is "spin." Although spin is an intrinsically relativistic concept, it is necessary to include it in nonrelativistic quantum mechanics. The spin angular momentum operator is designated by the symbol $\hat{S}$ so that for spin angular momentum $\hat{J}^2 \rightarrow \hat{S}^2$ and $\hat{J}_z \rightarrow \hat{S}_z$ and the components of $\hat{S}$ obey the commutation rule

$$\left[\hat{S}_i, \hat{S}_j\right] = i\hbar \hat{S}_k \epsilon_{ijk} \tag{8.8}$$

Unlike orbital angular momentum L there are no spatial coordinates associated with spin. As will be seen in Problem 2, angular momentum eigenvalues can be only integral or half-integral multiples of $\hbar$. The most common spin encountered is that of a spin-$\frac{1}{2}$ particle, the spin of an electron. For spin-$\frac{1}{2}$ there are only two possible values of the eigenvalues of $\hat{S}_z$, $\pm\frac{1}{2}\hbar$, so the eigenvalues of $\hat{S}^2$ are $\frac{3}{4}\hbar^2$ [see Eq. (8.4)]. It is convenient to use matrix algebra when dealing with spin-$\frac{1}{2}$ calculations. The matrices are

$$\hat{S}^2 = \frac{3}{4}\hbar^2 \begin{pmatrix} 1 & 0 \\ 0 & 1 \end{pmatrix} \; ; \hat{S}_z = \frac{1}{2}\hbar \begin{pmatrix} 1 & 0 \\ 0 & -1 \end{pmatrix}$$

$$\hat{S}_x = \frac{1}{2}\hbar \begin{pmatrix} 0 & 1 \\ 1 & 0 \end{pmatrix} \; ; \hat{S}_y = \frac{1}{2}\hbar \begin{pmatrix} 0 & -i \\ i & 0 \end{pmatrix} \tag{8.9}$$

where, as is customary, $\hat{S}_z$ is chosen to have the simultaneous eigenket with $\hat{S}^2$. Note that the matrix representations of the two operators $\hat{S}^2$ and $\hat{S}_z$ are both diagonal. For simplicity, the *Pauli spin matrices* $(\hat{\sigma}_x, \hat{\sigma}_y, \hat{\sigma}_z)$ are often used for spin-$\frac{1}{2}$ calculations. These are the same as the $\hat{S}_i$ matrices above with the $\hbar/2$ omitted:

$$\hat{\sigma}_x = \begin{pmatrix} 0 & 1 \\ 1 & 0 \end{pmatrix}; \quad \hat{\sigma}_y = \begin{pmatrix} 0 & -i \\ i & 0 \end{pmatrix}; \quad \hat{\sigma}_z = \begin{pmatrix} 1 & 0 \\ 0 & -1 \end{pmatrix} \tag{8.10}$$

The simultaneous eigenkets of $\hat{S}^2$ and $\hat{S}_z$ are designated $|\alpha\rangle$ and $|\beta\rangle$ and have eigenvalues $(3/4)\hbar^2$ and $\pm\hbar/2$, respectively. In matrix notation the eigenkets are

$$|\alpha\rangle = \begin{pmatrix} 1 \\ 0 \end{pmatrix} \quad \text{and} \quad |\beta\rangle = \begin{pmatrix} 0 \\ 1 \end{pmatrix} \tag{8.11}$$

and are referred to as spinors. The eigenkets of $\hat{S}_x$ are designated $|\alpha\rangle_x$ and $|\beta\rangle_x$ with analogous notation for the eigenkets of $\hat{S}_y$.

It often happens that there is more than one angular momentum present in a system. The most common case is when orbital angular momentum and spin angular momentum exist simultaneously, e.g. an atomic electron. For generality we designate the two angular momenta by $\hat{J}_1$ and $\hat{J}_2$, but there will also be a total angular momentum $\hat{J} = \hat{J}_1 + \hat{J}_2$. The quantum numbers associated with these operators are (j, m_j), (j_1, m_{j1}) and (j_2, m_{j2}), respectively, where we have used numerical subscripts to denote the particular angular momentum with which a z-component is associated.

The operators $\left(\hat{J}_1^2, \hat{J}_{1z}, \hat{J}_2^2, \hat{J}_{2z}\right)$ are mutually commuting [1]. Therefore, the eigenvalues associated with these operators may be simultaneously obtained. The simultaneous eigenkets of these four operators are designated

$$|j_1, m_{j1}; j_2, m_{j2}\rangle \tag{8.12}$$

This set of eigenkets is referred to as the "uncoupled set" and the set of commuting operators that produces them constitutes the "uncoupled representation."

The set of operators $\left(\hat{J}_1^2, \hat{J}_2^2; \hat{J}^2, \hat{J}_z\right)$ are also mutually commuting so the set of quantum numbers associated with them are good (simultaneously obtainable) and the eigenkets of these four operators are designated

$$|j_1, j_2; j, m_j\rangle \tag{8.13}$$

These are the "coupled set" and the operators that produced them constitute the "coupled representation." Because each set of eigenkets is a complete set, any arbitrary ket may be written as a linear combination of eigenkets from one of the sets. If the arbitrary ket is expanded in terms of the coupled kets, we obtain information about the probability of measuring the total angular momentum, its

z-component, and the individual angular momenta. No information can be obtained about the individual z-components though. On the other hand, we could expand the arbitrary ket in terms of the complete set of uncoupled kets and obtain information on the individual z-components, but none on the total angular momentum. This follows because $\left[\hat{J}_{1z}, \hat{J}^2 \right] \neq 0 \neq \left[\hat{J}_{2z}, \hat{J}^2 \right]$ (see Problem 9).

Additionally, we can expand an eigenket of one of the sets in terms of the eigenkets of the other complete set. For example, let us expand one of the coupled kets $|j_1, j_2; j, m_j\rangle$ on the uncoupled set. Starting with the identity

$$|j_1, j_2; j, m_j\rangle \equiv |j_1, j_2; j, m_j\rangle \tag{8.14}$$

we can operate on the right-hand side of Eq. (8.14) with the identity operator [1, 2], rearrange and obtain

$$|j_1, j_2; j, m_j\rangle = \left(\sum_{m_{j1}=-j_1}^{j_1} \sum_{m_{j2}=-j_2}^{j_2} |j_1, m_{j1}; j_2, m_{j2}\rangle \langle j_1, m_{j1}; j_2, m_{j2}| \right)$$

$$\times |j_1, j_2; j, m_j\rangle$$

$$= \left(\sum_{m_{j1}=-j_1}^{j_1} \sum_{m_{j2}=-j_2}^{j_2} \langle j_1, m_{j1}; j_2, m_{j2} | j_1, j_2; j, m_j\rangle \right)$$

$$\times |j_1, m_{j1}; j_2, m_{j2}\rangle \tag{8.15}$$

The expansion coefficients are the summations in the parentheses in Eq. (8.15). Of course, we could have expanded an uncoupled ket on the coupled set and obtained analogous results. These expansion coefficients are known as the Clebsch–Gordan or Vector Coupling coefficients, some of which are listed in Table S.

Problems

1. Use the ladder operators $\hat{J}_\pm$ to show the following for a system in an eigenstate $|jm\rangle$.

   (a) $\langle \hat{J}_x \rangle = \langle \hat{J}_y \rangle = 0$
   (b) $\langle \hat{J}_x^2 \rangle = \langle \hat{J}_y^2 \rangle = \frac{\hbar^2}{2} \left[j(j+1) - m^2 \right]$

Solution

(a) From the equations for $\hat{J}_{\pm}$, Eq. (8.5)

$$\hat{J}_x = \frac{1}{2}\left(\hat{J}_+ + \hat{J}_-\right) \text{ and } \hat{J}_y = \frac{1}{2i}\left(\hat{J}_+ - \hat{J}_-\right) \qquad (8.16)$$

so that

$$\left\langle \hat{J}_x \right\rangle = \frac{1}{2}\langle jm|\left(\hat{J}_+ + \hat{J}_-\right)|jm\rangle$$

$$= 0 \qquad (8.17)$$

and

$$\left\langle \hat{J}_y \right\rangle = \frac{1}{2i}\langle jm|\left(\hat{J}_+ - \hat{J}_-\right)|jm\rangle$$

$$= 0 \qquad (8.18)$$

because of the raising and lowering action of $\hat{J}_{\pm}$. In short, they raise and lower the bras and kets into orthogonality.

The method of showing that $\left\langle \hat{J}_x \right\rangle = \left\langle \hat{J}_y \right\rangle = 0$ using ladder operators is very simple, but there is yet another way to do this by employing the commutation relations, Eq. (8.2), that define an angular momentum. The two methods are entwined because the existence of the ladder operators depends upon the commutation relations, but it is worthwhile to go through the algebra. We begin by writing $\left\langle \hat{J}_x \right\rangle$ in terms of the commutator

$$\left\langle \hat{J}_x \right\rangle = \frac{1}{i\hbar}\langle jm|\left[\hat{J}_y, \hat{J}_z\right]|jm\rangle$$

$$= \frac{1}{i\hbar}\left\{\langle jm|\hat{J}_y\hat{J}_z|jm\rangle - \langle jm|\hat{J}_z\hat{J}_y|jm\rangle\right\}$$

$$= \frac{m}{i\hbar}\left\{\langle jm|\hat{J}_y|jm\rangle - \langle jm|\hat{J}_y|jm\rangle\right\}$$

$$\equiv 0 \qquad (8.19)$$

It is clear that the same method will produce $\left\langle \hat{J}_y \right\rangle = 0$.

Let us consider this problem from a physical viewpoint and pretend that the angular momentum under consideration is orbital angular momentum (which we can visualize). We have chosen $\hat{L}_z$ to be the component for which we can simultaneously know the eigenvalue $m\hbar$ and the eigenvalue of the total angular momentum $\sqrt{\ell(\ell+1)}\hbar$. This choice precludes knowledge of the eigenvalues of $\hat{L}_x$ and $\hat{L}_y$: these components can have any value between

$-j$ and j with equal probability so their expectation values must vanish. As we will see in part (b) of this problem though, this is not true of the expectation values of the squares of these operators.

(b) We write the square of the total angular momentum as the sum of the squares of its individual components.

$$\hat{J}^2 = \hat{J}_x^2 + \hat{J}_y^2 + \hat{J}_z^2 \tag{8.20}$$

Rearranging and taking the average values we have

$$\left\langle \hat{J}_x^2 \right\rangle + \left\langle \hat{J}_y^2 \right\rangle = \left\langle \hat{J}^2 \right\rangle - \left\langle \hat{J}_z^2 \right\rangle \tag{8.21}$$

But, by symmetry $\left\langle \hat{J}_x^2 \right\rangle = \left\langle \hat{J}_y^2 \right\rangle$ because the choice of these axes is arbitrary. We will, however, prove this assertion below.

Remembering that the system is in an eigenstate $|jm\rangle$ we have

$$\left\langle \hat{J}_x^2 \right\rangle = \left\langle \hat{J}_y^2 \right\rangle = \frac{1}{2} \left[\langle jm| \hat{J}^2 |jm\rangle - \langle jm| \hat{J}_z^2 |jm\rangle \right]$$

$$= \frac{\hbar^2}{2} \left[j(j+1) - m^2 \right] \tag{8.22}$$

Now prove that $\left\langle \hat{J}_x^2 \right\rangle = \left\langle \hat{J}_y^2 \right\rangle$. Whenever possible we appeal to the ladder operators and invoke Eq. (8.16). We write $\left\langle \hat{J}_x^2 \right\rangle$ and $\left\langle \hat{J}_y^2 \right\rangle$ as

$$\left\langle \hat{J}_x^2 \right\rangle = \left(\frac{1}{2} \right)^2 \langle jm| \left(\hat{J}_+ + \hat{J}_- \right)^2 |jm\rangle$$

$$\left\langle \hat{J}_y^2 \right\rangle = \left(\frac{1}{2i} \right)^2 \langle jm| \left(\hat{J}_+ - \hat{J}_- \right)^2 |jm\rangle \tag{8.23}$$

The first of these equations becomes

$$\left\langle \hat{J}_x^2 \right\rangle = \left(\frac{1}{4} \right) \langle jm| \left(\hat{J}_+^2 + \hat{J}_+\hat{J}_- + \hat{J}_-\hat{J}_+ + \hat{J}_-^2 \right) |jm\rangle$$

$$= \left(\frac{1}{4} \right) \langle jm| \left(\hat{J}_+\hat{J}_- + \hat{J}_-\hat{J}_+ \right) |jm\rangle \tag{8.24}$$

The second is

$$\left\langle \hat{J}_y^2 \right\rangle = \left(-\frac{1}{4} \right) \langle jm| \left(\hat{J}_+^2 - \hat{J}_+\hat{J}_- - \hat{J}_-\hat{J}_+ + \hat{J}_-^2 \right) |jm\rangle$$

$$= \left(\frac{1}{4} \right) \langle jm| \left(\hat{J}_+\hat{J}_- + \hat{J}_-\hat{J}_+ \right) |jm\rangle \tag{8.25}$$

Equations (8.24) and (8.25) are identical, proving that the expectation values of the squares of the x- and y- coordinates of $\hat{J}$ must be equal.

As long as we have Eqs. (8.24) and (8.25) we may as well evaluate them (one of them) because it must yield the same answer as the easily obtained result in Eq. (8.22). Applying Eqs. (8.6) and (8.7) to each of the terms in Eq. (8.24) we have

$$
\begin{aligned}
\hat{J}_- \hat{J}_+ |jm\rangle &= \hbar \sqrt{(j-m)(j+m+1)} \hat{J}_- |j(m+1)\rangle \\
&= \hbar^2 \sqrt{(j-m)(j+m+1)} \sqrt{(j+m+1)(j-m)} |jm\rangle \\
&= \hbar^2 (j-m)(j+m+1) |jm\rangle \\
&= \hbar^2 \left(j^2 + j - m^2 - m \right) |jm\rangle
\end{aligned} \tag{8.26}
$$

and

$$
\begin{aligned}
\hat{J}_+ \hat{J}_- |jm\rangle &= \hbar \sqrt{(j+m)(j-m+1)} \hat{J}_+ |j(m-1)\rangle \\
&= \hbar^2 \sqrt{(j+m)(j-m+1)} \sqrt{(j-m+1)(j+m)} |jm\rangle \\
&= \hbar^2 (j+m)(j-m+1) |jm\rangle \\
&\quad \hbar^2 \left(j^2 + j - m^2 + m \right) |jm\rangle
\end{aligned} \tag{8.27}
$$

Inserting these terms into Eq. (8.24) [or Eq. (8.25)] we have

$$
\begin{aligned}
\left\langle \hat{J}_x^2 \right\rangle &= \left(\frac{1}{4} \right) \langle jm| \left(\hat{J}_+ \hat{J}_- + \hat{J}_- \hat{J}_+ \right) |jm\rangle \\
&= \left(\frac{\hbar^2}{4} \right) 2 \left(j^2 + j - m^2 \right) \\
&= \left(\frac{\hbar^2}{2} \right) \left[j(j+1) - m^2 \right]
\end{aligned} \tag{8.28}
$$

Equation (8.28) is identical with the more easily obtained Eq. (8.22), but it is comforting to obtain the same result in a seemingly different way, albeit using more effort. Note that the ease with which Eq. (8.22) was obtained depended upon using the symmetry of the problem to simplify our calculation.

2. This problem once again emphasizes the utility of the ladder operators and shows that angular momentum can take on only integral or half-integral values of $\hbar$.

(a) Show the following relations.

$$
\left[\hat{J}_z, \hat{J}_\pm \right] = \pm \hbar \hat{J}_\pm \; ; \; \left[\hat{J}^2, \hat{J}_\pm \right] = 0 \tag{8.29}
$$

and

$$\hat{J}_{\mp}\hat{J}_{\pm} = \left(\hat{J}_x \mp i\hat{J}_y\right)\left(\hat{J}_x \pm i\hat{J}_y\right) \tag{8.30}$$

(b) Use the ladder operators and the above results to show that, in general, the eigenvalues of $\hat{J}$ and $\hat{J}_z$ have eigenvalues j and m which can only be integers or half-integers. Moreover, the quantum number j must be positive while m can range from $+j$ to $-j$.

Solution

(a) Writing $\hat{J}_{\pm}$ in terms of its components and using the fundamental commutation relations that define angular momentum, Eq. (8.5), we have

$$\left[\hat{J}_z, \hat{J}_{\pm}\right] = \left[\hat{J}_z, \hat{J}_x\right] \pm i\left[\hat{J}_z, \hat{J}_y\right]$$
$$= i\hbar\hat{J}_y \pm i\left(-i\hbar\hat{J}_x\right) = (\hbar)\left(\pm i\hat{J}_x + \hat{J}_y\right)$$
$$= \pm\hbar\left(i\hat{J}_x \pm \hat{J}_y\right) = \pm\hbar\hat{J}_{\pm} \tag{8.31}$$

and

$$\hat{J}_{\mp}\hat{J}_{\pm} = \left(\hat{J}_x \mp i\hat{J}_y\right)\left(\hat{J}_x \pm i\hat{J}_y\right)$$
$$= \hat{J}_x^2 \pm i\hat{J}_x\hat{J}_y \mp i\hat{J}_y\hat{J}_x + \hat{J}_y^2$$
$$= \hat{J}^2 - \hat{J}_z^2 \mp i\left[\hat{J}_x, \hat{J}_y\right]$$
$$= \hat{J}^2 - \hat{J}_z^2 \mp \hbar\hat{J}_z \tag{8.32}$$

where we have added and subtracted $\hat{J}_z^2$ in the third line of the last equation.

(b) In view of Eqs. (8.6) and (8.7) it is clear that the actions of $\hat{J}_+$ and $\hat{J}_-$ are to raise and lower the values of the quantum number m by unity when operating on the eigenkets $|jm\rangle$ of $\hat{J}^2$ and $\hat{J}_z$. Let us assume that we have operated on an eigenket with $\hat{J}^2$. We have

$$\hat{J}^2 |jm\rangle = \left(\hat{J}_x^2 + \hat{J}_y^2 + \hat{J}_z^2\right)|jm\rangle \tag{8.33}$$

or, from Eq. (8.4), we have

$$\left(\hat{J}_x^2 + \hat{J}_y^2\right)|jm\rangle = \left[j\left(j+1\right) - m^2\right]\hbar^2 |jm\rangle \tag{8.34}$$

Because $\left(\hat{J}_x^2 + \hat{J}_y^2 \right)$ is manifestly positive (or zero) we must have

$$j\,(j+1) \geq m^2 \tag{8.35}$$

and there must be a maximum value of m, call it $m_{\max}$, beyond which the eigenstate $|jm_{\max}\rangle$ cannot be raised by $\hat{J}_+$. That is

$$\hat{J}_+ |jm_{\max}\rangle = 0 \tag{8.36}$$

If we now operate on the left of the last equation with $\hat{J}_-$ and apply Eq. (8.32), we obtain

$$\left(\hat{J}^2 - \hat{J}_z^2 - \hbar\hat{J}_z \right) |jm_{\max}\rangle = 0$$

$$\left[j\,(j+1)\,\hbar^2 - m_{\max}^2\hbar^2 - \hbar m_{\max} \right] |jm_{\max}\rangle = 0 \tag{8.37}$$

Solving this equation for $j\,(j+1)\,\hbar^2$ we have

$$j\,(j+1)\,\hbar^2 = (m_{\max}\hbar)\,(m_{\max}\hbar + 1) \tag{8.38}$$

There must also be a minimum value of the quantum number m so

$$\hat{J}_- |jm_{\min}\rangle = 0 \tag{8.39}$$

We may now apply $\hat{J}_+$ to this last equation again using Eq. (8.32) we obtain

$$j\,(j+1)\,\hbar^2 = (m_{\min}\hbar)\,(m_{\min}\hbar - 1) \tag{8.40}$$

Comparing Eqs. (8.38) and (8.40) we must have

$$(m_{\min}\hbar)\,(m_{\min}\hbar - 1) = (m_{\max}\hbar)\,(m_{\max}\hbar + 1)$$

$$m_{\min}^2\hbar - m_{\min} = m_{\max}^2\hbar + m_{\max} \tag{8.41}$$

Thus,

$$m_{\max} = -m_{\min} \tag{8.42}$$

Equation (8.42) shows that successive values of the quantum number m must differ by unity. Therefore, m must be either an integer or a half-integer (try it). Moreover, because $j\,(j+1) \geq m_{\max}^2$ [Eq. (8.35)] we must have $j \geq 0$.

3. The normalized angular part of a certain wave function for a particle is given in spherical coordinates by

$$\psi\,(\theta, \phi) = \frac{1}{\sqrt{14}} \cdot \frac{1}{\sqrt{4\pi}} \left[1 + 2\sqrt{3}\cos\theta + 3\sqrt{\frac{5}{4}}\,(3\cos^2\theta - 1) \right] \tag{8.43}$$

(a) If a measurement is made of the total angular momentum, what are the possible values that could be measured and with what probabilities for each?

(b) If a measurement is made of the z-component of the angular momentum, what are the possible values that could be measured and with what probabilities for each?

(c) What are the expectation values of the z-component of the orbital angular momentum $\left\langle \hat{L}_z \right\rangle$ and of the square of the total orbital angular momentum $\left\langle \hat{L}^2 \right\rangle$? The spherical harmonics given in Appendix R should be helpful.

Solution

(a) We must first write $\psi\,(\theta, \phi)$ in terms of the spherical harmonics. Inspection of Table R.1 shows that the first term is proportional to $Y_{00}\,(\theta, \phi)$, the second to $Y_{10}\,(\theta, \phi)$, and the third to $Y_{20}\,(\theta, \phi)$. Multiplying the factor $1/\sqrt{4\pi}$ into the square brackets we have

$$\psi\,(\theta, \phi) = \frac{1}{\sqrt{14}} \cdot \left[\frac{1}{\sqrt{4\pi}} + 2\sqrt{\frac{3}{4\pi}} \cos\theta + 3\sqrt{\frac{5}{16\pi}} \left(3\cos^2\theta - 1\right) \right]$$

$$= \frac{1}{\sqrt{14}} \cdot \left[Y_{00}\,(\theta, \phi) + 2Y_{10}\,(\theta, \phi) + 3Y_{20}\,(\theta, \phi) \right] \qquad (8.44)$$

Therefore, the only possible values of ℓ are 0, 1, and 2. The total angular momentum possibilities are

$$\sqrt{\ell\,(\ell + 1)}\hbar = 0, \ \sqrt{2}\hbar, \ \sqrt{6}\hbar \qquad (8.45)$$

The probabilities are $1/14$, $4/14$, and $9/14$, respectively, reflective of the admixture of the spherical harmonics.

(b) Because $m = 0$ in each of the spherical harmonics that constitute $\psi\,(\theta, \phi)$ the only possible z-component of the angular momentum for the system described by this wave function is zero.

(c) Clearly $\left\langle \hat{L}_z \right\rangle = 0$ because of the result of part (b).

To calculate $\left\langle \hat{L}^2 \right\rangle$ we must compute the indicated expectation value. Remembering that the spherical harmonics are orthonormal we have

$$\left\langle \hat{L}^2 \right\rangle = \left\langle \psi\,(\theta, \phi) \middle| \hat{L}^2 \middle| \psi\,(\theta, \phi) \right\rangle$$

$$= \frac{\hbar^2}{14} \left(1^2 \cdot 0 + 2^2 \cdot 2 + 3^2 \cdot 6 \right)$$

$$\approx 4.6\hbar^2 \qquad (8.46)$$

The admixture is heavily weighted toward $\ell = 2$ for which $\ell(\ell+1)\hbar^2 = 6\hbar^2$ [see Eq. (8.44)]. Therefore, while $\langle \hat{L}^2 \rangle$ is expected to be less than $6\hbar^2$ it is considerably greater than $2\hbar^2$, the value for $\ell = 1$.

4. This problem is in the same spirit as Problem 6 of Chap. 5. It is known that the intrinsic angular momentum, the spin, of an electron is $\hbar/2$. Assume that the electron is a uniform solid sphere spinning at a frequency ω, with total mass m_e and radius r_e, the classical radius of the electron (see Problem 5 of Chap. 5). Find the speed of a point on the surface of the sphere $v_s = \omega r_e$ and show that it exceeds the speed of light.

Solution

Equating the spin angular momentum, $\hbar/2$, to the mechanical angular momentum of the spinning sphere we have

$$\frac{\hbar}{2} = I\omega \qquad (8.47)$$

where $I = (2/5)\, m_e r_e^2$, the moment of inertia of the solid sphere. In Eq. (8.47) it is assumed that the spin angular momentum and the mechanical angular momentum are parallel so we do not concern ourselves with vectors for this order of magnitude calculation. Substituting for I and ω in Eq. (8.47) we have

$$\frac{\hbar}{2} = \left(\frac{2}{5} m_e r_e^2\right) \frac{v_s}{r_e} \qquad (8.48)$$

Solving for v_s we obtain

$$
\begin{aligned}
v_s &= \frac{5}{4} \frac{\hbar}{m_e r_e} \\
&= \frac{5}{4} \frac{\hbar}{m_e} \left[\left(\frac{4\pi\epsilon_0}{e^2}\right) m_e c^2 \right] \\
&= \frac{5}{4} (m_e)(\hbar) \left(\frac{4\pi\epsilon_0}{e^2}\right) c^2 \qquad (8.49)
\end{aligned}
$$

where we have replaced r_e with its value deduced in Eq. (5.33) of Chap. 5. In atomic units (a.u.) all quantities in parentheses in Eq. (8.49) are unity except the speed of light c which is 137 (see Table C.1). This result is reminiscent of that obtained in Problem 6 of Chap. 5. We see that in a.u. v_s is

$$v_s = \frac{5}{4} 137^2$$

$$\approx 170 c \qquad (8.50)$$

about one-half the estimate obtained Problem 6 of Chap. 5. It, however, confirms that the notion of the electron being a spinning sphere of finite dimension is unrealistic.

5. An electron has total angular momentum quantum number $s = \frac{1}{2}$.

   (a) A measurement is made of the x-component of the spin. What are the possible results of this measurement?

   (b) If the system is in the normalized state represented by the spinor

$$|\chi\rangle = \frac{1}{5\sqrt{2}} \begin{pmatrix} 1 \\ 7 \end{pmatrix}$$ (8.51)

on the basis set of $\hat{S}_z$ find the probability of measuring the z-component of the spin to be $-\frac{1}{2}\hbar$?

Solution

   (a) Electrons are spin $\frac{1}{2}$ particles so the only value of any component that can be measured is $\pm\frac{1}{2}\hbar$.

   (b) The normalized spinor $|\chi\rangle$ represents the spin wave function of the electron which in Eq. (8.51) is written in terms of $|\alpha\rangle$ and $|\beta\rangle$ the eigenkets of $\hat{S}_z$ [see Eq. (8.11)].

$$|\chi\rangle = \frac{1}{5\sqrt{2}} \left[\begin{pmatrix} 1 \\ 0 \end{pmatrix} + 7 \begin{pmatrix} 0 \\ 1 \end{pmatrix} \right]$$

$$= \frac{1}{5\sqrt{2}} [|\alpha\rangle + 7 |\beta\rangle]$$ (8.52)

where $|\alpha\rangle$ and $|\beta\rangle$ are the spin-up and spin-down eigenkets of $\hat{S}_z$. Thus, the probability of measuring the spin angular momentum to be $-\frac{1}{2}\hbar$ is given by the absolute square of the coefficient of $|\beta\rangle$ which is

$$\frac{7^2}{\left(5\sqrt{2}\right)^2} = \frac{49}{50}$$

Because the admixture of $|\chi\rangle$ is so heavily weighted toward $|\beta\rangle$ the eigenvalue $-\frac{1}{2}\hbar$ will be measured most often, in fact 98 times out of 100.

6. Use the Pauli spin matrices to find $|\alpha\rangle_x$ and $|\beta\rangle_x$, the spin up and spin down eigenkets of $\hat{S}_x$, in terms of $|\alpha\rangle$ and $|\beta\rangle$, the spin up and spin down eigenkets of $\hat{S}_z$. Do the same for $|\alpha\rangle_y$ and $|\beta\rangle_y$, the spin up and spin down eigenkets of $\hat{S}_y$.

Solution

The eigenvalue equation for the operator $\hat{\sigma}_x$ is

$$\begin{pmatrix} 0 & 1 \\ 1 & 0 \end{pmatrix} \begin{pmatrix} a \\ b \end{pmatrix} = \lambda \begin{pmatrix} a \\ b \end{pmatrix} \tag{8.53}$$

where λ represents the eigenvalues of $\hat{\sigma}_x$. We know, however, that, $\lambda = \pm 1$ because all components of spin angular momenta are on an equal footing. By convention $\hat{S}_z$ is chosen to be the one that commutes with $\hat{S}^2$.

Multiplying the matrices on the left side of this equation and equating matrix elements we have

$$b = \pm a \quad \text{and} \quad a = \pm b \tag{8.54}$$

Thus, the matrix elements of the eigenkets either have the same sign or the opposite sign. The eigenkets are to be normalized so the one for which the signs of the elements are the same is

$$|\alpha\rangle_x = \frac{1}{\sqrt{2}} \begin{pmatrix} 1 \\ 1 \end{pmatrix}$$

$$= \frac{1}{\sqrt{2}} |\alpha\rangle + \frac{1}{\sqrt{2}} |\beta\rangle \tag{8.55}$$

The one for which the elements have different signs is

$$|\beta\rangle_x = \frac{1}{\sqrt{2}} \begin{pmatrix} 1 \\ -1 \end{pmatrix}$$

$$= \frac{1}{\sqrt{2}} |\alpha\rangle - \frac{1}{\sqrt{2}} |\beta\rangle \tag{8.56}$$

The subscripts on $|\alpha\rangle_x$ and $|\beta\rangle_x$ in Eqs. (8.55) and (8.56) indicate that they correspond to spin up and spin down with respect to $\hat{S}_x$. The eigenkets of $\hat{S}_y$ can be obtained in an analogous manner.

$$\begin{pmatrix} 0 & -i \\ i & 0 \end{pmatrix} \begin{pmatrix} a \\ b \end{pmatrix} = \lambda \begin{pmatrix} a \\ b \end{pmatrix} \tag{8.57}$$

so

$$-ib = \pm a \quad \text{and} \quad ia = \pm b \tag{8.58}$$

and we have

$$|\alpha\rangle_y = \frac{1}{\sqrt{2}} \begin{pmatrix} 1 \\ i \end{pmatrix}$$

$$= \frac{1}{\sqrt{2}} |\alpha\rangle + \frac{i}{\sqrt{2}} |\beta\rangle \tag{8.59}$$

and

$$|\beta\rangle_y = \frac{1}{\sqrt{2}} \begin{pmatrix} 1 \\ -i \end{pmatrix}$$

$$= \frac{1}{\sqrt{2}} |\alpha\rangle - \frac{i}{\sqrt{2}} |\beta\rangle \qquad (8.60)$$

7. Use the spin ladder operators to show that

$$\hat{S}_x |\alpha\rangle = \frac{1}{2}\hbar |\beta\rangle$$

$$\hat{S}_x |\beta\rangle = \frac{1}{2}\hbar |\alpha\rangle$$

$$\hat{S}_y |\alpha\rangle = \frac{i}{2}\hbar |\beta\rangle$$

$$\hat{S}_y |\beta\rangle = -\frac{i}{2}\hbar |\alpha\rangle \qquad (8.61)$$

Solution

The x- and y-components of the spin operators are given in terms of the ladder operators as

$$\hat{S}_x = \frac{1}{2}\left(\hat{S}_+ + \hat{S}_-\right) \quad \text{and} \quad \hat{S}_y = \frac{1}{2i}\left(\hat{S}_+ - \hat{S}_-\right) \qquad (8.62)$$

Thus

$$\hat{S}_- |\alpha\rangle = \hbar\sqrt{(s + m_s)(s - m_s + 1)} |\beta\rangle$$

$$= \hbar\sqrt{\left(\frac{1}{2} + \frac{1}{2}\right)} |\beta\rangle$$

$$= \hbar |\beta\rangle \qquad (8.63)$$

Similarly

$$\hat{S}_+ |\beta\rangle = \hbar\sqrt{(s - m_s)(s + m_s + 1)} |\alpha\rangle$$

$$= \hbar\sqrt{\left[\frac{1}{2} - \left(-\frac{1}{2}\right)\right]} |\alpha\rangle$$

$$= \hbar |\alpha\rangle \qquad (8.64)$$

so

$$\hat{S}_x \,|\alpha\rangle = \frac{1}{2}\left(\hat{S}_+ \,|\alpha\rangle + \hat{S}_- \,|\alpha\rangle\right)$$

$$= \frac{1}{2}\left(0 + \hbar \,|\beta\rangle\right)$$

$$= \frac{1}{2}\hbar \,|\beta\rangle \tag{8.65}$$

$$\hat{S}_x \,|\beta\rangle = \frac{1}{2}\left(\hat{S}_+ \,|\beta\rangle + \hat{S}_- \,|\beta\rangle\right)$$

$$= \frac{1}{2}\left(\hbar \,|\alpha\rangle + 0\right)$$

$$= \frac{1}{2}\hbar \,|\alpha\rangle \tag{8.66}$$

$$\hat{S}_y \,|\alpha\rangle = \frac{1}{2i}\left(\hat{S}_+ \,|\alpha\rangle - \hat{S}_- \,|\alpha\rangle\right)$$

$$= \frac{1}{2i}\left(0 - \hbar \,|\beta\rangle\right)$$

$$= \frac{i}{2}\hbar \,|\beta\rangle \tag{8.67}$$

$$\hat{S}_y \,|\beta\rangle = \frac{1}{2i}\left(\hat{S}_+ \,|\beta\rangle - \hat{S}_- \,|\beta\rangle\right)$$

$$= \frac{1}{2i}\left(\hbar \,|\alpha\rangle - 0\right)$$

$$= -\frac{i}{2}\hbar \,|\alpha\rangle \tag{8.68}$$

8. At time $t = 0$, an electron is in the spin up state in the x-direction, which we designate by the ket $|\alpha\rangle_x$. The electron is now placed in a constant magnetic field in the positive z-direction. The Hamiltonian is given by

$$\hat{H} = -\boldsymbol{\mu} \bullet \boldsymbol{B} \tag{8.69}$$

where $\hat{\boldsymbol{\mu}} = g\hat{\boldsymbol{S}}$ is the magnetic moment of the electron, which is proportional to the spin $\hat{\boldsymbol{S}}$; g is a constant of proportionality.

(a) Find $|\chi(t)\rangle$, the time dependent state vector that describes this electron for $t > 0$ in terms of $|\alpha\rangle$ and $|\beta\rangle$, the spin up and spin down eigenkets of $\hat{S}_z$.

(b) Find the time dependent probability that the electron is in the $|\alpha\rangle_x$ state if a measurement is made of the x-component of the electrons's spin for any time after $t = 0$.

Solution

(a) From Eq. (8.55) we know $|\alpha\rangle_x$ in terms of $|\alpha\rangle$ and $|\beta\rangle$, so that

$$|\chi(t=0)\rangle = |\alpha\rangle_x = \frac{1}{\sqrt{2}}|\alpha\rangle + \frac{1}{\sqrt{2}}|\beta\rangle \tag{8.70}$$

To find the time dependence of this state vector we must multiply each term by the appropriate exponential time factor which means that we must have the eigenenergies. We must thus solve the TISE with the Hamiltonian given in Eq. (8.69), which may be written

$$\hat{H} = -\boldsymbol{\mu}\bullet\boldsymbol{B} = -g\hat{S}_z B \tag{8.71}$$

because $\boldsymbol{B}$ is in the z-direction. Thus, the eigenvalue equation is

$$\hat{H}|\chi(t=0)\rangle = -gB\hat{S}_z|\chi(t=0)\rangle$$
$$= \mp\frac{gB\hbar}{2}|\chi(t=0)\rangle \tag{8.72}$$

Note that the minus sign corresponds to spin up and the plus to spin down. With $\omega = gB/2$ we may now write $|\chi(t)\rangle$ as

$$|\chi(t)\rangle = \frac{1}{\sqrt{2}}|\alpha\rangle e^{i\omega t} + \frac{1}{\sqrt{2}}|\beta\rangle e^{-i\omega t} \tag{8.73}$$

(b) To find the probability that the electron spin will be measured to be in the $|\alpha\rangle_x$ state we must re-cast Eq. (8.73) in terms of the kets $|\alpha\rangle_x$ and $|\beta\rangle_x$. This is easily done by solving Eqs. (8.59) and (8.60) of Problem 6 of this chapter for $|\alpha\rangle_x$ and $|\beta\rangle_x$ in terms of $|\alpha\rangle_x$ and $|\beta\rangle_x$. The result is

$$|\alpha\rangle = \frac{1}{\sqrt{2}}|\alpha\rangle_x + \frac{1}{\sqrt{2}}|\beta\rangle_x$$
$$|\beta\rangle = \frac{1}{\sqrt{2}}|\alpha\rangle_x - \frac{1}{\sqrt{2}}|\beta\rangle_x \tag{8.74}$$

Inserting these into Eq. (8.73) we have

$$|\chi(t)\rangle = \frac{1}{2}|\alpha\rangle_x\left(e^{i\omega t}+e^{-i\omega t}\right) - \frac{1}{2}|\beta\rangle_x\left(e^{i\omega t}-e^{-i\omega t}\right)$$
$$= |\alpha\rangle_x\cos\omega t - |\beta\rangle_x\sin\omega t \tag{8.75}$$

The probability that the system is in the state $|\alpha\rangle_x$ is given by the square of the expansion coefficient, in particular

$$\text{Probability in } |\alpha\rangle_x = \cos^2\omega t \text{ where } \omega = gB/2 \tag{8.76}$$

Note that $|\chi(t)\rangle$ is normalized since $\cos^2\omega t + \sin^2\omega t = 1$.

9. Show that $\left[\hat{J}_{1z}, \hat{J}^2\right] \neq 0 \neq \left[\hat{J}_{2z}, \hat{J}^2\right]$, which show that neither $\hat{J}_{1z}$ nor $\hat{J}_{2z}$ can be specified simultaneously with $\hat{J}.^2$

Solution

We will work it for $\left[\hat{J}_{1z}, \hat{J}^2\right]$.

$$\left[\hat{J}_{1z}, \hat{J}^2\right] = \left[\hat{J}_{1z}, \left(\hat{J}_x^2 + \hat{J}_y^2 + \hat{J}_z^2\right)\right]$$
$$= \left[\hat{J}_{1z}, \hat{J}_x^2\right] + \left[\hat{J}_{1z}, \hat{J}_y^2\right] + \left[\hat{J}_{1z}, \hat{J}_z^2\right]$$
$$= \left[\hat{J}_{1z}, \left(\hat{J}_{1x} + \hat{J}_{2x}\right)^2\right] + \left[\hat{J}_{1z}, \left(\hat{J}_{1y} + \hat{J}_{2y}\right)^2\right] + 0 \qquad (8.77)$$

Because $\left[\hat{J}_{1u}, \hat{J}_{2u}\right] \equiv 0$ where $u = x, y, z$ and $\left[\hat{J}_{1z}, \hat{J}_{1z}\right] \equiv 0$ we have

$$\left[\hat{J}_{1z}, \hat{J}^2\right] = \left[\hat{J}_{1z}, \left(\hat{J}_{1x}^2 + 2\hat{J}_{1x}\hat{J}_{2x}\right)\right] + \left[\hat{J}_{1z}, \left(\hat{J}_{1y}^2 + 2\hat{J}_{1y}\hat{J}_{2y}\right)\right]$$
$$= \left[\hat{J}_{1z}, \left(\hat{J}_{1x}^2 + \hat{J}_{1y}^2\right)\right] + 2\left[\hat{J}_{1z}, \hat{J}_{1x}\right]\hat{J}_{2x} + 2\left[\hat{J}_{1z}, \hat{J}_{1y}\right]\hat{J}_{2y}$$
$$= \left[\hat{J}_{1z}, \left(\hat{J}_1^2 - \hat{J}_{1z}^2\right)\right] + 2i\hbar\left(\hat{J}_{1y}\hat{J}_{2x} - \hat{J}_{1x}\hat{J}_{2y}\right)$$
$$\neq 0 \qquad (8.78)$$

10. In an effort to demystify "Clebsch-Gordanry" you are asked in this problem to construct the table of Clebsch–Gordan coefficients for coupling two angular momenta $j_1 = 1$ and $j_1 = 1/2$. This is not as difficult as it might at first appear. Begin by writing the coupled kets as linear combinations of the uncoupled kets and then cleverly apply the ladder operators to the top and bottom of the ladder states. Note that you are being asked to construct Table S.3.
[Hint: First count and enumerate the different states in each representation.]

Solution

Following the hint we first enumerate the states using the quantum numbers in either representation. Of course, we had better get the same number of states no matter which representation we choose so we will use both as a check. Let us use the coupled quantum numbers first, $j = 3/2$ and $j = 1/2$. For $j = 3/2$ there are four values of m_j: $3/2, 1/2, -1/2, -3/2$. For $j = 1/2$

there are two values of m_j: $1/2, -1/2$. Thus, there are a total of six states. In the uncoupled representation the quantum numbers are $j_1 = 1$ and $j_2 = 1/2$ with $m_{j1} = 1, 0, -1$. Because each m_{j1} state can have $m_{j2} = \pm 1/2$ there are (again) six states.

Now we correlate the coupled and uncoupled states. First the easy ones, the top and bottom of the ladder states. It is clear that the only way to make up $m_j = \pm 3/2$ is when the two individual z-components, m_{j1} and m_{j2}, have the same sign. For convenience we recall the notation.

$$\text{Coupled: } \left|1, \tfrac{1}{2}; j, m_j\right\rangle \quad ; \quad \text{Uncoupled: } \left|1, m_\ell; \tfrac{1}{2}, m_s\right\rangle \tag{8.79}$$

Now, the top and bottom of the ladder states are

$$\left|1, \tfrac{1}{2}; \tfrac{3}{2}, \pm\tfrac{3}{2}\right\rangle_c = \left|1, \pm 1; \tfrac{1}{2}, \pm\tfrac{1}{2}\right\rangle_u \tag{8.80}$$

The other four states must be made up of linear combinations of two of the states of the other representation. The correlations are listed in Table 8.1.

From this enumeration of states we can see that our table of Clebsch–Gordan coefficients will have only ten non-zero entries, two each for four of the states and one for each of the end of the ladder states.

Our task is to evaluate the constants, the C_i. Inasmuch as $\left|1, \tfrac{1}{2}; \tfrac{3}{2}, \tfrac{1}{2}\right\rangle_c$ can be a linear combination of only $\left|1, 1; \tfrac{1}{2}, -\tfrac{1}{2}\right\rangle_c$ and $\left|1, 1; \tfrac{1}{2}, -\tfrac{1}{2}\right\rangle_u$ we write

$$\left|1, \tfrac{1}{2}; \tfrac{3}{2}, \tfrac{1}{2}\right\rangle_c = C_1 \left|1, 1; \tfrac{1}{2}, -\tfrac{1}{2}\right\rangle_u + C_2 \left|1, 0; \tfrac{1}{2}, \tfrac{1}{2}\right\rangle_u \tag{8.81}$$

Table 8.1 Coupled and uncoupled states for an electron with orbital angular momentum ℓ

Coupled $\left|1, \tfrac{1}{2}; j, m_j\right\rangle$	Uncoupled $\left|1, m_\ell; \tfrac{1}{2}, m_s\right\rangle$
$\left|1, \tfrac{1}{2}; \tfrac{3}{2}, \tfrac{3}{2}\right\rangle$	$\left|1, 1; \tfrac{1}{2}, \tfrac{1}{2}\right\rangle$
$\left|1, \tfrac{1}{2}; \tfrac{3}{2}, \tfrac{1}{2}\right\rangle$	$\left|1, 1; \tfrac{1}{2}, -\tfrac{1}{2}\right\rangle$
	$\left|1, 0; \tfrac{1}{2}, \tfrac{1}{2}\right\rangle$
$\left|1, \tfrac{1}{2}; \tfrac{3}{2}, -\tfrac{1}{2}\right\rangle$	$\left|1, -1; \tfrac{1}{2}, \tfrac{1}{2}\right\rangle$
	$\left|1, 0; \tfrac{1}{2}, -\tfrac{1}{2}\right\rangle$
$\left|1, \tfrac{1}{2}; \tfrac{3}{2}, -\tfrac{3}{2}\right\rangle$	$\left|1, -1; \tfrac{1}{2}, -\tfrac{1}{2}\right\rangle$
$\left|1, \tfrac{1}{2}; \tfrac{1}{2}, \tfrac{1}{2}\right\rangle$	$\left|1, 0; \tfrac{1}{2}, \tfrac{1}{2}\right\rangle$
	$\left|1, 1; \tfrac{1}{2}, -\tfrac{1}{2}\right\rangle$
$\left|1, \tfrac{1}{2}; \tfrac{1}{2}, -\tfrac{1}{2}\right\rangle$	$\left|1, -1; \tfrac{1}{2}, \tfrac{1}{2}\right\rangle$
	$\left|1, 0; \tfrac{1}{2}, -\tfrac{1}{2}\right\rangle$

The left-hand column is a listing of the possible coupled states. The right-hand column contains the uncoupled states that correlate with the adjacent coupled states. Therefore, in the right-hand column, some uncoupled states are listed more than once

Table 8.2 Clebsch–Gordan coefficients for $j_1 = \ell = 1$ and $j_2 = 1/2$

$j_1 = 1; j_2 = 1/2$		$j = 3/2$	$j = 1/2$					
m_ℓ	m_s	3/2	1/2	−1/2	−3/2	1/2	−1/2
1	1/2	1					
1	−1/2		C_1			C_2	
0	1/2		C_3			C_4	
0	−1/2			C_5			C_6
−1	1/2			C_7			C_8
−1	−1/2				1		

The state described in Eq. (8.81) can, however, be easily obtained by applying the lowering operator, Eq. (8.7), to the top of the ladder ket in each representation in Eq. (8.80). This technique gives the Clebsch–Gordan coefficients, the C_i listed in Table 8.2.
We have

$$\hat{J}_- \left|1, \tfrac{1}{2}; \tfrac{3}{2}, \tfrac{3}{2}\right\rangle_c = \left(\hat{L}_- + \hat{S}_-\right) \left|1, 1; \tfrac{1}{2}, \tfrac{1}{2}\right\rangle_u$$

$$\sqrt{3}\left|1, \tfrac{1}{2}; \tfrac{3}{2}, \tfrac{1}{2}\right\rangle_c = \hat{L}_- \left|1, 1; \tfrac{1}{2}, \tfrac{1}{2}\right\rangle_u + \hat{S}_- \left|1, 1; \tfrac{1}{2}, \tfrac{1}{2}\right\rangle_u$$

$$= \sqrt{2}\left|1, 0; \tfrac{1}{2}, \tfrac{1}{2}\right\rangle_u + \left|1, 1; \tfrac{1}{2}, -\tfrac{1}{2}\right\rangle_u \qquad (8.82)$$

Comparing Eq. (8.82) with Table 8.2 we see that $C_1 = 1/\sqrt{3}$ and $C_2 = \sqrt{2/3}$ so we have the first non-trivial entry in our Clebsch–Gordan table. To obtain the next one down we continue the procedure. We can, however, save some effort by applying the raising operator to the bottom of the ladder state thus requiring only minimal work for the computation. The final result is Table S.3.

11. A system is known to be in a particular coupled eigenstate $\left|j_1, j_2; j, m_j\right\rangle_c$ that is known to be a linear combination of the uncoupled eigenstates $\left|j_1, m_{j1}; j_2, m_{j2}\right\rangle$ as

$$\left|j_1, j_2; j, m_j\right\rangle_c = C_1 \left|1, 0; \tfrac{1}{2}, \tfrac{1}{2}\right\rangle_u + C_2 \left|1, 1; \tfrac{1}{2}, -\tfrac{1}{2}\right\rangle_u \qquad (8.83)$$

where C_1 and C_2 are real constants of opposite sign.

(a) Find the specific coupled ket in Eq. (8.83). That is, find C_1 and C_2 in Eq. (8.83).

(b) If a measurement is made of the z-component of the individual angular momenta, the quantum numbers m_{j1} and m_{j2}, what are the possible values of the measurement and with what probabilities will these values be measured?

Solution

(a) This problem is an exercise in combining the various quantum numbers in the two different representations, coupled and uncoupled. After deducing the correct relationship between the coupled and uncoupled kets in Eq. (8.83), determination of the Clebsch–Gordan coefficients C_1 and C_2 is an exercise in selecting the correct table in Appendix S.

First we note that because the kets on the rhs are uncoupled kets the values of the individual angular momenta are $j_1 = 1$ and $j_2 = \frac{1}{2}$. Thus, the coupled ket must be of the form $\left|1, \frac{1}{2}; j, m_j\right\rangle_c$ so for convenience we re-write Eq. (8.83) as

$$\left|1, \tfrac{1}{2}; j, m_j\right\rangle_c = C_1 \left|1, 0; \tfrac{1}{2}, \tfrac{1}{2}\right\rangle_u + C_2 \left|1, 1; \tfrac{1}{2}, -\tfrac{1}{2}\right\rangle_u \qquad (8.84)$$

Now, to determine j and m_j we see that for the first ket on the rhs $m_{j1}+m_{j2} = 0 + \frac{1}{2} = \frac{1}{2}$. For the second ket on the rhs $m_{j1} + m_{j2} = 1 + \left(-\frac{1}{2}\right) = \frac{1}{2}$. This tells us that the quantum number m_j in the coupled ket on the lhs is $m_j = m_{j1} + m_{j2} = \frac{1}{2}$. The total angular momentum quantum number can, however, be either $j = j_1 + j_2 = \frac{3}{2}$ or $j = j_1 - j_2 = \frac{1}{2}$. Therefore, the two possible coupled kets are

$$\left|j_1, j_2; j, m_j\right\rangle_{c1} = \left|1, \tfrac{1}{2}; \tfrac{1}{2}, \tfrac{1}{2}\right\rangle_{c1}$$

or

$$\left|j_1, j_2; j, m_j\right\rangle_{c2} = \left|1, \tfrac{1}{2}; \tfrac{3}{2}, \tfrac{1}{2}\right\rangle_{c2} \qquad (8.85)$$

Because $j_1 = 1$ and $j_2 = \frac{1}{2}$ we require Table S.3 to decide between the two coupled kets in Eq. (8.85). We see that for $j = \frac{1}{2}$ one of the Clebsch–Gordan constants is negative, but for $j = \frac{3}{2}$ both are positive and it was stated in the problem that C_1 and C_2 are of opposite sign. We conclude, therefore, that $j = \frac{1}{2}$ and the ket that we seek is $\left|1, \frac{1}{2}; \frac{1}{2}, \frac{1}{2}\right\rangle$. Moreover, we see that for $m_{j1} = 0$, $m_{j2} = \frac{1}{2}$ the CG coefficient is $-\sqrt{1/3}$ while for $m_{j1} = 1$, $m_{j2} = -\frac{1}{2}$ the CG coefficient is $\sqrt{2/3}$. Therefore, the coupled eigenket is

$$\left|1, \tfrac{1}{2}; \tfrac{3}{2}, \tfrac{1}{2}\right\rangle_c = -\frac{1}{\sqrt{3}} \left|1, 0; \tfrac{1}{2}, \tfrac{1}{2}\right\rangle_u + \sqrt{\tfrac{2}{3}} \left|1, 1; \tfrac{1}{2}, -\tfrac{1}{2}\right\rangle_u \qquad (8.86)$$

(b) From the uncoupled kets on the rhs of Eq. (8.83) we see that the only possible values of the quantum numbers m_{j1} and m_{j2} that could be measured are $\left(0, \frac{1}{2}\right)$ and $\left(1, -\frac{1}{2}\right)$. Of course, this was known at the outset because it is essentially given in Eq. (8.83). What is not given are the probabilities of measuring these couplets. These probabilities are simply the squares of the CG coefficients, so the probabilities are $1/3$ and $2/3$, respectively. It is worthwhile to recall that because $\left[\hat{J}_{1z}, \hat{J}_{2z}\right] = 0$, m_{j1} and m_{j2} can be measured simultaneously.

12. The Hamiltonian for the interaction between the spins of the electron and the
proton in a H-atom, the hyperfine interaction is

$$\hat{H}_{HF} = \frac{2\kappa}{\hbar^2}\hat{S}_1\cdot\hat{S}_2 \tag{8.87}$$

where $\hat{S}_1$ and $\hat{S}_2$ represent the spins of the electron and proton and κ is a positive
constant. $\hat{S}_1$ and $\hat{S}_2$ can represent either the electron or the proton since the
Hamiltonian is symmetric in these operators. Using the coupled representation
and the notation $|SM\rangle$ for the coupled kets find the energy eigenvalues and the
eigenkets for the hyperfine interaction.

Solution

Noting that

$$\hat{S}^2 = \left(\hat{S}_1 + \hat{S}_2\right)\cdot\left(\hat{S}_1 + \hat{S}_2\right)$$
$$= \hat{S}_1^2 + 2\hat{S}_1\cdot\hat{S}_2 + \hat{S}_2^2 \tag{8.88}$$

we can solve for $\hat{S}_1\cdot\hat{S}_2$ so that

$$\hat{H}_{HF} = \frac{\kappa}{\hbar^2}\left(\hat{S}^2 - \hat{S}_1^2 - \hat{S}_2^2\right) \tag{8.89}$$

Because Eq. (8.89) contains only the squares of the total and individual angular
momenta, it is clear that the coupled kets, the $|SM\rangle$ are the eigenkets of $\hat{H}_{HF}$.
Additionally, there is no operator in $\hat{H}_{HF}$ that involves M so this quantum
number remains constant. Therefore, the states $|1M\rangle$ are degenerate, in fact
threefold degenerate ($M = 0, \pm1$), and are known as triplet states. The
remaining coupled state, the $|00\rangle$ state, is non-degenerate and is known as a
singlet state. To find the energy eigenvalues we apply $\hat{H}_{HF}$ to the coupled kets.
For the triplet we have

$$\hat{H}_{HF}|1M\rangle = \frac{\kappa}{\hbar^2}\left(\hat{S}^2 - \hat{S}_1^2 - \hat{S}_2^2\right)|1M\rangle$$
$$= \kappa\left[1(1+1) - \frac{1}{2}\left(\frac{1}{2}+1\right) - \frac{1}{2}\left(\frac{1}{2}+1\right)\right]|1M\rangle$$
$$= \frac{\kappa}{2}|1M\rangle \tag{8.90}$$

For the singlet it is

$$\hat{H}_{HF} |00\rangle = \frac{\kappa}{\hbar^2} \left(\hat{S}^2 - \hat{S}_1^2 - \hat{S}_2^2\right) |00\rangle$$

$$= \kappa \left[0 - \frac{1}{2}\left(\frac{1}{2}+1\right) - \frac{1}{2}\left(\frac{1}{2}+1\right)\right]|1M\rangle$$

$$= -\frac{3\kappa}{2}|1M\rangle \tag{8.91}$$

13. As noted in Problem 12 of this chapter the Hamiltonian for the interaction between the spins of the electron and the proton in a H-atom, the hyperfine interaction is

$$\hat{H}_{HF} = \frac{2\kappa}{\hbar^2}\hat{S}_1 \cdot \hat{S}_2 \tag{8.92}$$

where $\hat{S}_1$ and $\hat{S}_2$ represent the spins of the electron and proton and κ is a positive constant.

(a) Show that this Hamiltonian can be written in terms of the uncoupled spin ladder operators as

$$\hat{H}_{HF} = \frac{\kappa}{\hbar^2}\left(\hat{S}_{1+}\hat{S}_{2-} + \hat{S}_{1-}\hat{S}_{2+}\right) + \frac{2\kappa}{\hbar^2}\hat{S}_{1z}\hat{S}_{2z} \tag{8.93}$$

(b) Write the singlet and triplet states, $|1M\rangle$ and $|00\rangle$, in terms of the uncoupled states $|m_{s1}, m_{s2}\rangle$.

(c) Using this result show that the eigenvalue of the triplet and singlet coupled kets are $-3\kappa/2$ and $\kappa/2$, respectively.

Solution

(a)

$$\hat{H}_{HF} = \frac{2\kappa}{\hbar^2}\hat{S}_1 \cdot \hat{S}_2$$

$$= \frac{2\kappa}{\hbar^2}\left(\hat{S}_{1x}\hat{S}_{2x} + \hat{S}_{1y}\hat{S}_{2y} + \hat{S}_{1z}\hat{S}_{2z}\right)$$

$$= \frac{\kappa}{2\hbar^2}\left[\left(\hat{S}_{1+} + \hat{S}_{1-}\right)\left(\hat{S}_{2+} + \hat{S}_{2-}\right) - \left(\hat{S}_{1+} - \hat{S}_{1-}\right)\left(\hat{S}_{2+} - \hat{S}_{2-}\right)\right]$$

$$+ \frac{2\kappa}{\hbar^2}\hat{S}_{1z}\hat{S}_{2z}$$

$$= \frac{\kappa}{\hbar^2}\left(\hat{S}_{1+}\hat{S}_{2-} + \hat{S}_{1-}\hat{S}_{2+}\right) + \frac{2\kappa}{\hbar^2}\hat{S}_{1z}\hat{S}_{2z} \tag{8.94}$$

(b) To write the coupled states as linear combinations of the uncoupled states we, as usual, appeal to the tables of CG coefficients. In this relatively simple case we require Table S.2 which, for convenience, we reproduce here.

S	$m_{s2} = 1/2$	$m_{s2} = -1/2$
1	$\sqrt{(1+M)/2}$	$\sqrt{(1-M)/2}$
0	$-1/\sqrt{2}$	$1/\sqrt{2}$

Using the table we find

$$S = 1, M = 1, m_{s2} = 1/2 \Rightarrow m_{s1} = 1/2 \Rightarrow |11\rangle = \left|\tfrac{1}{2}, \tfrac{1}{2}\right\rangle$$

$$S = 1, M = 0, m_{s2} = \pm 1/2 \Rightarrow m_{s1} = \mp 1/2 \Rightarrow |10\rangle$$

$$= \frac{1}{\sqrt{2}} \left(\left|-\tfrac{1}{2}, \tfrac{1}{2}\right\rangle + \left|\tfrac{1}{2}, -\tfrac{1}{2}\right\rangle \right)$$

$$S = 1, M = -1, m_{s2} = -1/2 \Rightarrow m_{s1} = -1/2 \Rightarrow |1-1\rangle = \left|-\tfrac{1}{2}, -\tfrac{1}{2}\right\rangle$$

$$S = 0, M = 0, m_{s2} = \mp 1/2 \Rightarrow m_{s1} = \pm 1/2 \Rightarrow |00\rangle$$

$$= -\frac{1}{\sqrt{2}} \left(\left|-\tfrac{1}{2}, \tfrac{1}{2}\right\rangle - \left|\tfrac{1}{2}, -\tfrac{1}{2}\right\rangle \right) \tag{8.95}$$

(c) We first evaluate $\hat{H}_{\mathrm{HF}} |00\rangle$ using the Hamiltonian as evaluated in Part (a) of this problem.

$$-\frac{\hbar^2}{\kappa} \hat{H}_{\mathrm{HF}} |00\rangle = \left[\left(\hat{S}_{1+}\hat{S}_{2-} + \hat{S}_{1-}\hat{S}_{2+} \right) + 2\hat{S}_{1z}\hat{S}_{2z} \right]$$

$$\times \frac{1}{\sqrt{2}} \left(\left|-\tfrac{1}{2}, \tfrac{1}{2}\right\rangle - \left|\tfrac{1}{2}, -\tfrac{1}{2}\right\rangle \right)$$

$$= \frac{1}{\sqrt{2}} \hbar^2 \left(\left|\tfrac{1}{2}, -\tfrac{1}{2}\right\rangle - \left|-\tfrac{1}{2}, \tfrac{1}{2}\right\rangle \right)$$

$$+ 2 \frac{1}{\sqrt{2}} \left(-\frac{\hbar^2}{4} \right) \left(\left|-\tfrac{1}{2}, \tfrac{1}{2}\right\rangle - \left|\tfrac{1}{2}, -\tfrac{1}{2}\right\rangle \right)$$

$$= \frac{1}{\sqrt{2}} \hbar^2 [\left(\left|\tfrac{1}{2}, -\tfrac{1}{2}\right\rangle - \left|-\tfrac{1}{2}, \tfrac{1}{2}\right\rangle \right)$$

$$+ \frac{1}{2} \left(\left|\tfrac{1}{2}, -\tfrac{1}{2}\right\rangle - \left|-\tfrac{1}{2}, \tfrac{1}{2}\right\rangle \right)]$$

$$= \frac{3\hbar^2}{2} \left[-\frac{1}{\sqrt{2}} \left(- \left|\tfrac{1}{2}, -\tfrac{1}{2}\right\rangle + \left|-\tfrac{1}{2}, \tfrac{1}{2}\right\rangle \right) \right] \tag{8.96}$$

But the quantity in brackets is $|00\rangle$ so we have

$$-\frac{\hbar^2}{\kappa}\hat{H}_{\mathrm{HF}}|00\rangle = \frac{3\hbar^2}{2}|00\rangle \implies \hat{H}_{\mathrm{HF}}|00\rangle = -\frac{3\kappa}{2}|00\rangle \tag{8.97}$$

Now evaluate $\hat{H}_{\mathrm{HF}}|10\rangle$:

$$\frac{\hbar^2}{\kappa}\hat{H}_{\mathrm{HF}}|10\rangle = \left[\left(\hat{S}_{1+}\hat{S}_{2-} + \hat{S}_{1-}\hat{S}_{2+}\right) + 2\hat{S}_{1z}\hat{S}_{2z}\right]$$

$$\times \frac{1}{\sqrt{2}}\left(|-\tfrac{1}{2},\tfrac{1}{2}\rangle + |\tfrac{1}{2},-\tfrac{1}{2}\rangle\right)$$

$$= \frac{1}{\sqrt{2}}\hbar^2\left(|\tfrac{1}{2},-\tfrac{1}{2}\rangle + |-\tfrac{1}{2},\tfrac{1}{2}\rangle\right)$$

$$+ 2\frac{1}{\sqrt{2}}\left(-\frac{\hbar^2}{4}\right)\left(|-\tfrac{1}{2},\tfrac{1}{2}\rangle + |\tfrac{1}{2},-\tfrac{1}{2}\rangle\right)$$

$$= \frac{1}{\sqrt{2}}\hbar^2[\left(|\tfrac{1}{2},-\tfrac{1}{2}\rangle + |-\tfrac{1}{2},\tfrac{1}{2}\rangle\right)$$

$$- \frac{1}{2}\left(|\tfrac{1}{2},-\tfrac{1}{2}\rangle + |-\tfrac{1}{2},\tfrac{1}{2}\rangle\right)]$$

$$= \frac{\hbar^2}{2}\left[\frac{1}{\sqrt{2}}\left(-|\tfrac{1}{2},-\tfrac{1}{2}\rangle + |-\tfrac{1}{2},\tfrac{1}{2}\rangle\right)\right] \tag{8.98}$$

But the quantity in brackets is $|10\rangle$ so we have

$$\frac{\hbar^2}{\kappa}\hat{H}_{\mathrm{HF}}|10\rangle = \frac{\hbar^2}{2}|10\rangle \implies \hat{H}_{\mathrm{HF}}|10\rangle = \frac{\kappa}{2}|10\rangle \tag{8.99}$$

We could have made it easy on ourselves and worked it out for the bottom or top of the ladder triplet state. For example,

$$\hat{H}_{\mathrm{HF}}|11\rangle = \hat{H}_{\mathrm{HF}}|\tfrac{1}{2},\tfrac{1}{2}\rangle$$

$$= \left[\frac{\kappa}{\hbar^2}\left(\hat{S}_{1+}\hat{S}_{2-} + \hat{S}_{1-}\hat{S}_{2+}\right) + \frac{2\kappa}{\hbar^2}\hat{S}_{1z}\hat{S}_{2z}\right]|\tfrac{1}{2},\tfrac{1}{2}\rangle$$

$$= \left[\frac{\kappa}{\hbar^2}(0+0) + \frac{2\kappa}{\hbar^2}\left(\frac{\hbar^2}{4}\right)|\tfrac{1}{2},\tfrac{1}{2}\rangle\right]$$

$$= \frac{\kappa}{2}|\tfrac{1}{2},\tfrac{1}{2}\rangle = \frac{\kappa}{2}|11\rangle \tag{8.100}$$

These Problems 12 and 13 illustrate how a judicious choice of basis set can greatly simplify a problem. While the use of the uncoupled set for obtaining the energy eigenvalues and the eigenkets was not too laborious, use of the coupled kets led almost immediately to these quantities.

14. This problem treats elastic scattering of quantum particles from a hard sphere of radius R; it is the quantum mechanical analogue of Problem 12 of Chap. 1, where the particle mass m is much less than the target sphere mass. The solution shows that incorporating the wave nature of the scattered particles, taken to be electrons or light atoms (not BBs!), leads to a substantially different result for the low-energy cross-section than that obtained in the classical treatment. Figure 8.1 is a schematic depiction of the problem.

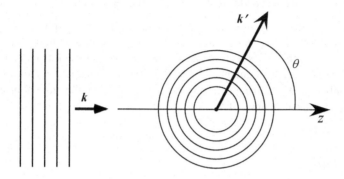

Fig. 8.1 Problem 14

A plane wave with wave vector k and kinetic energy E directed along the z-axis represents the particle flux incident on a spherical target. The magnitudes of the energy and the wave vector are related by

$$k = \frac{\sqrt{2mE}}{\hbar} \tag{8.101}$$

The wave vector after the collision is designated k'. In the general central potential case, the scattered flux intensity varies with angle θ. No energy is transferred, so $k' = k$ and both the incident and scattering flux intensities are symmetric around the z-axis. The incoming plane wave and outgoing scattered wave can be expanded in terms of spherical waves, $\exp{(ikr)}/r$, and Legendre polynomials, $P_\ell (\cos \theta)$, where ℓ is the angular momentum quantum number of the particular expansion term. The problem is simplified if only the $\ell = 0$ terms are significant; this occurs in the low collision energy limit. In this $\ell = 0$ limit (known as s-wave scattering) the wave function that represents the outgoing particles depends only on the radial coordinate r, and outside the range of the scattering potential $\psi (r)$ can be written [2]

$$\psi_{\ell=0} (r) = \frac{1}{2ikr} \left(e^{-ikr} - e^{ikr}\right) - f_{\ell=0} (k) \frac{e^{ikr}}{r} \tag{8.102}$$

where $f_{\ell=0}(k)$ is the scattering amplitude. It is the coefficient of the term that represents the outgoing spherical wave, e^{ikr}/r, which is the solution of the scattering problem. This coefficient is related to the differential scattering cross section $d\sigma/d\Omega$ by

$$\frac{d\sigma}{d\Omega} = |f_{\ell=0}(k)|^2 \tag{8.103}$$

Equation (8.102) is valid for any central potential that falls off faster than $1/r$.

(a) Obtain the scattering amplitude $f_{\ell=0}(k)$ for a hard sphere potential of radius R. Use Eq. (8.103) to find the total cross section σ in the limit $k \to 0$. Compare this cross section result with the classical result from Problem 12.
(b) Use a classical argument to define the term "low energy collision," such that only $\ell = 0$ scattering is important. Remember that the angular momentum is quantized in units of $\hbar$.

Solution

(a) For the hard-sphere the wave function $\psi(r)$ must vanish at $r = R$. Setting $\psi(R) = 0$ in Eq. (8.102) and rearranging, one obtains

$$f_{\ell=0}(k) = \frac{1}{2ik} \left(e^{-ikR} - e^{ikR} \right) e^{-ikR}$$

$$= \frac{1}{k} e^{-ikR} \sin kR \tag{8.104}$$

so

$$|f_{\ell=0}(k)|^2 = \frac{\sin^2 kR}{k^2} \tag{8.105}$$

Taking the limit as $k \to 0$ leads to

$$\lim_{k \to 0} |f_{\ell=0}(k)|^2 = R^2 \tag{8.106}$$

To find the total cross section σ we integrate Eq. (8.106) over all Ω and obtain

$$\sigma = 4\pi R^2 \tag{8.107}$$

This value of σ is four times larger than the geometric classical cross section obtained in Problem 12 of Chap. 1. This result illustrates the interference phenomena associated with the wave nature of quantum particles.

The fact that diffraction of the incoming particles occurs at the surface of the hard sphere causes more particles (four times more) to be scattered than in the classical case for which no such diffraction occurs.

(b) The approximation to the total cross section in Eq. (8.107) is accurate to within roughly 10 % for $kR \lesssim 0.6$ via Eq. (8.105). For a given R this defines a limiting k. The validity of Eq. (8.106) by itself does not guarantee that higher angular momenta (e.g., $\ell = 1$) are not important. The classical angular momentum for a collision that just grazes the hard sphere (see Fig. 1.13) is given by

$$L_{\text{classical}} = R\sqrt{2mE} \qquad (8.108)$$

and the quantum angular momentum for $\ell = 1$ is $\hbar$. Now we argue that if $R\sqrt{2mE} < \hbar$, a classical collision corresponding to $\ell = 1$ with collision energy E is not possible, so only $\ell = 0$ collisions are important. This sets an energy E_{max} for $\ell = 0$ collisions to dominate:

$$E_{\text{max}} < \frac{\hbar^2}{2mR^2} \qquad (8.109)$$

To particularize, we can assume a hard sphere target of radius $R = 0.5$ nm, or $R \simeq 10$ a.u. (see Appendix C), being impacted by a beam of electrons ($m = 1$ a.u.). In atomic units $\hbar = 1$, so Eq. (8.109) yields $E_{\text{max}} < 0.005$ a.u. (or 0.14 eV). If the irradiating particles were neutrons ($m = 1839$ a.u.) or the sphere radius 0.5 μm, s-wave scattering would lose dominance at very low collision energy.

Chapter 9
Indistinguishable Particles

The defining characteristic of indistinguishable particles is that their interchange cannot be detected. In this chapter we will concentrate on two-particle systems because they are relatively simple and they contain the essential physics. For a two-particle system we designate a state by $|a\,b\rangle$ where a and b represent the states (both spatial and spin states) of each of the particles. The positions of a and b in the ket $|a\,b\rangle$ designate particle 1 or particle 2, respectively, under the assumption that they can be labelled. Thus, the ket $|a\,b\rangle$ represents a state in which particle 1 is in state a and particle 2 is in state b. The two particles are, however, indistinguishable so that a ket such as $|a\,b\rangle$ is unacceptable because the probability distribution is not invariant under particle exchange. To accomplish this invariance the kets representing a given state must be a linear combination of $|a\,b\rangle$ and $|ba\rangle$ that is either symmetric or antisymmetric under particle interchange to insure that the probability distribution is invariant. The acceptable symmetric and antisymmetric kets are designated $|\psi_+\rangle$ and $|\psi_-\rangle$, respectively, and are

$$|\psi_+\rangle = \frac{1}{\sqrt{2}}\left(|a\,b\rangle + |b\,a\rangle\right) \qquad (9.1)$$

and

$$|\psi_-\rangle = \frac{1}{\sqrt{2}}\left(|a\,b\rangle - |b\,a\rangle\right) \qquad (9.2)$$

For these eigenfunctions exchange of all particle coordinates leads to the original eigenfunction or the negative of the original eigenfunction.

© Springer International Publishing AG 2017
J.D. Kelley, J.J. Leventhal, *Problems in Classical
and Quantum Mechanics*, DOI 10.1007/978-3-319-46664-4_9

There are two types of indistinguishable particles, fermions (half-integral spin) and bosons (integral spin).

Fermions are characterized by antisymmetric state vectors and bosons by symmetric state vectors.

Examples of fermions are electrons, protons, and neutrons; examples of bosons are photons and α-particles (bare helium nuclei).

Problems

1. For two identical bosons

    (a) find the average positions of each of the particles. Do the same for a pair of fermions.
    (b) Find average of the squares of the separations between the bosons. Do the same for the fermion pair. Contrast the results with those obtained for distinguishable particles.

Solution

(a) According to Eqs. (9.1) and (9.2) the allowable eigenfunctions for identical particles may be written

$$|\psi_\pm\rangle = \frac{1}{\sqrt{2}}\left(|a\,b\rangle \pm |b\,a\rangle\right) \tag{9.3}$$

The average value of the position of, for example, particle 1 is

$$\langle x_1\rangle = \langle \psi_\pm| x_1 |\psi_\pm\rangle$$

$$= \frac{1}{2}\left(\langle a\,b| \pm \langle b\,a|\right) x_1 \left(|a\,b\rangle \pm |b\,a\rangle\right)$$

$$= \frac{1}{2}\left[\langle a\,b| x_1 |a\,b\rangle \pm \langle a\,b| x_1 |b\,a\rangle \pm \langle b\,a| x_1 |a\,b\rangle + \langle b\,a| x_1 |b\,a\rangle\right]$$

$$= \frac{1}{2}\left[\langle x_1\rangle_a + \langle x_1\rangle_b\right] \tag{9.4}$$

where $\langle x_1\rangle_a$ is the average of x_1 when the particle is in state a; similarly for $\langle x_1\rangle_b$. The terms with $\pm$ vanish because of orthogonality of the states a and b. Clearly $\langle x_1\rangle \equiv \langle x_2\rangle$.

Let us understand the meaning of each term in Eq. (9.4). Because the particles are identical, each can be in either state a or state b so $\langle x_1\rangle_a$ is simply the average value of the position of particle 1 when it is in state a. The term $\langle x_1 \rangle_b$ represents the average value of the position of the particle when in state b. It is not unexpected that $\langle x_1 \rangle$, the average value of x for particle 1 when it is in either state, a or b is the average of $\langle x_a \rangle_a$ and $\langle x_a \rangle_b$. Note that the values of $\langle x_1 \rangle$ and $\langle x_2 \rangle$ are the same for bosons and fermions.

(b) The average of the separation between the two particles is

$$\langle (x_1 - x_2)^2 \rangle_\pm = \langle \psi_\pm | (x_1 - x_2)^2 | \psi_\pm \rangle$$

$$= \langle \psi_\pm | (x_1^2 - 2x_1 x_2 + x_2^2)^2 | \psi_\pm \rangle$$

$$= \langle x_1^2 \rangle + \langle x_2^2 \rangle - 2 \langle \psi_\pm | (x_1 x_2) | \psi_\pm \rangle \qquad (9.5)$$

Now, inserting the kets $|\psi_\pm\rangle$ from Eqs. (9.1) and (9.2) let us examine the last term in Eq. (9.5).

$$\langle \psi_\pm | (x_1 x_2) | \psi_\pm \rangle = \langle (\langle a\, b| \pm \langle b\, a|) | (x_1 x_2) | (|a\, b\rangle \pm |b\, a\rangle) \rangle$$

$$= \langle a\, b| (x_1 x_2) |a\, b\rangle \pm \langle a\, b| (x_1 x_2) |b\, a\rangle$$

$$\pm \langle b\, a| (x_1 x_2) |a\, b\rangle + \langle b\, a| (x_1 x_2) | b\, a\rangle$$

$$= \langle x_1 \rangle_a \langle x_2 \rangle_b \pm \langle a| x_1 |b\rangle \langle b| x_2 |a\rangle$$

$$\pm \langle b| x_1 |a\rangle \langle a| x_2 |b\rangle + \langle x_1 \rangle_b \langle x_2 \rangle_a \qquad (9.6)$$

The matrix elements are the same regardless of whether we use x_1 or x_2 so we simply use x. Equation (9.6) becomes

$$\langle (x_1 - x_2)^2 \rangle_\pm = \langle x^2 \rangle_a + \langle x^2 \rangle_b - 2 \langle x \rangle_b \langle x \rangle_a \mp 2 \langle a| x |b\rangle^2 \qquad (9.7)$$

where we have taken advantage of the fact that $\langle b| x |a\rangle = \langle a| x |b\rangle$ in the last term.

Equation (9.7) contains some very important physics. If the particles were distinguishable, then the last term would not be present. The $\mp$ sign shows that the average separation between the two indistinguishable particles is smaller for bosons and larger for fermions. We see then that bosons like to congregate while fermions are antisocial. This effect results from the so-called exchange force, which is not a force in the usual sense. The effect is a consequence of the symmetry requirements on the eigenfunctions. The exchange force is responsible for the ordering of the periodic table as well as many other effects such as boson condensation.

2. Two identical non-interacting spin-$\frac{1}{2}$ particles (fermions) of mass m are subject to a one-dimensional harmonic oscillator potential. Determine the ground state and first excited state kets and their corresponding energies

Solution

The Hamiltonian is given by

$$\hat{H}(x_1, x_2) = \left(\frac{p_1^2}{2m} + \frac{1}{2}m\omega^2 x_1^2 \right) + \left(\frac{p_2^2}{2m} + \frac{1}{2}m\omega^2 x_2^2 \right)$$

$$= \hat{H}(x_1) + \hat{H}(x_2) \tag{9.8}$$

Because the Hamiltonian is separable as indicated in Eq. (9.8) the eigenfunctions will be the products of the eigenfunction of the individual Hamiltonians and the energy eigenvalues will be the sums of the individual eigenvalues. Thus, the spatial part of the eigenfunctions may be written

$$|n_1 n_2\rangle_x = |n_1\rangle |n_2\rangle \tag{9.9}$$

The energy eigenvalues are

$$E(n_1, n_2) = (n_1 + n_2 + 1)\hbar\omega \tag{9.10}$$

To construct the total eigenfunctions we must include the spin eigenfunctions. The coupled kets are designated $|SM\rangle_{\text{spin}}$ where $S = 0, 1$. For two spin-$\frac{1}{2}$ particles there are four states, the singlet state $|00\rangle_{\text{spin}}$ and three $|1M\rangle_{\text{spin}}$ states (the "triplet" states). In terms of the uncoupled kets $|m_{s1} m_{s2}\rangle$ these four states are

$$|00\rangle_{\text{spin}} = -\frac{1}{\sqrt{2}} |-\tfrac{1}{2}\tfrac{1}{2}\rangle + \frac{1}{\sqrt{2}} |\tfrac{1}{2} - \tfrac{1}{2}\rangle$$

$$|11\rangle_{\text{spin}} = |\tfrac{1}{2}, \tfrac{1}{2}\rangle$$

$$|10\rangle_{\text{spin}} = \frac{1}{\sqrt{2}} |-\tfrac{1}{2}\tfrac{1}{2}\rangle + \frac{1}{\sqrt{2}} |\tfrac{1}{2} - \tfrac{1}{2}\rangle$$

$$|1 - 1\rangle_{\text{spin}} = |-\tfrac{1}{2} - \tfrac{1}{2}\rangle \tag{9.11}$$

Note that under exchange of particles the singlet spin state is antisymmetric and the triplet spin states are all symmetric.

Ground state: The spatial part of the ground state is the symmetric state $|00\rangle$ with energy

$$E_{00} = \hbar\omega \tag{9.12}$$

Because these two fermions are in a symmetric spatial state the spins must be in the antisymmetric (singlet) state, clearly $|00\rangle_{\text{spin}}$ (see Eq. (9.11)). Thus, the ket representing the ground state of the system $|\psi\rangle_{00}$ is

$$|\psi\rangle_{00} = |00\rangle_x |00\rangle_{\text{spin}} \tag{9.13}$$

First excited state:
The first excited state is more complicated because the states are fourfold degenerate. The spatial parts of the total eigenkets are $|01\rangle_x$ and $|10\rangle_x$ each of which has the same energy eigenvalue

$$E_{01} = E_{10} = 2\hbar\omega \tag{9.14}$$

These eigenkets are, however, neither symmetric nor antisymmetric so they must be symmetrized. Inasmuch as they are degenerate, linear combinations of them are also eigenfunctions so that using Eqs. (9.1) and (9.2) we have

$$|\xi\rangle_s = \frac{1}{\sqrt{2}} (|10\rangle_x + |01\rangle_x)$$

$$|\xi\rangle_a = \frac{1}{\sqrt{2}} (|10\rangle_x - |01\rangle_x) \tag{9.15}$$

where ξ is used to designate the spatial part of the eigenkets. The total kets for the first excited states are

$$|\xi\rangle_s |00\rangle_{\text{spin}} = \frac{1}{\sqrt{2}} (|10\rangle_x + |01\rangle_x) |00\rangle_{\text{spin}}$$

$$|\xi\rangle_a |11\rangle_{\text{spin}} = \frac{1}{\sqrt{2}} (|10\rangle_x - |01\rangle_x) |11\rangle_{\text{spin}}$$

$$|\xi\rangle_a |10\rangle_{\text{spin}} = \frac{1}{\sqrt{2}} (|10\rangle_x - |01\rangle_x) |10\rangle_{\text{spin}}$$

$$|\xi\rangle_a |1-1\rangle_{\text{spin}} = \frac{1}{\sqrt{2}} (|10\rangle_x - |01\rangle_x) |1-1\rangle_{\text{spin}} \tag{9.16}$$

Chapter 10
Bound States in Three Dimensions

Three-dimensional problems in quantum mechanics almost always involve central potentials of the type that were dealt with in classical mechanics in Chap. 3. These central potential problems are usually solved using spherical coordinates. When the TISE can be separated in an additional coordinate system an additional symmetry exists (beyond the obvious isotropic symmetry associated with a central potential). This additional symmetry signals an "accidental degeneracy" [1], a degeneracy beyond the $(2\ell + 1)$-fold degeneracy associated with the isotropy of the central potential [1]. In this chapter we concentrate on central potentials with a small dose of Cartesian potential problems just to show that there are other coordinate systems used for three-dimensional problems in quantum mechanics.

For quantum mechanical central potential problems the symmetry of the potential dictates that the eigenfunctions are given by

$$\psi_{n\ell m}(r, \theta, \phi) = R_{n\ell}(r) Y_{\ell m}(\theta, \phi) \tag{10.1}$$

where the $Y_{\ell m}(\theta, \phi)$ are the spherical harmonics (see Appendix R and [1]) and $R_{n\ell}(r)$ is the solution to the radial part of the separated TISE which is

$$\left[-\frac{\hbar^2}{2mr^2} \frac{d}{dr} \left(r^2 \frac{d}{dr} \right) + \frac{\ell(\ell + 1)\hbar^2}{2mr^2} + U(r) \right] R_{n\ell}(r) = E_{n\ell} R_{n\ell}(r) \tag{10.2}$$

Making the substitution $u_{n\ell}(r) = r R_{n\ell}(r)$ Eq. (10.2) is sometimes re-written in the form

$$-\frac{\hbar^2}{2m} \frac{d^2 u_{n\ell}(r)}{dr^2} + \left\{ \frac{\ell(\ell + 1)\hbar^2}{2mr^2} + U(r) \right\} u_{n\ell}(r) = E_{n\ell} u_{n\ell}(r) \tag{10.3}$$

© Springer International Publishing AG 2017
J.D. Kelley, J.J. Leventhal, *Problems in Classical
and Quantum Mechanics*, DOI 10.1007/978-3-319-46664-4_10

Using Eq. (10.3) and the trial solution $u(r) = r^s$ the behavior of $u(r)$ and of $R(r)$ as $r \to 0$ can be deduced. The result is

$$\lim_{r\to 0} u(r) \sim r^{\ell+1} \quad \Rightarrow \quad \lim_{r\to 0} R(r) \sim r^\ell \tag{10.4}$$

Additionally, again using Eq. (10.3), the behavior as $r \to \infty$ for bound states $[E < U(\infty)]$ is determined to be

$$\lim_{r\to\infty} u(r) \propto e^{-\kappa r} \quad \Rightarrow \quad \lim_{r\to\infty} R(r) \propto \frac{e^{-\kappa r}}{r} \tag{10.5}$$

where $\kappa = \sqrt{-2mE/\hbar^2}$.

Problems

1. A particle of mass m is trapped inside a cube, each side of which has length L. The walls of the cube are impenetrable. It is a three-dimensional L-box.

   (a) Find the energy eigenfunctions in Cartesian coordinates.
   (b) Find the energy eigenvalues and compare the ground state energy with the ground state energy of the one-dimensional L-box?
   (c) What are the degeneracies of the first three energy levels?

Solution

(a) Inside the cube the Hamiltonian is

$$\hat{H}(x, y, z) = \frac{p_x^2}{2m} + \frac{p_y^2}{2m} + \frac{p_z^2}{2m}$$

$$= \hat{H}_x(x) + \hat{H}_y(y) + \hat{H}_z(z) \tag{10.6}$$

where each Hamiltonian $\hat{H}_{x_i}(x_i)$ is a one-dimensional L-box Hamiltonian in each of the three coordinates. Because the Hamiltonian is the sum of three Hamiltonians, each containing only a single coordinate, and the boundary conditions are the same as those for an L-box, the solution to the three-dimensional TISE is a product of the wave functions for the individual Hamiltonians. Using quantum numbers $n_x, n_y, n_z = 1, 2, 3, \ldots$ we have

$$\psi(x, y, z) = \left(\sqrt{\frac{2}{L}}\right)^3 \sin\left(\frac{n_x \pi}{L}x\right) \sin\left(\frac{n_y \pi}{L}y\right) \sin\left(\frac{n_z \pi}{L}z\right) \tag{10.7}$$

(b) Again because the Hamiltonian is the sum of three Hamiltonians, each containing only a single coordinate, the eigenvalues are the sums of the individual eigenvalues. Therefore

$$E_{n_x n_y n_z} = \frac{\pi^2 \hbar^2}{2mL^2} \left(n_x^2 + n_y^2 + n_z^2 \right) \tag{10.8}$$

Inasmuch as there are three quantum numbers, each with a minimum value $n_i = 1$, it is clear from Eq. (10.8) that the ground state energy of the three-dimensional L-box is three times that of a comparable one-dimensional L-box.

(c) The sets of quantum numbers for the ground state and the first excited state are

$$\text{ground state:} \quad (1, 1, 1)$$

$$\text{1st excited state:} \quad (1, 1, 2) ; (1, 2, 1) ; (2, 1, 1)$$

so the degeneracies are $g_1 = 1$ and $g_2 = 3$. For the second excited state we have

$$\text{2nd excited state:} \quad (1, 2, 2) ; (2, 1, 2) ; (2, 2, 1)$$

so that $g_3 = 3$ as well.

Notice that $g_2 = g_3$ because in each triplet of quantum numbers for a given energy level two of the three quantum numbers are identical, for example (a, a, b). If the quantum numbers are all different (a, b, c), the degeneracy is 6 because the number of permutations of 3 distinct objects is 3!, and this is the highest degeneracy that can occur for any level of a cubic box.

2. Consider a 3-dimensional L-box as in Problem 1 of this chapter. Assume the box is large enough so that the energy levels are closely spaced and states with high values of n_x, n_y, and n_z are occupied. In this case it is useful to think about the number of states present in a narrow energy band ΔE rather than a single specific state (n_x, n_y, n_z). When the energy levels are closely spaced, we can treat the n_i's as continuous variables, and we define a quantity

$$n = \sqrt{n_x^2 + n_y^2 + n_z^2} \tag{10.9}$$

The number of states N with n-values from zero to a specific value n_0 is

$$N = \left(\frac{1}{8} \right) \left(\frac{4}{3} \pi n_0^3 \right) \tag{10.10}$$

which is simply the volume of a sphere in n-space divided by 8 to account for the fact that (n_x, n_y, n_z) are all positive. The *density of states* $g(E)$, which is the number of states in an energy interval dE, is defined as

$$g(E) = \frac{dN}{dE} \tag{10.11}$$

Find an explicit expression for the density of states for a particle in an L-box [use Eq. (10.8)].

Solution

From Eq. (10.8),

$$n = \frac{1}{2\sqrt{\beta}}\sqrt{E} \tag{10.12}$$

where

$$\beta = \frac{\pi^2 \hbar^2}{2mL^2} \tag{10.13}$$

Rewrite Eq. (10.11) as

$$g(E) = \frac{dN}{dn} \cdot \frac{dn}{dE} \tag{10.14}$$

and use Eqs. (10.10) and (10.12) to obtain the density of states

$$g(E) = \left(\frac{\pi}{4\beta^{3/2}}\right) E^{1/2} \tag{10.15}$$

The density of states is essentially the energy degeneracy in the continuous limit where the energy states are very closely spaced. The density of states is used in treating radiation (photon modes in an enclosure) and in statistical physics.

3. Consider a diatomic molecule as a pair of atoms with masses m_1 and m_2 connected by a rigid massless rod of length r. The moment of inertia is $I = \mu r^2$ where $\mu = m_1 m_2 / (m_1 + m_2)$. This problem is referred to as the "rigid rotor" and may be considered the simplest of all central potential problems. Find the energy eigenfunctions, the energy eigenvalues and the degree of degeneracy for this rigid rotor.

Solution

The rod is rigid so the distance between the masses is fixed. There is therefore no potential energy and the Hamiltonian consists of only kinetic energy $\hat{T}$. This kinetic energy is purely rotational and can be written in terms of the total angular momentum operator $\hat{L}$ as

$$\hat{T} = \hat{H} = \frac{\hat{L}^2}{2I} \tag{10.16}$$

We need go no further in solving the TISE because we have already solved it. We know that the Hamiltonian operator is proportional to $\hat{L}^2$ so the eigenfunctions must be the spherical harmonics, the $Y_{\ell m}(\theta, \phi)$. Moreover we know that

$$\hat{H} Y_{\ell m}(\theta, \phi) = \frac{1}{2I} \hat{L}^2 Y_{\ell m}(\theta, \phi)$$

$$= \left[\frac{1}{2\mu r^2} \ell (\ell + 1) \hbar^2 \right] Y_{\ell m}(\theta, \phi)$$

$$= E_\ell Y_{\ell m}(\theta, \phi) \tag{10.17}$$

so the eigenvalues must be

$$E_\ell = \frac{\ell (\ell + 1) \hbar^2}{2\mu r^2} \tag{10.18}$$

Because the eigenvalues do not depend upon the quantum number m this system has the usual degeneracy associated with a central potential. Physically, the central nature of the potential means that there is no preferred direction in space.

4. Another simple central potential is the three-dimensional analog of the one-dimensional infinitely deep square well. The potential energy function is thus

$$U(r) = 0 \quad r < a$$

$$= \infty \quad r > a \tag{10.19}$$

Find the eigenenergies and the energy eigenfunctions for the $\ell = 0$ states. Note the difficulty of finding the same quantities for $\ell \neq 0$.

Solution

Clearly, $R_{n\ell}(r) = 0$ for $r > a$. For $r < a$ the radial TISE for $u_{n\ell} = rR_{n\ell}(r)$ is

$$-\frac{\hbar^2}{2m} \frac{d^2 u_{n\ell}(r)}{dr^2} + \left[\frac{\ell (\ell + 1) \hbar^2}{2mr^2} \right] u_{n\ell}(r) = E_{n\ell} u_{n\ell}(r) \tag{10.20}$$

For $\ell = 0$, however, this equation is

$$\frac{d^2 u_{n0}(r)}{dr^2} + k^2 u_{n0}(r) = 0 \quad \text{where} \quad k^2 = \frac{2mE_{n0}}{\hbar^2} \tag{10.21}$$

We see that we have already solved this equation, which is the same as that encountered in the solution of the L-box (see Appendix M). The well-known solutions are

$$u_{n0}(r) = A_0 \sin(kr) + B_0 \cos(kr) \tag{10.22}$$

where the subscript zero on the constants of integration reminds us that this solution is peculiar to the $\ell = 0$ case.

Inasmuch as $\cos(kr) \to 1$ as $r \to 0$ we must set $B_0 = 0$. In addition, the boundary condition that $R_{n0}(a) = 0$ demands that $ka = n\pi$ so that

$$E_{n0} = \frac{n^2 \pi^2 \hbar^2}{2ma^2} \tag{10.23}$$

which is the same result at that for an L-box or an a-box (see Appendix M). Moreover, we already know the normalization constant A_{n0} so we may write the complete $\ell = 0$ eigenfunctions. They are

$$\psi_{n00}(r, \theta, \phi) = Y_{00}(\theta, \phi) \sqrt{\frac{2}{a}} \frac{\sin(k_{n0}r)}{r}$$

$$= \sqrt{\frac{1}{2\pi a}} \frac{\sin(k_{n0}r)}{r} \tag{10.24}$$

These radial eigenfunctions are the spherical Bessel functions of order zero [1]. Note that in Eq. (10.24) for the eigenfunctions, k now has a subscript $n0$ because it corresponds to the energy eigenvalues E_{n0}.

This problem was relatively simple only because we chose to solve it for the $\ell = 0$ case. If $\ell \neq 0$, then the radial TISE is considerably more complicated and the solutions are higher order spherical Bessel functions.

5. While Problem 4 of this chapter is likely too elementary to be found on a PhD qualifying exam and variations of it for $\ell > 0$ are too complicated because they require knowledge of spherical Bessel functions, it is possible to construct a problem using this theme which might be acceptable on such an exam. Consider the shell potential

$$U(r) = \infty \quad r < a$$

$$= 0 \quad a < r < b$$

$$= \infty \quad r > b \tag{10.25}$$

We ask the same questions as those asked for Problem 4, namely, find the eigenenergies and the energy eigenfunctions for the $\ell = 0$ states. Do not bother to normalize the eigenfunction.

Solution

For $r < a$ and $r > b$, $R(r) = 0$. For the region in the shell $a < r < b$ the radial TISE for $u = rR(r)$ and $\ell = 0$ is

$$\frac{d^2u(r)}{dr^2} + k^2u(r) = 0 \quad \text{where} \quad k^2 = \frac{2mE}{\hbar^2} \tag{10.26}$$

the solution to which is again

$$u(r) = A_0 \sin(kr) + B_0 \cos(kr) \tag{10.27}$$

The boundary conditions are $R(a) = 0 = R(b)$ so that $u(a) = 0 = u(b)$. Inserting these boundary condition into Eq. (10.26) leads to two equations and two unknowns for A_0 and B_0.

$$\sin(ka)A_0 + \cos(ka)B_0 = 0$$

$$\sin(kb)A_0 + \cos(kb)B_0 = 0 \tag{10.28}$$

Inasmuch as these are homogeneous equations the only non-trivial solution is the one obtained by setting the determinant of the coefficients equal to zero. Thus

$$\sin(ka)\cos(kb) - \cos(ka)\sin(kb) = 0 \tag{10.29}$$

which, using Eq. (E.1), is

$$\sin k(b - a) = 0 \tag{10.30}$$

Reminiscent of square well problems already worked, Eq. (10.30) demands that $k(b - a) = n\pi$ so that

$$\frac{2mE}{\hbar^2}(b - a)^2 = n^2\pi^2 \tag{10.31}$$

Solving for E_{n0} as in Problem 4 of this chapter we have

$$E_{n0} = \frac{n^2\pi^2\hbar^2}{2m(b - a)^2} \tag{10.32}$$

Additionally, from Eq. (10.28) we see that A_0 and B_0 are related by

$$B_0 = -\tan(ka)A_0 \tag{10.33}$$

so that, from Eq. (10.27) the radial energy eigenfunction is

$$R_{n0}(r) = \frac{u(r)}{r} = \frac{A_0}{r}\left[\sin(k_{n0}r) - \tan(k_{n0}a)\cos(k_{n0}r)\right] \tag{10.34}$$

Because we are dealing with only $\ell = 0$ states the angular part of the eigenfunctions is $Y_{00}(\theta, \phi)$.

6. Find the probability that the electron in the ground state of a H-atom will be found in the classically forbidden region? Use atomic units (see Appendix C) in which the Coulomb potential is $-1/r$, the ground state eigenfunction is

$$\psi_{100}(r, \theta, \phi) = \frac{1}{\sqrt{\pi}}e^{-r} \tag{10.35}$$

and the energy eigenvalues are $E_n = -1/(2n^2)$ for all values of n.

Solution

For the ground state $n = 1$, $\ell = 0$ so we must first find the value of r_c the classical turning point for the effective potential with $\ell = 0$, i.e. the pure Coulomb potential as shown in Fig. 10.1.

Fig. 10.1
Problem 6—solution

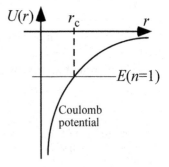

The classical turning point occurs when the kinetic energy is zero. Therefore, the total energy, which is the energy eigenvalue, must equal the potential energy and

$$-\frac{1}{2n^2} = -\frac{1}{r_c} \implies r_c = 2n^2 \tag{10.36}$$

Therefore, the probability $P_{\text{forbidden}}$ of finding the electron in the classical forbidden region for the ground state is

$$P_{\text{forbidden}} = \int_{r_c=2}^{\infty} dr \left(\frac{1}{\sqrt{\pi}} e^{-r} \right)^2 r^2 \int_0^{\pi} \sin\theta d\theta \int_0^{2\pi} d\phi$$

$$= \frac{4\pi}{\pi} \int_{r_c=2}^{\infty} r^2 e^{-2r} dr \qquad (10.37)$$

Using the integral in Eq. (G.1) we have

$$P_{\text{forbidden}} = 4 \left. \frac{e^{-2r}}{-2} \left(r^2 + \frac{2r}{2} + \frac{2}{2^2} \right) \right|_{\infty}^{2}$$

$$= 2e^{-4} \left(2^2 + \frac{2 \cdot 2}{2} + \frac{1}{2} \right)$$

$$= 13e^{-4}$$

$$\approx 0.24 \qquad (10.38)$$

It is fairly surprising that the electron spends nearly one quarter of its time in the classically forbidden region.

7. Find the expectation value $\langle r^{-2} \rangle$ for the ground state of a H-atom. The normalized radial eigenfunction for a H-atom in the ground state is (see Table T.2)

$$R_{10}(r) = 2 \left(\frac{1}{a_0} \right)^{3/2} e^{-r/a_0} \qquad (10.39)$$

where a_0 is the Bohr radius.

Solution

The expectation value of r^{-2} for the ground state of hydrogen is (see Table R.1 for $Y_{00}(\theta, \phi)$)

$$\left\langle \frac{1}{r^2} \right\rangle_{100} = \int_{\text{all space}} Y_{00}^*(\theta, \phi) R_{10}^*(r) \left(\frac{1}{r^2} \right) Y_{00}(\theta, \phi) R_{10}(r) \, dV$$

$$= \left\{ \int_{\text{all } \Omega} [Y_{00}(\theta, \phi)]^2 \, d\Omega \right\} \int_0^{\infty} \frac{[R_{10}(r)]^2}{r^2} r^2 dr$$

$$= \{1\} \cdot \left\{ 4 \left[\left(\frac{1}{a_0} \right)^3 \right] \int_0^\infty e^{-2r/a_0} dr \right\}$$

$$= 4 \left(\frac{1}{a_0} \right)^3 \frac{a_0}{2} \left[e^{-2r/a_0} \right]_\infty^0$$

$$= 2 \left(\frac{1}{a_0} \right)^2 \tag{10.40}$$

It is satisfying that our computed value of $\langle r^{-2} \rangle$ has units $1/\,m^2$.

8. Kramer's relation permits calculation of $\langle r^s \rangle$ for arbitrary values of n and ℓ, and integral values of s (see the last entry in Table T.3). This relation is

$$\frac{(s+1)}{n^2} \langle r^s \rangle - (2s+1) a_0 \langle r^{s-1} \rangle + \frac{s}{4} a_0^2 \left[(2\ell+1)^2 - s^2 \right] \langle r^{s-2} \rangle = 0 \tag{10.41}$$

(a) Using Kramer's relation evaluate $\langle r^{-1} \rangle$ and $\langle r \rangle$ in that order.
(b) Using Kramer's relation evaluate $\langle r^{-3} \rangle$ for the ground state. [Hint: To avoid having to evaluate $\langle r^{-2} \rangle$ use Table T.3.]

Solution

(a) Choosing $s = 0$ in Eq. (10.41) immediately leads to $\langle r^{-1} \rangle$ because, conveniently, the last term vanishes for $s = 0$. We have

$$\frac{1}{n^2} - a_0 \langle r^{-1} \rangle = 0 \Rightarrow \langle r^{-1} \rangle = \frac{1}{n^2 a_0} \tag{10.42}$$

To evaluate $\langle r \rangle$ we choose $s = 1$ and obtain

$$\frac{2}{n^2} \langle r \rangle - 3 a_0 + \frac{a_0^2}{4} \left[(2\ell+1)^2 - 1 \right] \langle r^{-1} \rangle = 0 \tag{10.43}$$

It is clear from Eq. (10.43) that $\langle r^{-1} \rangle$ is required to evaluate $\langle r \rangle$. This illustrates the bootstrapping necessary to use Kramer's relation (except for $s = 0$). Solving Eq. (10.43) for $\langle r \rangle$ and using the result in Eq. (10.42) we have

$$\langle r \rangle = \frac{n^2}{2} \left[3 a_0 - \frac{a_0^2}{4} \left[(2\ell+1)^2 - 1 \right] \left(\frac{1}{n^2 a_0} \right) \right]$$

$$= \left[\frac{3n^2}{2} - \frac{1}{4} \left(4\ell^2 + 4\ell \right) \right] a_0$$

$$= \left[3n^2 - \ell \left(\ell+1 \right) \right] \frac{a_0}{2} \tag{10.44}$$

(b) To evaluate $\langle r^{-3} \rangle$ we use $s = -1$ in Kramer's relation to obtain

$$a_0 \langle r^{-2} \rangle - \frac{1}{4} a_0^2 \left[(2\ell + 1)^2 - 1 \right] \langle r^{-3} \rangle = 0 \qquad (10.45)$$

Although we have computed $\langle r^{-1} \rangle$, this result is not helpful for other negative values of s. We require $\langle r^{-2} \rangle$ to obtain $\langle r^{-3} \rangle$. Using the hint in the statement of the problem we have from Table T.3

$$\langle r^{-2} \rangle = \frac{1}{a_0^2} \left\{ \frac{1}{n^3 \left(\ell + \frac{1}{2} \right)} \right\} \qquad (10.46)$$

Solving Eq. (10.45) for $\langle r^{-3} \rangle$ we have

$$\langle r^{-3} \rangle = \frac{4a_0}{a_0^2 \left[(2\ell + 1)^2 - 1 \right]} \langle r^{-2} \rangle$$

$$= \frac{4a_0}{a_0^2 \left[4\ell^2 + 4\ell \right]} \cdot \frac{1}{a_0^2} \left\{ \frac{1}{n^3 \left(\ell + \frac{1}{2} \right)} \right\}$$

$$= \frac{1}{(na_0)^3 \left[\ell \left(\ell + 1 \right) \right] \left(\ell + \frac{1}{2} \right)} \qquad (10.47)$$

This problem illustrates the usefulness of the Kramer relation as well as its limitations. Fortunately there are several ways to obtain the value of $\langle r^{-2} \rangle$ that is required to obtain $\langle r^{-3} \rangle$ (see, for example, [2]).

9. A particle of mass m is subjected to an unknown central potential $U(r)$ that vanishes as $r \to \infty$. A particular (unnormalized) eigenfunction in spherical coordinates is $\psi(r, \theta, \phi) = e^{-\beta r}$ where β is a real positive constant. Use atomic units (see Appendix C).

(a) Does this state have definite angular momentum? If so, why and what is the total angular momentum and the z-component of the angular momentum of this state?

(b) Is the state bound or free? What is the energy eigenvalue of this state? Hint: Because $\lim_{r \to \infty} U(r) = 0$ all the energy at $r = \infty$ is kinetic energy. In a.u. the kinetic energy operator is given by

$$\hat{T} = -\frac{1}{2} \nabla^2$$

$$= -\frac{1}{2} \left\{ \frac{1}{r^2} \frac{\partial}{\partial r} \left(r^2 \frac{\partial}{\partial r} \right) + \right.$$

$$\left. \frac{1}{r^2} \left[\frac{1}{\sin \theta} \frac{\partial}{\partial \theta} \left(\sin \theta \frac{\partial}{\partial \theta} \right) + \frac{1}{\sin^2 \theta} \frac{\partial^2}{\partial \phi^2} \right] \right\} \qquad (10.48)$$

(c) What is the potential energy?

Solution

(a) Because the given function is known to be an eigenfunction for a central potential it must have definite angular momentum and z-component of angular momentum. The angular momentum eigenfunction is clearly $Y_{00}(\theta,\phi) = \sqrt{1/4\pi}$ because it is the only spherical harmonic with no angular dependence [see Eq. (10.1)]. This means that we must have $\ell = 0$ and $m = 0$ and the total angular momentum is $\sqrt{\ell(\ell+1)} = 0$.

(b) If we operated on $\psi(r,\theta,\phi)$ with $\hat{H} = \hat{T} + U(r)$, we would get the eigenvalue. We do not, however, know $U(r)$ other than that it vanishes as $r \to \infty$. Therefore, we operate on $\psi(r,\theta,\phi)$ with $\hat{H}$ and take the limit as $r \to \infty$. Because $\lim_{r\to\infty} [U(r)\psi(r,\theta,\phi)] = 0$, the eigenvalue will be the eigenvalue of $\hat{T}\psi(r,\theta,\phi)$ as $r \to \infty$ which is the total energy. Using $\hat{T} = -\frac{1}{2}\nabla^2$ from the hint

$$\lim_{r\to\infty} \hat{T}\psi(r,\theta,\phi) = -\frac{1}{2}\lim_{r\to\infty} \nabla^2\psi(r,\theta,\phi)$$
$$= E\psi(r,\theta,\phi) \tag{10.49}$$

We have then

$$\nabla^2\psi(r,\theta,\phi) = [\frac{1}{r^2}\frac{\partial}{\partial r}\left(r^2\frac{\partial}{\partial r}\right)]e^{-\beta r}$$

$$= -\beta\frac{1}{r^2}\frac{\partial}{\partial r}[r^2 e^{-\beta r}]$$

$$= -\beta\frac{1}{r^2}\left(2r - \beta r^2\right)e^{-\beta r}$$

$$= \left(-\frac{2\beta}{r} + \beta^2\right)\psi(r,\theta,\phi) \tag{10.50}$$

Taking the limit as $r \to \infty$ and multiplying by $-\frac{1}{2}$ we see that the eigenvalue of $\hat{T}$ at ∞ is $(-1/2)\beta^2$ which is a manifestly negative number meaning that at $r = \infty$ the particle is in the classically forbidden region and thus this energy represents a bound state.

Actually, the bound nature of the state was evident at the outset because

$$\lim_{r\to\infty} \psi(r,\theta,\phi) = 0 \tag{10.51}$$

The derivation above was worthwhile, however, because it provided both the nature of the state and its eigenvalue.

(c) Because $\hat{H} = E = \hat{T}(r) + U(r)$

$$\hat{H}\psi(r,\theta,\phi) = E\psi(r,\theta,\phi) = \hat{T}(r)\psi(r,\theta,\phi) + U(r)\psi(r,\theta,\phi) \tag{10.52}$$

Therefore

$$U\left(r\right)\psi\left(r,\theta,\phi\right)=E\psi\left(r,\theta,\phi\right)-\hat{T}\left(r\right)\psi\left(r,\theta,\phi\right) \tag{10.53}$$

so we simply subtract $\hat{T}\psi\left(r,\theta,\phi\right)\left[\text{not}\lim_{r\to\infty}T\psi\left(r,\theta,\phi\right)\right]$ from the total energy found above. Remembering to include the $-\frac{1}{2}$ that makes the eigenvalue of ∇^2 the kinetic energy we have

$$\begin{aligned}U\left(r\right)&=-\frac{1}{2}\beta^2-\left[\left(-\frac{1}{2}\right)\left(-\frac{2\beta}{r}+\beta^2\right)\right]\\&=-\frac{1}{2}\beta^2-\left[\frac{\beta}{r}-\frac{\beta^2}{2}\right]\\&=-\frac{1}{2}\beta^2-\frac{\beta}{r}+\frac{\beta^2}{2}\\&=-\frac{\beta}{r}\end{aligned} \tag{10.54}$$

which is the Coulomb potential. We should have anticipated the Coulomb potential because $\psi\left(r,\theta,\phi\right)=e^{-\beta r}$ is the (unnormalized) ground state eigenfunction of the H-atom.

10. The effective potential for the H-atom in atomic units for which $e=\hbar=m_e=1$ (see Appendix C) is

$$U_{\text{eff}}\left(r\right)=-\frac{1}{r}+\frac{\ell\left(\ell+1\right)}{2r^2} \tag{10.55}$$

where ℓ is the usual angular momentum quantum number. In the same units, the quantized energy is

$$E_n=-\frac{1}{2n^2} \tag{10.56}$$

(a) Find the position of the minimum of $U_{\text{eff}}\left(r\right)$, call it r_0, and also find $U_{\text{eff}}\left(r_0\right)$. Sketch $U_{\text{eff}}\left(r\right)$ vs. r for several values of ℓ (say three), including $\ell=0$.
(b) By comparing $U_{\text{eff}}\left(r_0\right)$ with nearby values of the H-atom eigenvalues E_n determine the relationship between the magnitudes of ℓ and n. Namely, for a given n, what is the restriction on ℓ. This problem provides a graphical way of determining the relationship between principal quantum number n and the angular momentum quantum number ℓ for the H-atom.

Solution

(a) Taking the derivative and setting equal to zero leads to

$$\left.\frac{dU_{\text{eff}}(r)}{dr}\right|_{r=r_0} = \frac{1}{r_0^2} - \frac{\ell(\ell+1)}{r_0^3} = 0$$

$$\Longrightarrow r_0 = \ell(\ell+1) \tag{10.57}$$

The second derivative is

$$\left.\frac{d^2U_{\text{eff}}(r)}{dr^2}\right|_{r=\ell(\ell+1)} = \frac{-2}{r^3} + \frac{3\ell(\ell+1)}{r^4}$$

$$= \frac{-2}{[\ell(\ell+1)]^3} + \frac{3\ell(\ell+1)}{[\ell(\ell+1)]^4} > 0 \tag{10.58}$$

Therefore $r_0 = \ell(\ell+1)$ is a minimum. Moreover, as ℓ increases r_0 increases. Also

$$U_{\text{eff}}(r_0, \ell) = -\frac{1}{\ell(\ell+1)} + \frac{\ell(\ell+1)}{2[\ell(\ell+1)]^2} = -\frac{1}{2\ell(\ell+1)} \tag{10.59}$$

Thus, as ℓ increases, both the position of the minimum and the minimum value of $U_{\text{eff}}(r_0, \ell)$ increase as shown in Fig. 10.2.

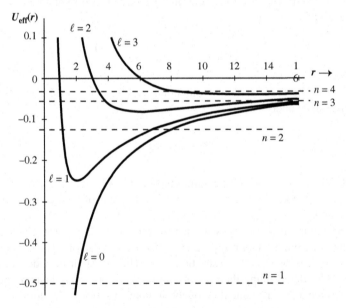

Fig. 10.2 Problem 10a—solution. The *dashed lines* represent the hydrogen atom energy eigenvalues

(b) Using Fig. 10.2 we focus on two adjacent energies and their relation to $U_{\text{eff}}(r, \ell)$. Although any two adjacent energies could be chosen, we elect to examine the relationship of $E_{n=2}$ and $E_{n=3}$ to the effective potential that falls between them, $U_{\text{eff}}(r, \ell = 2)$. It is seen that $E_{n=2}$ is not compatible with $U_{\text{eff}}(r, \ell = 2)$ because the energy $E_{n=2}$ does not intersect this effective potential. Notice that $E_{n=2}$ *is* compatible with both $\ell = 0$ and $\ell = 1$ and $E_{n=3}$ is compatible with $\ell = 0, 1, 2$.

This behavior is characteristic of all *adjacent* energy levels E_{n-1} and E_n. The energy corresponding to the smaller of the two principal quantum numbers is always lower than the minimum value of the effective potential, $U_{\text{eff}}(r_0, \ell = n)$. Therefore, for compatibility between n and ℓ we must have

$$E_n \geq U_{\text{eff}}(r_0, \ell) \tag{10.60}$$

or

$$-\frac{1}{2n^2} \geq -\frac{1}{2\ell(\ell + 1)} \tag{10.61}$$

Solving this inequality leads to

$$n^2 \geq \ell(\ell + 1) \tag{10.62}$$

Because n and ℓ are integers we must have $n > \ell$. The relationship between n and ℓ in Eq. (10.62) is usually stated as:

$$\ell \leq n - 1 \tag{10.63}$$

Equation (10.63) can also be obtained directly from the solution of the TISE for the H-atom [1].

11. In Problem 10 of this chapter it was determined that the relation between the principal quantum number n and the azimuthal (angular momentum) quantum number ℓ is

$$\ell \leq n - 1 \tag{10.64}$$

The energy levels of the H-atom in atomic units are

$$E_n = -\frac{1}{2n^2} \tag{10.65}$$

Because E_n depends only upon n it is clear from Table T.1 that the states of the H-atom are degenerate because there can be several values of ℓ and m for a given n. Find the *degree of degeneracy* of the H-atom g_H (do not consider spin). That is, determine how many states that have a given value of n have the same energy.

Solution

To determine g_H we must sum all ℓ-states for a given n over all their possible sub-states. Using the relationships given in Table T.1 we can calculate the total number of states for a given principal quantum number. For a given ℓ there are $2\ell + 1$ possible values of m. Therefore, we must sum these values for every possible value of ℓ for a given n, that is, from $\ell = 0$ to $\ell = n - 1$. We have then

$$g_H = \sum_{\ell=0}^{n-1} (2\ell + 1) \tag{10.66}$$

First Method

We use Gauss's trick, which provides a formula for summing the first j integers [1]. This formula is

$$\sum_{j=0}^{M} j = \frac{M(M+1)}{2} \tag{10.67}$$

Thus,

$$g_H = 2\sum_{\ell=0}^{n-1} \ell + \sum_{\ell=0}^{n-1} 1$$

$$= 2 \cdot \frac{(n-1)n}{2} + n$$

$$= n^2 \tag{10.68}$$

Second Method (This is Actually a Derivation of Gauss' Trick)

We first write the terms of Eq. (10.66) as follows:

$$\sum_{\ell=0}^{n-1} (2\ell + 1) = 1 + 3 + 5 + \cdots (2n - 3) + (2n - 1) \tag{10.69}$$

Now re-write Eq. (10.69) with the terms in reverse order.

$$\sum_{\ell=0}^{n-1} (2\ell + 1) = (2n - 1) + (2n - 3) + \cdots 5 + 3 + 1 \tag{10.70}$$

We now add Eqs. (10.69) and (10.70) term by term starting from the equal signs to obtain

$$2\sum_{\ell=0}^{n-1}(2\ell+1)=2n+2n+2n+\cdots2n+2n+2n \tag{10.71}$$

Inasmuch as there are n terms on the right-hand side (we started counting at $\ell=0$) we may write

$$2\sum_{\ell=0}^{n-1}(2\ell+1)=n(2n) \tag{10.72}$$

or

$$g_H=\sum_{\ell=0}^{n-1}(2\ell+1)=n^2 \tag{10.73}$$

This second method is really no different from the first method because it is the same trick of adding the re-written sums that are used to establish Gauss' trick. The origin of the trick and thus its appellation are interesting themselves [1].

12. While the H-atom energy eigenvalues depend only upon the principal quantum number n, other atoms that resemble hydrogen atoms have nearly hydrogenic energy level formulas. This is especially true of the alkali atoms, which have a single valence electron outside a closed shell rare gas configuration. At the end of the nineteenth century Rydberg found that the energies of many atoms, especially the alkali atoms, could be described by a hydrogen-like formula that was ℓ-dependent. In atomic units (see Appendix C)

$$E_{n\ell}=-\frac{1}{2(n-\delta_\ell)^2} \tag{10.74}$$

where δ_ℓ is called the quantum defect and is dependent upon the angular momentum of the valence electron. Notice that this formula reduces to the H-atom energy if $\delta_\ell=0$. In this model the atom is viewed as having a single valence electron under the influence of a spherical ball of charge consisting of the point nucleus of charge $+Ze$ surrounded by $(Z-1)$ electrons. If the angular momentum of the valence electron is high, it stays away from the ionic core and the system behaves in a very hydrogenic manner. If the angular momentum is low, the valence electron encounters the electron cloud and deviates from hydrogenic behavior.

The effects of core penetration by the valence electron may be approximated by assuming that the potential as seen by the valence electron is, in atomic units,

$$U(r)=-\frac{1}{r}-\frac{b}{r^2} \tag{10.75}$$

Of course, the first term is simply the Coulomb potential while the second term accounts for the increasing attraction of the valence electron by the nucleus as the valence electron penetrates the electron cloud that shields the nuclear charge. Multi-electron atoms for which the energy is given by Eq. (10.74) are called Rydberg atoms.

Now for the problem: Find an approximate expression for the quantum defect δ_ℓ in terms of b and ℓ.

Solution

The key point here is that the non-Coulombic term in the potential $U(r)$ has the same dependence on r as does the centrifugal term in the effective potential. We may thus write the effective potential in the form

$$U_{\text{eff}}(r) = -\frac{1}{r} - \frac{b}{r^2} + \frac{\ell(\ell+1)}{2r^2}$$
$$= -\frac{1}{r} + \frac{[\ell(\ell+1) - 2b]}{2r^2} \tag{10.76}$$

We already know the solution to the eigenvalue problem with an effective potential in the form of Eq. (10.76). We merely identify the numerator in the second term with $\ell(\ell+1)$ in the solution of the pure hydrogen problem. We therefore let

$$\ell'(\ell'+1) = \ell(\ell+1) - 2b \tag{10.77}$$

At this point it is necessary to recall the relation between n and ℓ that led to the solution of the radial Schrödinger equation for the H-atom [1]. It was necessary to terminate an assumed infinite series to avoid non-physical conditions. We had

$$n = n_r + (\ell+1) \tag{10.78}$$

where n_r is the index of the last term in the series. Incidentally, this relationship provides an analytical method of arriving at the conclusion arrived in Problem 10 of this chapter, Eq. (10.63).

The energy of the H-atom in terms of n_r is

$$E_H = -\frac{1}{2[n_r + (\ell+1)]^2} \tag{10.79}$$

By analogy with the H-atom the Rydberg atom energy is

$$E_R = -\frac{1}{2\left[n_r + (\ell' + 1)\right]^2}$$

$$= \frac{1}{2\left[n_r + (\ell + 1) + \ell' - \ell\right]^2}$$

$$= \frac{1}{2\left[n - (\ell - \ell')\right]^2} \tag{10.80}$$

Comparing Eq. (10.80) with Eq. (10.74) we see that

$$\delta_\ell = \left(\ell - \ell'\right) \tag{10.81}$$

Now we must eliminate ℓ' from Eq. (10.81) using Eq. (10.77)

$$\left(\ell'^2 + \ell'\right) - \left(\ell^2 + \ell\right) = -2b \tag{10.82}$$

which is a quadratic equation for ℓ'.

$$\ell'^2 + \ell' + [2b - \ell(\ell + 1)] = 0 \tag{10.83}$$

Then

$$\ell' = -\frac{1}{2} \pm \frac{1}{2}\sqrt{1 - 4 \cdot [2b - \ell(\ell + 1)]}$$

$$= -\frac{1}{2} \pm \frac{1}{2}\sqrt{1 - 8b + 4\ell(\ell + 1)}$$

$$= -\frac{1}{2} \pm \sqrt{\ell^2 + \ell + \frac{1}{4} - 2b}$$

$$= -\frac{1}{2} \pm \sqrt{\left(\ell + \frac{1}{2}\right)^2 - 2b} \tag{10.84}$$

and

$$\delta_\ell = \ell - \left[-\frac{1}{2} \pm \sqrt{\left(\ell + \frac{1}{2}\right)^2 - 2b}\right]$$

$$= \left(\ell + \frac{1}{2}\right) \mp \sqrt{\left(\ell + \frac{1}{2}\right)^2 - 2b} \tag{10.85}$$

Now we must consider the appearance of the $\pm$ sign in Eq. (10.85). There cannot be two values of the quantum defect. We can use a limiting case to decide which root is physically meaningful. Suppose $b = 0$. If this were the case we would have a pure H-atom potential [Eq. (10.76)] and $\delta_\ell \equiv 0$ which can only occur with the minus sign in Eq. (10.85). The plus sign gives an extraneous root so we can ignore it and move on.

Now, let us consider the properties of δ_ℓ. Assuming that $\ell > b$ (because we are interested in high angular momentum states) we expand the expression for δ_ℓ using the binomial theorem.

$$
\begin{aligned}
\delta_\ell &= \left(\ell + \frac{1}{2}\right) - \sqrt{\left(\ell + \frac{1}{2}\right)^2 - 2b} \\
&= \left(\ell + \frac{1}{2}\right)\left\{1 - \left[1 - \frac{2b}{\left(\ell + \frac{1}{2}\right)^2}\right]^{1/2}\right\} \\
&\approx \left(\ell + \frac{1}{2}\right)\left[\frac{b}{\left(\ell + \frac{1}{2}\right)^2} \cdots\right] \\
&= \frac{b}{\left(\ell + \frac{1}{2}\right)}
\end{aligned}
\tag{10.86}
$$

Two important conclusions can be drawn from Eq. (10.86).

1. This relation between δ_ℓ and ℓ is consistent with our premise that higher angular momentum states are more nearly hydrogenic than states having lower angular momentum inasmuch as the valence electron "stays away" from the electron core that causes the non-hydrogenic effects.
2. The accidental degeneracy of the H-atom is broken by the presence of the electronic core because the energy depends upon the quantum defect which depends upon the angular momentum, a parameter of which the true hydrogen energy is independent.

Chapter 11
Approximation Methods

Most practical problems in quantum mechanics do not lead to Schrödinger equations that allow exact solution. For this reason approximation methods are necessary. In this chapter we present some problems that illustrate the most frequently used of these methods.

11.1 The WKB Approximation

This approximation technique is developed by expanding the TISE for a single independent variable in powers of $\hbar$. Retaining the first two terms in the expansion (the first term is the classical limit, $\hbar = 0$; the second term is linear in $\hbar$) one eventually obtains expressions for the wave function and energy. The details are discussed at length in most textbooks [1].

For a potential function $U(x)$ with one minimum and classical turning points at $x = a$ and $x = b$, the wave function for $U(x) < E$ (the classically allowed region between a and b) is given by

$$\psi(x) = \frac{B}{\sqrt{k(x)}} \exp\left[i \int_a^x k(x)\,dx\right] + \frac{C}{\sqrt{k(x)}} \exp\left[-i \int_a^x k(x)\,dx\right] \qquad (11.1)$$

where B and C are constants; k is the classical momentum divided by $\hbar$.

$$k(x) = \frac{1}{\hbar}\sqrt{2m\left[E - U(x)\right]} \qquad (11.2)$$

© Springer International Publishing AG 2017
J.D. Kelley, J.J. Leventhal, *Problems in Classical
and Quantum Mechanics*, DOI 10.1007/978-3-319-46664-4_11

In the classically forbidden regions $x < a$ and $x > b$, where $U(x) > E$, the wave functions are

$$\psi(x) = \frac{A}{\sqrt{k(x)}} \exp\left[\int_x^a \kappa(x)\, dx\right] \quad x < a$$

$$\psi(x) = \frac{D}{\sqrt{k(x)}} \exp\left[-\int_b^x \kappa(x)\, dx\right] \quad x < a \tag{11.3}$$

with

$$\kappa(x) = \frac{1}{\hbar}\sqrt{2m[U(x) - E]} \tag{11.4}$$

Finally, if the potential permits bound states, the discrete energies are found from

$$\int_a^b p(x)\, dx = \left(n + \frac{1}{2}\right)\pi\hbar \tag{11.5}$$

where $p(x)$ is the classical momentum

$$p(x) = \hbar k(x) = \sqrt{2m[E - U(x)]} \tag{11.6}$$

It is also possible to use the WKB approximation to estimate the transmission coefficient through a barrier. The approximation to the transmission coefficient is

$$T_{\text{WKB}} = \exp\left[-\frac{2}{\hbar}\int_a^b \sqrt{2m[U(x) - E]}dx\right] \tag{11.7}$$

Problems

1. Use the WKB approximation to find the quantized energies for an L-box. Compare the result with the exact solution. What happens as $n \to \infty$?

Solution

Applying Eq. (11.5) to the L-box we have

$$\int_a^b p(x)\, dx = \int_0^L \sqrt{2mE}dx = L\sqrt{2mE}$$

$$= \left(n + \frac{1}{2}\right)\pi\hbar \tag{11.8}$$

which leads to

$$E_n = \left(n + \frac{1}{2}\right)^2 \frac{\pi^2\hbar^2}{2mL^2} \tag{11.9}$$

This value of the quantized energy is very nearly the same as the exact solution that is obtained by solving the TISE (see Appendix M). This exact energy is

$$E_n = \frac{n^2\pi^2\hbar^2}{2mL^2} \tag{11.10}$$

so it is evident that as $n \to \infty$, which is the region of validity of the WKB approximation, the approximate energy expression approaches the exact one.

2. Use the WKB approximation to find the energy levels for a harmonic oscillator. How do the results compare with the exact solution?

Solution

For the harmonic oscillator the momentum is given by

$$p(x) = \sqrt{2m(E - Kx^2)} \quad \text{with} \quad K = \frac{1}{2}m\omega^2$$

At a given energy E the classical turning points of the harmonic oscillator occur when the classical momentum vanishes. Thus, the classical turning points for $x < 0$ and $x > 0$ are

$$x_a = -\sqrt{\frac{E}{K}} \quad \text{and} \quad x_b = \sqrt{\frac{E}{K}} \tag{11.11}$$

From Eq. (11.5) we have

$$\int_{x_a}^{x_b} p(x)\,dx = \left(n + \frac{1}{2}\right)\pi\hbar$$

$$= 2\sqrt{2mK} \int_0^{x_b} \sqrt{(E/K - x^2)}\,dx$$

$$= 2\sqrt{2mK} \int_0^{x_b} \sqrt{(x_b^2 - x^2)}\,dx \tag{11.12}$$

To evaluate the integral we employ the integration formula of Eq. (G.5).

$$\int_0^{x_b} p(x)\, dx = 2\sqrt{2mK} \left[\frac{x\sqrt{x_b^2 - x^2}}{2} + \frac{x_b^2}{2} \sin^{-1} \frac{x}{x_b} \right]_0^{x_b}$$

$$= 2\sqrt{2mK} \left[\frac{x_b^2}{2} \cdot \frac{\pi}{2} \right]$$

$$= \frac{\pi}{2} \sqrt{2mK} \cdot \frac{E}{K} = \pi \cdot \sqrt{\frac{m}{2K}} \cdot E$$

$$= \pi \cdot \frac{1}{\omega} \cdot E \tag{11.13}$$

Inserting this result into Eq. (11.5) we have

$$\pi \cdot \frac{1}{\omega} \cdot E_n = \left(n + \frac{1}{2} \right) \pi \hbar \Rightarrow E_n = \left(n + \frac{1}{2} \right) \hbar \omega \tag{11.14}$$

which is the same as the exact result.

3. A particle of mass m and kinetic energy $E < U_0$ is incident from the left on a one-dimensional parabolic potential barrier given by

$$U(x) = U_0 \left(1 - \frac{x^2}{x_0^2} \right) \tag{11.15}$$

as shown in Fig. 11.1. Use the WKB approximation to find the approximate transmission rate through the barrier as a function of $E < U_0$.

Fig. 11.1 Problem 3

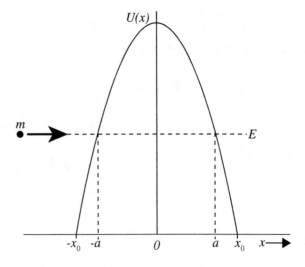

Solution

The WKB approximation to the transmission coefficient T is given in Eq. (11.7). Inserting the parameters of this problem we have

$$T_{\text{WKB}} = \exp\left[-\frac{2}{\hbar}\int_{-a}^{a}\sqrt{2m\left[U\left(x\right) - E\right]}dx\right] \tag{11.16}$$

where $x = a$ when $U\left(x\right) = E$ (see Fig. 11.1) so that

$$E = U_0\left(1 - \frac{a^2}{x_0^2}\right) \Rightarrow a^2 = x_0^2\left(1 - \frac{E}{U_0}\right) \tag{11.17}$$

Then

$$T_{\text{WKB}} = \exp\left[-\frac{2\sqrt{2m}}{\hbar}\int_{-a}^{a}\left(U_0 - U_0\frac{x^2}{x_0^2} - U_0 + U_0\frac{a^2}{x_0^2}\right)^{1/2}dx\right]$$

$$= \exp\left[-\frac{2a\sqrt{2mU_0}}{\hbar x_0}\int_{-a}^{a}\sqrt{1 - \frac{x^2}{a^2}}dx\right] \tag{11.18}$$

We let $x/a = u$ to put Eq. (11.18) in a form consistent with the integral given in Eq. (G.6), which is

$$\int_{-1}^{1}\sqrt{1 - u^2}du = \frac{\pi}{2} \tag{11.19}$$

Using Eq. (11.17) along with the new variable u converts Eq. (11.18) to

$$T_{\text{WKB}} = \exp\left[-x_0^2\left(1 - \frac{E}{U_0}\right)\frac{2\sqrt{2mU_0}}{\hbar x_0}\int_{-1}^{1}\sqrt{1 - u^2}du\right] \tag{11.20}$$

which, using Eq. (11.19), becomes

$$T_{\text{WKB}} = \exp\left[-x_0\frac{\pi\left(U_0 - E\right)}{\hbar}\sqrt{\frac{2m}{U_0}}\right] \tag{11.21}$$

The first thing we notice about Eq. (11.21) is that when $E = U_0$ the transmission coefficient is unity. Just to be sure that our answer makes sense let us check the units. Inasmuch as we are dealing with an exponential the exponent must be unitless.

$$-x_0 \frac{\pi\,(U_0 - E)}{\hbar} \sqrt{\frac{2m}{U_0}} = \frac{m \cdot \sqrt{kg \cdot J}}{(J \cdot s)}$$

$$= \frac{m \cdot \sqrt{kg}}{s \cdot \sqrt{\frac{kg \cdot m^2}{s^2}}} \tag{11.22}$$

which is unitless. This does not prove that the answer is correct, but it gives us some confidence.

11.2 The Variational Method

The principle behind the variational method is simple to understand. We start with the assumption that the Hamiltonian has a set of eigenfunctions and associated eigenvalues so that we can write

$$\hat{H}\,|\phi_n\rangle = E_n\,|\phi_n\rangle \tag{11.23}$$

where the $|\phi_n\rangle$ are assumed to be the exact orthonormal eigenkets and the E_n the corresponding energy eigenvalues. Neither are known; the only thing known in Eq. (11.23) is the Hamiltonian.

We begin by guessing a normalized trial ground-state eigenfunction, which we designate by $|\psi\rangle$. We should choose a trial wave function that matches the boundary conditions of the unknown exact function and that has any known characteristics, e.g. no nodes for a ground state, and appropriate symmetry. We can expand $|\psi\rangle$ in terms of the complete, but unknown, set of orthonormal eigenkets $|\phi_n\rangle$.

$$|\psi\rangle = \sum_{n=0}^{\infty} c_n\,|\phi_n\rangle \tag{11.24}$$

The expectation value of the energy in the state represented by $|\psi\rangle$ is

$$\langle E\rangle = \langle\psi|\hat{H}|\psi\rangle$$

$$= \sum_{n=0}^{\infty} |c_n|^2\,\langle\phi_n|\hat{H}|\phi_n\rangle$$

$$= \sum_{n=0}^{\infty} |c_n|^2\,E_n \tag{11.25}$$

The value of the ground state energy, E_0, is lower than any of the other eigenvalues. Therefore, if in Eq. (11.25) we replace all of the E_n by E_0 we obtain the inequality

$$\langle E \rangle \geq E_0 \sum_{n=0}^{\infty} |c_n|^2$$

$$\geq E_0 \tag{11.26}$$

where, because $|\psi\rangle$ is normalized, the summation in Eq. (11.26) is unity. In coordinate space this inequality is

$$\langle E \rangle = \int \psi^*(r) \hat{H} \psi(r) \, dV$$

$$\geq E_0 \tag{11.27}$$

so, given *any* trial wave function, we are assured that the integral in Eq. (11.27) will be greater than the true ground state energy.

In typical applications, parameters are incorporated into the trial wave function. After computing the integral in Eq. (11.27), $|\psi\rangle$ is then minimized with respect to these parameters to obtain the lowest energy possible for the chosen functional form. No matter how many parameters are included in the trial wave function $\langle E \rangle$ will be greater than the ground state energy, although with modern computers and sophisticated functional forms one can come very close to the actual value.

It is sometimes possible to obtain useful variational approximations to the first excited state provided symmetry precludes the ground state from consideration. For example, if a one-dimensional potential well is an even function and has more than one bound state, the ground state eigenfunction will be an even function and the first excited state eigenfunction will be an odd function (see Sect. 6.2). If the trial function is chosen to be odd, the variational method will converge toward the first excited state energy. In a sense the first excited state may be thought of as the ground state for odd symmetry.

Problems

1. Estimate the ground state energy of a harmonic oscillator using the variational method with the trial wave function $\psi(x) = A e^{-\beta^2 x^2}$. Because this trial wave function is of the correct form of the actual ground state eigenfunction we expect to obtain the exact ground state energy after minimizing $\langle E \rangle$ with respect to the parameter β. Naturally we expect that the value of β that minimizes $\langle E \rangle$ will be $\beta = \alpha/\sqrt{2}$ where $\alpha = \sqrt{m\omega/\hbar}$. Show all of this.

Solution

First normalize using Eq. (G.3).

$$\langle \psi \,|\, \psi \rangle = 1 = A^2 \int_{-\infty}^{\infty} e^{-2\beta^2 x^2} dx$$

$$= A^2 \sqrt{\frac{\pi}{2\beta^2}} \tag{11.28}$$

so

$$A^2 = \sqrt{\frac{2\beta^2}{\pi}} \tag{11.29}$$

Then

$$\langle E \rangle = \langle \psi \,|\, \hat{H} \,|\, \psi \rangle$$

$$= \sqrt{\frac{2\beta^2}{\pi}} \int_{-\infty}^{\infty} e^{-\beta^2 x^2} \left[-\frac{\hbar^2}{2m} \frac{d^2}{dx^2} + \frac{1}{2} m\omega^2 \right] e^{-\beta^2 x^2} dx$$

$$= \sqrt{\frac{2\beta^2}{\pi}} \left(\frac{1}{2} m\omega^2 - \frac{2\hbar^2 \beta^4}{m} \right) \int_{-\infty}^{\infty} x^2 e^{-2\beta^2 x^2} dx + \frac{\beta^2 \hbar^2}{m} \int_{-\infty}^{\infty} e^{-2\beta^2 x^2} dx$$

$$= \sqrt{\frac{2\beta^2}{\pi}} \left(\frac{1}{2} m\omega^2 - \frac{2\hbar^2 \beta^4}{m} \right) \frac{1}{4\beta^2} \sqrt{\frac{\pi}{2\beta^2}} + \frac{\beta^2 \hbar^2}{m} \sqrt{\frac{\pi}{2\beta^2}}$$

$$= \frac{m\omega^2}{8\beta^2} + \frac{\beta^2 \hbar^2}{2m} \tag{11.30}$$

Because β occurs only as β^2 we may as well minimize with respect to β^2 to find β_0, the value of β that minimizes $\langle E \rangle$.

$$\frac{d \langle E \rangle}{d\beta^2} = -\frac{m\omega^2}{8\beta_0^4} + \frac{\hbar^2}{2m} = 0 \tag{11.31}$$

so

$$\beta_0^2 = \frac{m\omega}{2\hbar} \implies \beta_0 = \frac{1}{\sqrt{2}} \alpha \tag{11.32}$$

Therefore, we know that the energy "estimate" will yield the exact value, $\frac{1}{2}\hbar\omega$. Moreover, the trial wave function does indeed reduce to the exact ground state eigenfunction

$$\psi(x) = \left(\frac{2\beta_0^2}{\pi}\right)^{1/4} e^{-\beta^2 x^2} \rightarrow \psi_0(x) = \left(\frac{\alpha^2}{\pi}\right)^{1/4} e^{-\alpha^2 x^2/2} \qquad (11.33)$$

2. Estimate the energy of the first excited state of a harmonic oscillator using the variational method and the trial wave function $\psi(x) = Af(x)e^{-\beta^2 x^2}$ where $A =$ constant and β is an adjustable parameter. It is up to you to choose $f(x)$. You may choose any function you wish, but it is strongly advised that you choose the simplest possible function that will meet the necessary conditions, in this case an odd function. Recall that $\alpha = \sqrt{m\omega/\hbar}$.

Solution

The first excited state energy can be calculated using the variational method because the eigenfunctions of the harmonic oscillator potential have definite parity. Therefore, the ground state is even and the first excited state is odd. The first excited state is thus the "ground state" of the odd eigenfunctions. We choose $f(x) = x$ because the first excited state eigenfunction must have only one node and $f(x) = x$ is the simplest function that makes the given trial wave function odd.

We require the integral given in Eq. (G.4) which may be re-written as

$$\int_{-\infty}^{\infty} x^m e^{-ax^2} dx = \frac{\Gamma[(m+1)/2]}{a^{(m+1)/2}} \qquad (11.34)$$

First we normalize by finding A.

$$|A|^2 \int_{-\infty}^{\infty} x^2 e^{-2\beta^2 x^2} dx = 1$$

$$|A|^2 \frac{\Gamma(3/2)}{\left(2\beta^2\right)^{3/2}} = 1$$

$$|A|^2 \frac{\sqrt{\pi}}{2} \cdot \frac{1}{2^{3/2}\beta^3} = 1 \Longrightarrow |A|^2 = \frac{2^{5/2}\beta^3}{\sqrt{\pi}} \qquad (11.35)$$

We must first evaluate $\langle E \rangle$.

$$\langle E \rangle = \langle \psi | \hat{H} | \psi \rangle$$
$$= \langle T \rangle + \langle U \rangle \qquad (11.36)$$

where $|\psi\rangle$ represents the trial wave function.

We evaluate $\langle T \rangle$ and $\langle U \rangle$ separately.

$$\langle T \rangle = |A|^2 \int_{-\infty}^{\infty} xe^{-\beta^2 x^2} \left[-\frac{\hbar^2}{2m} \frac{d^2}{dx^2} \right] xe^{-\beta^2 x^2} dx \tag{11.37}$$

Part of the integrand is

$$\left[\frac{d^2}{dx^2} \right] xe^{-\beta^2 x^2} = \frac{d}{dx} \left(1 - 2\beta^2 x^2 \right) e^{-\beta^2 x^2}$$

$$= -2\beta^2 xe^{-\beta^2 x^2} - 2\beta^2 \frac{d}{dx} \left(x^2 e^{-\beta^2 x^2} \right)$$

$$= -2\beta^2 xe^{-\beta^2 x^2} - 2\beta^2 \left(2x - 2\beta^2 x^3 \right) e^{-\beta^2 x^2}$$

$$= \left(-6\beta^2 x + 4\beta^4 x^3 \right) e^{-\beta^2 x^2} \tag{11.38}$$

so

$$\langle T \rangle = |A|^2 \int_{-\infty}^{\infty} xe^{-\beta^2 x^2} \left[-\frac{\hbar^2}{2m} \left(-6\beta^2 x + 4\beta^4 x^3 \right) \right] e^{-\beta^2 x^2} dx$$

$$= |A|^2 \int_{-\infty}^{\infty} \left[3\frac{\hbar^2}{m} \beta^2 x - 2\frac{\hbar^2}{m} \beta^4 x^3 \right] xe^{-2\beta^2 x^2} dx$$

$$= |A|^2 \frac{\hbar^2 \beta^2}{m} \left[3 \int_{-\infty}^{\infty} x^2 e^{-2\beta^2 x^2} dx - 2\beta^2 \int_{-\infty}^{\infty} x^4 e^{-2\beta^2 x^2} dx \right]$$

$$= |A|^2 \left(\frac{\hbar^2 \beta^2}{m} \right) \left[3\frac{1}{|A|^2} - 2\beta^2 \frac{\Gamma(5/2)}{\left(2\beta^2 \right)^{5/2}} \right]$$

$$= |A|^2 \left(\frac{\hbar^2 \beta^2}{m} \right) \left[3\frac{1}{|A|^2} - 2\beta^2 \frac{3\sqrt{\pi}}{4} \frac{1}{2^{5/2}\beta^5} \right]$$

$$= 3 \left(\frac{\hbar^2 \beta^2}{m} \right) - \frac{3}{2} \left(\frac{\hbar^2 \beta^2}{m} \right) = \frac{3}{2} \left(\frac{\hbar^2 \beta^2}{m} \right) \tag{11.39}$$

Now, $\langle U \rangle$ is given by

$$\langle U \rangle = \frac{1}{2} m\omega^2 |A|^2 \int_{-\infty}^{\infty} x^4 e^{-2\beta^2 x^2} dx$$

$$= \frac{1}{2} \frac{m\omega^2}{\left(2\beta^2 \right)^{5/2}} \Gamma(5/2) |A|^2 = \frac{1}{2} \frac{m\omega^2}{2^{5/2}\beta^5} \cdot \frac{3\sqrt{\pi}}{4} \left(\frac{2^{5/2}\beta^3}{\sqrt{\pi}} \right)$$

$$= \frac{3}{8} \frac{m\omega^2}{\beta^2} \tag{11.40}$$

Then

$$\langle E \rangle = \frac{3}{2} \frac{\hbar^2 \beta^2}{m} + \frac{3}{8} \frac{m\omega^2}{\beta^2} \tag{11.41}$$

To minimize $\langle E \rangle$ with respect to the parameter β we differentiate Eq. (11.41) and set this derivative to zero. Because β occurs only as β^2 in Eq. (11.41) we may take the derivative with respect to β^2.

$$\frac{d\langle E \rangle}{d\beta^2} = \frac{3}{2}\frac{\hbar^2}{m} - \frac{3}{8}\frac{m\omega^2}{\beta_0^4} = 0 \tag{11.42}$$

Solving for β_0, the value of β that minimizes $\langle E \rangle$, we have

$$\beta_0^2 = \frac{1}{4}\frac{m\omega}{\hbar} = \frac{1}{2}\alpha^2 \tag{11.43}$$

Thus, the energy is given by

$$
\begin{aligned}
\langle E \rangle_{\beta=\beta_0} &= \frac{3}{2}\frac{\hbar^2}{m}\frac{1}{2}\alpha^2 + \frac{3}{8}m\omega^2\frac{2}{\alpha^2} \\
&= \frac{3}{2}\frac{\hbar^2}{m}\frac{1}{2}\frac{m\omega}{\hbar} + \frac{3}{8}m\omega^2\frac{2\hbar}{m\omega} \\
&= \frac{3}{4}\hbar\omega + \frac{3}{4}\hbar\omega = \frac{3}{2}\hbar\omega
\end{aligned} \tag{11.44}
$$

which, as expected, is the exact value. With $\beta = \beta_0$ the trial wave function is of course the exact eigenfunction of the first excited state, i.e.

$$
\begin{aligned}
\psi_1(x) &= \sqrt{\frac{2^{5/2}\beta^3}{\sqrt{\pi}}}xe^{-\beta^2 x^2} \\
&= \sqrt{\frac{\alpha}{2\sqrt{\pi}}}2[(\alpha x)]e^{-\alpha^2 x^2/2}
\end{aligned} \tag{11.45}
$$

3. A particle of mass m is trapped in a potential given by

$$
\begin{aligned}
U(x) &= Bx \quad x > 0 \\
&= \infty \quad\quad x \le 0
\end{aligned} \tag{11.46}
$$

(a) Using the variational method estimate the ground state energy using the trial wave function

$$\psi(x) = Axe^{-\beta x} \tag{11.47}$$

where A is a constant and β is an adjustable parameter.

(b) The exact energy of the ground state is (see, for example, [1])

$$E_1 = 2.338 \left(\frac{B^2 \hbar^2}{2m} \right)^{1/3} \tag{11.48}$$

Show that the energy calculated using the variational method is greater than the exact energy.

Solution

Note that the trial wave function was chosen so it meets the boundary condition $\psi(0) = \lim_{x \to \infty} \psi(x) = 0$.

(a) First find the value of A that normalizes $\psi(x)$. Using the integral given in Eq. (G.2) we have

$$|A|^2 \int_0^\infty |\psi(x)|^2 \, dx = 1$$

$$|A|^2 \int_0^\infty x^2 e^{-2\beta x} dx = 1$$

$$|A|^2 \frac{\Gamma(3)}{(2\beta)^3} = 1$$

$$|A|^2 = 4\beta^3 \tag{11.49}$$

Next we compute $\langle E \rangle = \langle \psi | \hat{H} | \psi \rangle$.

$$\langle E \rangle = 4\beta^3 \left(-\frac{\hbar^2}{2m} \right) \int_0^\infty x e^{-\beta x} \frac{d^2}{dx^2} \left(x e^{-\beta x} \right) dx$$

$$+ 4\beta^3 B \int_0^\infty x^3 e^{-2\beta x} dx$$

$$= \left(-\frac{2\hbar^2 \beta^3}{m} \right) \int_0^\infty x e^{-\beta x} \frac{d^2}{dx^2} \left(x e^{-\beta x} \right) dx$$

$$+ 4\beta^3 B \cdot \frac{\Gamma(4)}{(2\beta)^5}$$

$$= \left(-\frac{2\hbar^2 \beta^3}{m} \right) \int_0^\infty x e^{-\beta x} \frac{d}{dx} [x(-\beta) + 1] e^{-\beta x} dx$$

$$+ 4\beta^3 B \cdot \frac{3!}{(2\beta)^4}$$

$$= \left(\frac{2\hbar^2 \beta^3}{m} \right) \int_0^\infty x e^{-\beta x} \left[(-\beta)^2 x - \beta - \beta \right] e^{-\beta x} dx + \frac{3B}{2\beta}$$

$$= \left(-\frac{2\hbar^2 \beta^3}{m} \right) \beta \int_0^\infty \left[\beta x^2 - 2x \right] e^{-2\beta x} dx + \frac{3B}{2\beta}$$

$$= \left(-\frac{2\hbar^2 \beta^3}{m} \right) \left[\beta^2 \frac{\Gamma(3)}{(2\beta)^3} - 2\beta \frac{\Gamma(2)}{(2\beta)^2} \right] + \frac{3B}{2\beta}$$

$$= \left(-\frac{2\hbar^2 \beta^3}{m} \right) \left(\frac{1}{\beta} \right) \left[\frac{1}{4} - \frac{1}{2} \right] + \frac{3B}{2\beta}$$

$$= \frac{\hbar^2 \beta^2}{2m} + \frac{3B}{2\beta} \tag{11.50}$$

Now minimize $\langle E \rangle$ with respect to the parameter β and set to zero.

$$\frac{d \langle E \rangle}{d\beta} = \frac{\hbar^2 \beta}{m} - \frac{3B}{2\beta^2} = 0 \tag{11.51}$$

so

$$\beta^3 = \frac{3Bm}{2\hbar^2} \tag{11.52}$$

Therefore, the estimate of the energy using this trial wave function and the variational method is

$$\langle E \rangle = \frac{\hbar^2}{2m} \left(\frac{3Bm}{2\hbar^2} \right)^{2/3} + \frac{3B}{2} \left(\frac{2\hbar^2}{3Bm} \right)^{1/3}$$

$$= \frac{3^{5/3}}{2^{4/3}} \left(\frac{B^2 \hbar^2}{2m} \right)^{1/3} \tag{11.53}$$

(b) Comparing the variational answer $\langle E \rangle$ with the exact answer E_1 as given in Eq. (11.48) above, we have

$$\frac{\langle E \rangle}{E_1} = \frac{3^{5/3}}{2^{4/3}} \left(\frac{B^2 \hbar^2}{2m} \right)^{1/3} \cdot \frac{1}{2.338} \left(\frac{2m}{B^2 \hbar^2} \right)^{1/3}$$

$$= \frac{3^{5/3}}{2^{4/3}} \cdot \frac{1}{2.338} \approx 1.06 \tag{11.54}$$

so, indeed, the energy obtained using the variational method lies above the exact value.

4. Estimate the ground state energy and the first excited state energy of a particle of mass m trapped in a quartic potential given by

$$U(x) = Cx^4 \tag{11.55}$$

Use the variational method with appropriate harmonic oscillator eigenfunctions as the trial wave functions and use $\alpha = \sqrt{m\omega/\hbar}$ as the variable parameter.

Solution

The Hamiltonian for this problem is

$$\hat{H} = \frac{\hat{p}^2}{2m} + Cx^4 \tag{11.56}$$

We must first compute $\langle E \rangle = \langle n | \hat{H} | n \rangle$ where the bras and kets are harmonic oscillator eigenfunctions. We then minimize $\langle E \rangle$ using α as the variational parameter.

Before diving into the calculations let us map out the strategy. First, we use the matrix elements calculated in Chap. 7 to calculate $\langle E \rangle$. In fact, we have already calculated these expectation values in Problems 7 and 8 of Chap. 7, which we reproduce here.

$$\langle n | \hat{p}^2 | n \rangle = \left(n + \frac{1}{2} \right) \alpha^2 \hbar^2 \quad \text{and} \quad \langle n | x^4 | n \rangle = \frac{3}{4\alpha^4} \left(2n^2 + 2n + 1 \right) \tag{11.57}$$

Because the quartic potential is an even potential we can use the variational method to calculate upper bounds to both the ground state energy and the first excited state energy, $n = 0$ and $n = 1$. From parity considerations and the required number of nodes we see that $|0\rangle$ and $|1\rangle$ are the appropriate trial wave functions for the ground and first excited state, respectively.

In terms of α the $n = 0$ and $n = 1$ values of $\langle E \rangle$ are

$$\langle E_n^q (\alpha) \rangle = \langle n | \hat{H} | n \rangle = \frac{1}{2m} \langle n | \hat{p}^2 | n \rangle + C \langle n | x^4 | n \rangle$$

$$= \frac{\alpha^2 \hbar^2}{2m} \left(n + \frac{1}{2} \right) + \frac{3C}{4\alpha^4} \left(2n^2 + 2n + 1 \right) \tag{11.58}$$

Checking units as we proceed we note that in Eq. (11.58) both terms on the right-hand side have units of energy because α has units m^{-1} and C has units J/m^4.

Minimizing Eq. (11.58) with respect to α we have

$$\frac{dE_n^q(\alpha)}{d\alpha} = \left(n + \frac{1}{2}\right)\frac{\alpha\hbar^2}{m} - C\frac{3}{\alpha^5}\left(2n^2 + 2n + 1\right) \tag{11.59}$$

Setting the derivative equal to zero we obtain

$$\alpha_e^6 = \frac{\left(2n^2 + 2n + 1\right)}{\left(n + \frac{1}{2}\right)} \cdot \frac{3mC}{\hbar^2} \tag{11.60}$$

where α_e signifies the value of α that minimizes $\langle E_n^q(\alpha)\rangle$. Thus

$$\alpha_e^6\left(n = 0\right) = \frac{6mC}{\hbar^2} \quad \text{and} \quad \alpha_e^6\left(n = 1\right) = \frac{10mC}{\hbar^2} \tag{11.61}$$

Note that the values of α_e are different for the two states under consideration. We can calculate $E_0^q(\alpha_e)$ and $E_1^q(\alpha_e)$ by inserting α_e from Eq. (11.61) into Eq. (11.58).

For $n = 0$ we have

$$
\begin{aligned}
E_0^q &= \left[\frac{1}{4}\frac{\hbar^2}{m}\left(\frac{6mC}{\hbar^2}\right)^{1/3}\right] + \left[\frac{3C}{4}\left(\frac{\hbar^2}{6mC}\right)^{2/3}\right] \\
&= \left[\frac{1}{2^2}\frac{\hbar^{4/3}}{m^{2/3}}2^{1/3}3^{1/3}C^{1/3}\right] + \left[\frac{3C^{1/3}\hbar^{4/3}}{2^2 2^{2/3}3^{2/3}m^{2/3}}\right] \\
&= \frac{C^{1/3}\hbar^{4/3}}{m^{2/3}}\left(\frac{3^{1/3}}{2^{5/3}} + \frac{3^{1/3}}{2^{8/3}}\right) = \frac{C^{1/3}\hbar^{4/3}}{m^{2/3}}\left(\frac{3^{1/3}}{2^{5/3}} + \frac{3^{1/3}}{2^{5/3}2}\right) \\
&= \frac{C^{1/3}\hbar^{4/3}}{m^{2/3}}\frac{3^{1/3}}{2^{5/3}}\left(1 + \frac{1}{2}\right) = \left(\frac{C\hbar^4}{m^2}\right)^{1/3} \\
&= \frac{3^{4/3}}{2 \cdot 2^{5/3}}\left(\frac{C\hbar^4}{m^2}\right)^{1/3} \simeq 0.681\left(\frac{C\hbar^4}{m^2}\right)^{1/3} \tag{11.62}
\end{aligned}
$$

For $n = 1$ we have

$$
\begin{aligned}
E_1^q(\alpha) &= \frac{3}{4}\frac{\hbar^2}{m}\left(\frac{10mC}{\hbar^2}\right)^{1/3} + \frac{15C}{4}\left(\frac{\hbar^2}{10mC}\right)^{2/3} \\
&= \frac{3}{4}\frac{\hbar^2}{m}(10)^{1/3}\left(\frac{mC}{\hbar^2}\right)^{1/3} + \frac{15C}{4}\left(\frac{1}{10}\right)^{2/3}\left(\frac{\hbar^2}{mC}\right)^{2/3} \\
&= \frac{3}{4}\frac{\hbar^2}{m}[2^{1/3} \cdot 5^{1/3}\left(\frac{mC}{\hbar^2}\right)^{1/3}
\end{aligned}
$$

$$+5 \cdot C^{1/3} \left(\frac{1}{5}\right)^{2/3} \left(\frac{1}{2}\right)^{2/3} m^{1/3} \left(\frac{1}{\hbar^2}\right)^{1/3}]$$

$$= \frac{3}{4} \frac{\hbar^2}{m} \left(\frac{mC}{\hbar^2}\right)^{1/3} [2^{1/3} \cdot 5^{1/3} + 5^{1/3} \cdot 2^{-2/3}]$$

$$= 3 \cdot 5^{1/3} \frac{\hbar^2}{m} \left(\frac{mC}{\hbar^2}\right)^{1/3} \left[\frac{2^{1/3} + 2^{-2/3}}{2^2}\right]$$

$$= 3 \cdot 5^{1/3} \left[\frac{2^{1/3} + 2^{-2/3}}{2^2}\right] \left(\frac{C\hbar^4}{m^2}\right)^{1/3}$$

$$\simeq 1.282 \left(\frac{C\hbar^4}{m^2}\right)^{1/3} \tag{11.63}$$

Both $E_0^q(\alpha)$ and $E_1^q(\alpha)$ are proportional to $\left(C\hbar^4/m^2\right)^{1/3}$ which has units of energy.

$$\left(\frac{C\hbar^4}{m^2}\right)^{1/3} = \left[\left(\frac{J}{m^4}\right) \frac{(J \cdot s)^4}{kg^2}\right]^{1/3}$$

$$= \left[J^5 \cdot \left(\frac{kg \cdot m^2}{s^2}\right)^{-2}\right]^{1/3}$$

$$= J \tag{11.64}$$

This makes us happy. It is also comforting that $E_0^q(\alpha) < E_1^q(\alpha)$.

Note that the use of the matrix elements obtained using the harmonic oscillator ladder operators obviated the need to evaluate tedious integrals. This observation emphasizes the general rule:

- **For problems involving the harmonic oscillator first see if a solution using the ladder operators can be effected.**

11.3 Non-degenerate Perturbation Theory

Time independent perturbation theory is an often used approximation method (especially on examinations). It is most useful when the Hamiltonian for the system may be written as the sum of an "unperturbed" Hamiltonian for which the solution is known and a term that is much smaller, the perturbation. We begin with notation:

$$\hat{H} = \text{true Hamiltonian}$$

$$\hat{H}_0 = \text{unperturbed Hamiltonian}$$

$$\hat{H}_1 = \text{perturbing Hamiltonian.}$$

A subscript on an energy or a wave function (state vector) signifies the state. The superscript denotes the degree of approximation. No superscript refers to the exact energy or wave function. Thus, for example

$$E_n = \text{exact energy of the } n\text{th state}$$

$$E_n^{(0)} = \text{zeroth order approximation to } E_n$$

$$E_n^{(1)} = \text{first-order correction to } E_n^{(0)}$$

$$\psi_n(x) = \text{exact eigenfunction for the } n\text{th state (}x\text{-space)}$$

$$\psi_n^{(0)}(x) = \text{first-order approximation to } \psi_n(x)$$

$$\psi_n^{(1)}(x) = \text{first-order correction } \psi_n^{(0)}(x)$$

Synopsis of results:

First order:

$$E_n^{(1)} = \left\langle \psi_n^{(0)} \middle| \hat{H}_1 \middle| \psi_n^{(0)} \right\rangle \tag{11.65}$$

The first-order correction to the energy is the expectation value of the perturbing Hamiltonian on the unperturbed state.

$$\left| \psi_n^{(1)} \right\rangle = \sum_{k \neq n} \frac{\left\langle \psi_k^{(0)} \middle| \hat{H}_1 \middle| \psi_n^{(0)} \right\rangle}{\left(E_n^{(0)} - E_k^{(0)} \right)} \left| \psi_k^{(0)} \right\rangle \tag{11.66}$$

Second order:

$$E_n^{(2)} = \sum_{k \neq n} \frac{\left| \left\langle \psi_k^{(0)} \middle| \hat{H}_1 \middle| \psi_n^{(0)} \right\rangle \right|^2}{\left(E_n^{(0)} - E_k^{(0)} \right)} \tag{11.67}$$

Problems

1. Find the first-order correction to the energy of a hydrogen atom due to the gravitational attraction between the proton and the electron. Assume the electron is in a circular orbit the radius of which is the Bohr radius a_0. Find the ratio of this correction to the unperturbed energy $E_{n=1}^{(0)} = -13.6\,\text{eV}$.

Solution

The perturbing Hamiltonian is the gravitational potential between the electron and the proton. Because we assume a circular orbit it is a constant and is

$$H_1 (r = a_0) = \frac{-Gm_e m_p}{a_0} \tag{11.68}$$

where G is the gravitational constant; m_e and m_p are the masses of the electron and proton, respectively. The first-order correction to the energy $E_{n=1}^{(1)}$ is $\langle H_1 \rangle$. Because H_1 is a constant the computation is trivial.

$$E_{n=1}^{(1)} = \langle H_1 \rangle = \frac{-Gm_e m_p}{a_0}$$

$$\approx -1.24 \times 10^{-38}\, \text{eV} \tag{11.69}$$

The ratio of this quantity to $E_{n=1}^{(0)}$ is

$$\frac{E_{n=1}^{(1)}}{E_{n=1}^{(0)}} \approx \frac{1.24 \times 10^{-38}\, \text{eV}}{13.6\, \text{eV}} \approx 10^{-39} \tag{11.70}$$

This very small ratio shows that gravitational attraction between the electron and the proton is insignificant and has virtually no effect on electronic structure. This result is compatible with that obtained in Problem 2 of Chap. 5 where it was deduced that gravity had little effect on the binding of electrons to a nucleus.

2. A particle of mass m is trapped in an L-box potential (see Appendix M) given by

$$U(x) = 0 \quad 0 \le x \le L$$

$$= \infty \quad -\infty < x < 0; L < x < \infty \tag{11.71}$$

The L-box potential is perturbed by a δ-function

$$\hat{H}_1 (x) = \alpha \delta (x - L/2) \tag{11.72}$$

where α is a positive real constant. Find $E_n^{(1)}$, the first-order correction to the energy for all eigenstates of the L-box (Fig. 11.2). For what values of n does $E_n^{(1)}$ vanish and why?

Fig. 11.2 Problem 2

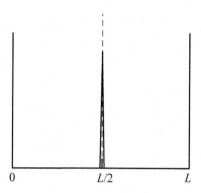

Solution

The eigenfunctions for an L-box are

$$\psi_n(x) = \sqrt{\frac{2}{L}} \sin\left(\frac{n\pi x}{L}\right) \quad 0 \leq x \leq L$$
$$= 0 \quad -\infty < x < 0; L < x < \infty \quad (11.73)$$

Therefore

$$E_n^{(1)} = \langle \psi_n | \hat{H}_1(x) | \psi_n \rangle$$
$$= \alpha \left(\frac{2}{L}\right) \int_0^L \delta(x - L/2) \sin^2\left(\frac{n\pi x}{L}\right) dx$$
$$= \left(\frac{2\alpha}{L}\right) \sin^2\left(\frac{n\pi}{2}\right)$$
$$\begin{cases} = 0 & n \text{ even} \\ = \left(\frac{2\alpha}{L}\right) & n \text{ odd} \end{cases} \quad (11.74)$$

The reason that there is no first-order correction to the energy for even values of n is that these eigenfunctions have nodes at $x = L/2$. Therefore, the even states never "feel" the perturbation because it exists only at $x = L/2$. $E_n^{(1)}$ must vanish for even n.

3. Find the first-order correction to the energy of an L-box

$$U(x) = 0 \quad 0 \leq x \leq L$$
$$= \infty \quad -\infty < x < 0; L < x < \infty \quad (11.75)$$

for a linear perturbation of the form

$$\hat{H}_1(x) = \left(\frac{U_0}{L}\right)x \tag{11.76}$$

and show the quantum state dependence of this correction. Note that this problem is equivalent to a charged particle in the L-box with an applied electric field (Fig. 11.3).

Fig. 11.3 Problem 3

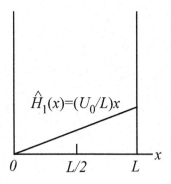

Solution

The first order perturbation to the energy $E_n^{(1)}$ for the nth state is

$$E_n^{(1)} = \left\langle \psi_n^{(0)} \middle| \hat{H}_1 \middle| \psi_n^{(0)} \right\rangle \tag{11.77}$$

For an L-box the normalized eigenfunctions are

$$\psi_n(x) = \sqrt{\frac{2}{L}} \sin\left(\frac{n\pi x}{L}\right) \qquad 0 \le x \le L$$

$$= 0 \qquad -\infty < x < 0; L < x < \infty \tag{11.78}$$

so

$$E_n^{(1)} = \frac{2U_0}{L} \int_0^L \frac{x}{L} \sin^2\left(\frac{n\pi x}{L}\right) dx \tag{11.79}$$

$$\text{Let } y = \frac{n\pi x}{L} \implies x = 0 \to y = 0 \text{ and } x = L \to y = n\pi \tag{11.80}$$

Then

$$E_n^{(1)} = \frac{2U_0}{L} \int_0^{n\pi} \left(\frac{y}{n\pi}\right) \sin^2 y \left(\frac{L}{n\pi} dy\right) = \frac{2U_0}{n^2\pi^2} \int_0^{n\pi} y \sin^2 y \, dy \qquad (11.81)$$

To perform this integral we use that given in Eq. (G.9) so

$$E_n^{(1)} = \frac{2U_0}{n^2\pi^2} \left\{ \frac{y^2}{4} - \frac{y \sin(2y)}{4} - \frac{\cos(2y)}{8} \right\}_0^{n\pi} \qquad (11.82)$$

The second term vanishes at both limits and the third term is the same at both limits so we have

$$E_n^{(1)} = \frac{2U_0}{n^2\pi^2} \frac{(n\pi)^2}{4} = \frac{1}{2}U_0 \qquad (11.83)$$

and the answer is independent of the quantum state. We could have saved ourselves the work of computing this integral with a bit of forethought. Let us examine the integral in Eq. (11.77) and re-write it as follows.

$$\begin{aligned} E_n^{(1)} &= \left\langle \psi_n^{(0)} \middle| \hat{H}_1 \middle| \psi_n^{(0)} \right\rangle \\ &= \left(\frac{U_0}{L}\right) \left\langle \psi_n^{(0)} \middle| x \middle| \psi_n^{(0)} \right\rangle \\ &= \left(\frac{U_0}{L}\right) \langle x \rangle \\ &= \left(\frac{U_0}{L}\right) \left(\frac{L}{2}\right) \\ &= \frac{U_0}{2} \end{aligned} \qquad (11.84)$$

This calculation was facilitated because it is obvious that $\langle x \rangle$, the average value of x for an L-box, is $L/2$. Both approaches were presented to emphasize that forethought about the problem can often save a lot of work.

4. A particle of mass m is confined to an L-box. The width of the box L is such that the rest mass energy $mc^2 \gg E_n^{(0)}$ for low values of n.

   (a) Find $E_n^{(1)}$ the first-order correction to the energy due to the relativistic kinetic energy of the particle. Write $E_n^{(1)}$ in terms of $E_n^{(0)}$, the nonrelativistic energy of the nth level of the L-box. (The total relativistic energy of a particle of mass m having momentum $\hat{p}$ is given by $E^2 = \hat{p}^2 c^2 + m^2 c^4$.)
   (b) Show that for a particle-in-a-box the nonrelativistic energy eigenfunctions are also eigenfunctions of the relativistic Hamiltonian.
   (c) Find the *exact* relativistic energy and show that for $mc^2 \gg E_n^{(0)}$ the solution reduces to the perturbation theory result.

Solution

(a) Inside the box the Hamiltonian is the kinetic energy. The relativistic kinetic energy is the total energy minus the rest energy mc^2. Therefore the relativistic Hamiltonian for a particle-in-a-box is

$$\hat{H} = \sqrt{\hat{p}^2c^2 + m^2c^4} - mc^2$$

$$= mc^2 \left(1 + \frac{\hat{p}^2}{m^2c^2}\right)^{1/2} - mc^2$$

$$\approx \frac{\hat{p}^2}{2m} - \frac{1}{8}\frac{\hat{p}^4}{m^3c^2} + \cdots \qquad (11.85)$$

The first term in Eq. (11.85) is the nonrelativistic kinetic energy, which we may take to be the unperturbed Hamiltonian $\hat{H}_0$. We therefore take the second term to be the perturbation Hamiltonian $\hat{H}_1$. Before jumping in and performing an integration that might be avoided we stop and examine $\hat{H}_1$ and note that it can be written in terms of $\hat{H}_0$ as follows.

$$\hat{H}_1 = -\frac{1}{8}\frac{\hat{p}^4}{m^3c^2}$$

$$= -\frac{1}{8m^3c^2}\left(2m\hat{H}_0\right)^2 \qquad (11.86)$$

We have thus managed to write $\hat{H}_1$ in terms of the Hamiltonian $\hat{H}_0$ which is a great simplification because the unperturbed eigenfunctions are now eigenfunctions of $\hat{H}_1$. This will save time in the calculation and simplify the computation, thus minimizing the possibility of error.

Designating the unperturbed kets by $|n\rangle$ the first-order correction to the energy is

$$E_n^{(1)} = \langle n| \hat{H}_1 |n\rangle$$

$$= -\frac{1}{2mc^2} \langle n| \hat{H}_0^2 |n\rangle$$

$$= -\frac{1}{2mc^2} \left(E_n^{(0)}\right)^2 \qquad (11.87)$$

where

$$E_n^{(0)} = \frac{n^2\pi^2\hbar^2}{2mL^2} \qquad (11.88)$$

Alternative Method (Brute Force)

We use direct integration with the spatial eigenfunctions to obtain $E_n^{(1)}$, in contrast to the simple method used above.

$$E_n^{(1)} = \langle n | \hat{H}_1 | n \rangle$$

$$= -\frac{1}{8} \frac{1}{m^3 c^2} \frac{2}{L} \int_0^L \sin\left(\frac{n\pi x}{L}\right) \hat{p}^4 \sin\left(\frac{n\pi x}{L}\right) dx$$

$$= -\frac{1}{4} \frac{1}{m^3 c^2 L} \int_0^L \sin\left(\frac{n\pi x}{L}\right) \left(\frac{\hbar}{i}\frac{d}{dx}\right)^4 \sin\left(\frac{n\pi x}{L}\right) dx$$

$$= -\frac{1}{4} \frac{\hbar^4}{m^3 c^2 L} \left(\frac{n\pi}{L}\right)^4 \int_0^L \sin^2\left(\frac{n\pi x}{L}\right) dx \tag{11.89}$$

To evaluate the integral we make the substitutions

$$y = (n\pi/L)x \Longrightarrow dx = (L/n\pi)\,dy$$
$$x = 0 \to y = 0;\ x = L \to y = n\pi \tag{11.90}$$

which lead to

$$E_n^{(1)} = -\frac{1}{4} \frac{\hbar^4}{m^3 c^2 L} \left(\frac{n\pi}{L}\right)^4 \left(\frac{L}{n\pi}\right) \int_0^{n\pi} \sin^2 y\, dy$$

$$= -\frac{1}{4} \frac{\hbar^4}{m^3 c^2 L} \left(\frac{n\pi}{L}\right)^4 \left(\frac{L}{n\pi}\right) \left(\frac{n\pi}{2}\right)$$

$$= -\frac{1}{2} \cdot \frac{1}{mc^2} \cdot \left(\frac{\pi^2 n^2 \hbar^2}{2mL^2}\right)^2$$

$$= -\frac{1}{2mc^2} \left(E_n^{(0)}\right)^2 \quad \text{as above} \tag{11.91}$$

(b) Rewriting the radical in the relativistic Hamiltonian, Eq. (11.85), provides a polynomial in the operator $\hat{p}^2$ as shown below.

$$\hat{H} = \left\{ mc^2 \left[1 + \frac{\hat{p}^2}{m^2 c^2}\right]^{1/2} - mc^2 \right\}$$

$$= \left\{ mc^2 \left[1 + \left(\frac{2}{mc^2}\right)\frac{\hat{p}^2}{2m}\right]^{1/2} - mc^2 \right\} \tag{11.92}$$

When the radical is expanded using the binomial theorem an infinite series in powers of $\hat{p}^2$ will be obtained. Inasmuch as the eigenfunctions of the unperturbed L-box, which we designate $|n\rangle$, are eigenfunctions of any operator that is proportional to $\hat{p}^2$ they will also be eigenfunctions of the relativistic Hamiltonian.

(c) To find the values of the relativistic Hamiltonian we simply operate on the eigenfunctions $|n\rangle$ with the relativistic Hamiltonian. This procedure amounts to expanding the radical, operating on $|n\rangle$ and then contracting the infinite series back to an expression within a radical. This is tantamount to replacing $\hat{p}^2/2m$ in Eq. (11.85) by $E_n^{(0)}$.

$$
\begin{aligned}
\hat{H}\,|n\rangle &= \left\{ mc^2 \left[1 + \frac{\hat{p}^2}{m^2 c^2} \right]^{1/2} - mc^2 \right\} |n\rangle \\
&= \left\{ mc^2 \left[1 + \left(\frac{2}{mc^2} \right) \frac{\hat{p}^2}{2m} \right]^{1/2} - mc^2 \right\} |n\rangle \\
&= \left\{ mc^2 \left[1 + \left(\frac{2}{mc^2} \right) E_n^{(0)} \right]^{1/2} - mc^2 \right\} |n\rangle
\end{aligned}
\tag{11.93}
$$

Therefore the exact relativistic energy eigenvalues are

$$
E_n^{(\mathrm{rel})} = mc^2 \left[1 + \left(\frac{2}{mc^2} \right) E_n^{(0)} \right]^{1/2} - mc^2
\tag{11.94}
$$

which, when expanded becomes

$$
\begin{aligned}
E_n^{(\mathrm{rel})} &\approx mc^2 \left\{ 1 + \frac{1}{2} \left[\left(\frac{2}{mc^2} \right) E_n^{(0)} \right] - \frac{1}{8} \left[\left(\frac{2}{mc^2} \right) E_n^{(0)} \right]^2 \right\} - mc^2 \\
&\approx E_n^{(0)} - \frac{1}{8} \left(\frac{2^2}{mc^2} \right) \left(E_n^{(0)} \right)^2 + \cdots \\
&= E_n^{(0)} - \frac{1}{2} \left(\frac{1}{mc^2} \right) \left(E_n^{(0)} \right)^2 + \cdots
\end{aligned}
\tag{11.95}
$$

Not surprisingly, this produces the same result for $E_n^{(1)}$ as that obtained in Eqs. (11.87) and (11.91).

5. An electron is subjected to a harmonic oscillator potential, $U(x)$ given by

$$
U(x) = \frac{1}{2} kx^2 = \frac{1}{2} m\omega^2 x^2
\tag{11.96}
$$

A constant electric field F is applied in the $-x$ direction.

(a) Determine the first nonvanishing correction to the harmonic oscillator energy using perturbation theory. Find the polarizability [1].
(b) Solve the problem exactly and show that the exact solution yields the result from perturbation theory.

Solution

(a) Treating the electric field as a perturbation we have

$$\hat{H}_1 = eFx \tag{11.97}$$

Note that for this potential energy the negative of the derivative yields a field in the $-x$ direction.

The first-order correction to the energy is given by

$$E_n^{(1)} = \langle n| \hat{H}_1 |n\rangle = eF \langle n| x |n\rangle \tag{11.98}$$

Clearly $E_n^{(1)} \equiv 0$ because the integrand in Eq. (11.98) is odd over symmetric limits. We must go to the second-order correction which is given by

$$E_n^{(2)} = \sum_{m\neq n} \frac{\left| \langle m| \hat{H}_1 |n\rangle \right|^2}{\left(E_n^{(0)} - E_m^{(0)} \right)}$$

$$= \sum_{m\neq n} \frac{|eF \langle m| x |n\rangle|^2}{\left(E_n^{(0)} - E_m^{(0)} \right)} \tag{11.99}$$

Next we evaluate the matrix elements $eF \langle m| x |n\rangle$ using the ladder operators. The action of the ladder operators is given by

$$\hat{a} |n\rangle = \sqrt{n} |n - 1\rangle$$
$$\hat{a}^\dagger |n\rangle = \sqrt{n + 1} |n + 1\rangle \tag{11.100}$$

Also, see Eq. (7.46),

$$\hat{x} = \frac{1}{\sqrt{2}\alpha} \left(\hat{a} + \hat{a}^\dagger \right) \; ; \; \alpha = \sqrt{\frac{m\omega}{\hbar}} \tag{11.101}$$

Therefore

$$eF \langle m| x |n\rangle = \frac{eF}{\sqrt{2}\alpha} \left(\sqrt{n}\delta_{m,n-1} + \sqrt{n + 1}\delta_{m,n+1} \right) \tag{11.102}$$

so that only two terms in Eq. (11.99) survive the summation. We have

$$
E_n^{(2)} = \sum_{m \neq n} \frac{|eF \langle m|x|n\rangle|^2}{\left(E_n^{(0)} - E_m^{(0)}\right)}
$$

$$
= \left(\frac{eF}{\sqrt{2\alpha}}\right)^2 \left[\frac{n}{(E_n - E_{n-1})} + \frac{n+1}{(E_n - E_{n+1})}\right]
$$

$$
= \frac{e^2 F^2}{2} \left(\frac{\hbar}{m\omega}\right) \left[\frac{n}{\hbar\omega} + \frac{n+1}{-\hbar\omega}\right]
$$

$$
= -\frac{e^2 F^2}{2m\omega^2} \tag{11.103}
$$

Note that this correction to the energy is independent of n so the same correction applies to all levels. The polarizability is

$$
\alpha = -\frac{d^2 E}{dF^2} = \frac{e^2}{m\omega^2} \tag{11.104}
$$

(b) To solve the problem exactly we note that the total potential energy may be rewritten by completing the square. It is still a parabola and the eigenfunctions and eigenvalues remain those of a harmonic oscillator. The total potential energy is

$$
U_T(x) = \frac{1}{2}m\omega^2 x^2 + eFx \tag{11.105}
$$

which, after completing the square, is

$$
U_T(x) = \frac{1}{2}m\omega^2 \left(x + \frac{eF}{m\omega^2}\right)^2 - \frac{e^2 F^2}{2m\omega^2} \tag{11.106}
$$

The minimum in the potential energy with the field on occurs at

$$
\frac{dU_T(x)}{dx} = m\omega^2 \left(x_{min} + \frac{eF}{m\omega^2}\right) = 0
$$

$$
\implies x_{min} = -\frac{eF}{m\omega^2} \tag{11.107}
$$

so the minimum value of $U_T(x)$ is

$$
U_T(x = x_{min}) = \frac{1}{2}m\omega^2 \left(-\frac{eF}{m\omega^2} + \frac{eF}{m\omega^2}\right)^2 - \frac{e^2 F^2}{2m\omega^2}
$$

$$
= -\frac{e^2 F^2}{2m\omega^2} \tag{11.108}
$$

Figure 11.4 shows the potential energy with and without the electric field on.

Fig. 11.4
Problem 5b—solution

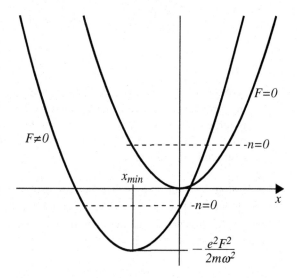

The shapes of both parabolas are the same as is easily seen by taking the second derivative of Eq. (11.105). Therefore the energy level spacing is the same with the field on or off. The total energy is merely displaced by $(e^2F^2)/(2m\omega^2)$. The energy eigenvalues are therefore

$$E_n = \left(n + \frac{1}{2}\right)\hbar\omega - \frac{e^2F^2}{2m\omega^2} \tag{11.109}$$

We can show this directly from the TISE using the same substitutions as above. The TISE is

$$\left[-\frac{\hbar^2}{2m}\frac{d^2}{dx^2} + U_T(x)\right]\psi(x) = E\psi(x) \tag{11.110}$$

Letting $y = x + eF/(m\omega^2)$ we have

$$\left[-\frac{\hbar^2}{2m}\frac{d^2}{dy^2} + \frac{1}{2}m\omega^2y^2 - \frac{e^2F^2}{2m\omega^2}\right]\psi(y) = E\psi(y) \tag{11.111}$$

or

$$\left[-\frac{\hbar^2}{2m}\frac{d^2}{dy^2} + \frac{1}{2}m\omega^2y^2\right]\psi(y) = \left(E + \frac{e^2F^2}{2m\omega^2}\right)\psi(y) \tag{11.112}$$

which is the TISE with energy eigenvalues given by Eq. (11.109).

$$E_n + \frac{e^2 F^2}{2m\omega^2} = \left(n + \frac{1}{2}\right) \hbar\omega$$

$$\Longrightarrow E_n = \left(n + \frac{1}{2}\right) \hbar\omega - \frac{e^2 F^2}{2m\omega^2} \qquad (11.113)$$

Note that the second-order correction to the energy derived in part (a) of this problem is the exact energy added to the system when the electric field is turned on. In this special case the perturbation treatment of the problem provides the exact answer so the magnitude of the "perturbation" F is not restricted to small values.

6. A H-atom is subjected to a repulsive δ-function potential at the origin of coordinates so that $\hat{H}_1 = U_0 \delta(r)$. Find $E_{n00}^{(1)}$, the first-order correction to the energy of *any* zero-angular momentum state. [Hint: $R_{n0}(r = 0) = 2(1/na_0)^{3/2}$.] What is the first-order correction to the energy if $\ell \neq 0$?

Solution

The first-order correction to the energy of any state $|n00\rangle$ is

$$E_{n00}^{(1)} = \langle n00| \hat{H}_1 |n00\rangle \qquad (11.114)$$

Because the perturbation is a δ-function it is easy to evaluate $E_{n00}^{(1)}$ using the corresponding integral. We require $\psi_{n00}(r)$, which is

$$\psi_{n00}(r) = Y_{00}(\theta, \phi) R_{n0}(r) \qquad (11.115)$$

Using the hint and the fact that $Y_{00}(\theta, \phi) = 1/\sqrt{4\pi}$ (see Appendix R) we have

$$\psi_{n00}(r = 0) = \sqrt{\frac{1}{4\pi}} \cdot \frac{2}{(na_0)^{3/2}} \qquad (11.116)$$

so that

$$E_{n00}^{(1)} = \int_{\text{all space}} |\psi_{n00}(r)|^2 \left[U_0 \delta(r)\right] d^3r$$

$$= U_0 |\psi_{n00}(r = 0)|^2$$

$$= \frac{U_0}{\pi n^3 a_0^3} \qquad (11.117)$$

We note that the integration in Eq. (11.117) can be made more transparent by evaluating $E_{n00}^{(1)}$ in Cartesian coordinates and using $\delta(\boldsymbol{r}) = \delta(x)\,\delta(y)\,\delta(z)$.

$$
\begin{aligned}
E_{n00}^{(1)} &= \int_{\text{all space}} |\psi_{n00}(x,y,z)|^2 \left[U_0\delta(x)\,\delta(y)\,\delta(z)\right] dx\,dy\,dz \\
&= U_0\,|\psi_{n00}(0,0,0)|^2 \\
&= \frac{U_0}{\pi n^3 a_0^3}
\end{aligned}
\tag{11.118}
$$

The first-order correction to the energy for $\ell \neq 0$ vanishes because the eigenfunctions for $\ell \neq 0$ are all zero at the origin. Although there is a quantum mechanical derivation of this fact [1] it is easy to remember by appealing to the classical definition of angular momentum $\boldsymbol{L} = \boldsymbol{r} \times \boldsymbol{p}$. It is clear that if $\boldsymbol{L} \neq 0$ then $\boldsymbol{r}$ cannot vanish.

7. Two H-atoms, each of mass m_p (ignore the electronic masses), are joined by a covalent bond to form the H_2 molecule. Small amplitude vibration of this molecule may be treated as harmonic oscillation with a force constant $k = \mu\omega^2$ where μ is the reduced mass of the nuclei and ω is the frequency of the vibration. The force constant k is obtained from the coefficient of the quadratic term in an expansion of the intermolecular potential.

(a) Find $E_0^{(1)}$ the first-order correction to the ground vibrational state energy $E_0^{(0)}$ due to the relativistic motion.

(b) Write $E_0^{(1)}$ in terms of $E_0^{(0)}$ to obtain the approximate ratio $\left|E_0^{(1)}/E_0^{(0)}\right|$ for the ground electronic state of the H_2 molecule in order to evaluate the validity of the use of perturbation theory. The energy separation between adjacent vibrational levels of the electronic ground state is $\hbar\omega \approx 0.54\,\text{eV}$ and $m_p c^2$ is $\sim 940\,\text{MeV}$.

Solution

(a) From Problem 4 of this chapter we know that the relativistic expansion of $\hat{T}$ the kinetic energy operator is [see Eq. (11.85)].

$$
\hat{T} \approx \frac{\hat{p}^2}{2\mu} - \frac{1}{8}\frac{\hat{p}^4}{\mu^3 c^2} + \cdots
\tag{11.119}
$$

Letting $\omega = \sqrt{k/m}$ the complete Hamiltonian is therefore

$$
\hat{H} = \hat{T} + \hat{V}
$$

$$= \frac{\hat{p}^2}{2m} - \frac{1}{8}\frac{\hat{p}^4}{\mu^3 c^2} + \cdots + \frac{1}{2}\mu\omega^2$$

$$\approx \hat{H}_0 - \frac{1}{8}\frac{\hat{p}^4}{\mu^3 c^2} \tag{11.120}$$

where $\hat{H}_0$ is the unperturbed harmonic oscillator Hamiltonian. The correction to the unperturbed vibrational ground state energy is simply $\langle 0| \hat{H}_1 |0\rangle$ where

$$\hat{H}_1 = -\frac{1}{8}\frac{\hat{p}^4}{\mu^3 c^2} \tag{11.121}$$

Now the question is what is the best (read: easiest) way to calculate $\langle 0| \hat{H}_1 |0\rangle$? Inasmuch as the commonly used eigenfunctions are those in coordinate space we can either convert them to momentum space or convert the momentum operator to its equivalent coordinate space operator. This conversion would then require *four* derivatives of the ground state eigenfunction. While neither method is difficult, they are both tedious enough that an error is likely. The most efficient approach is through the use of ladder operator as has been emphasized in Chap. 7.

Now, using Eq. (7.16) $\hat{H}_1$ may be written in terms of the ladder operators

$$\hat{H}_1 = -\frac{1}{8}\frac{1}{\mu^3 c^2}\left[\frac{\alpha\hbar}{\sqrt{2}}\left(\hat{a} - \hat{a}^\dagger\right)\right]^4 \tag{11.122}$$

and $E_0^{(1)} = \langle 0| \hat{H}_1 |0\rangle$ may be written [see Eq. (7.16)] as

$$E_0^{(1)} = \langle 0| \hat{H}_1 |0\rangle$$

$$= -\frac{1}{8\mu^3 c^2}\frac{\alpha^4\hbar^4}{4}\langle 0| \left(\hat{a} - \hat{a}^\dagger\right)^4 |0\rangle$$

$$= -\frac{\alpha^4\hbar^4}{32\mu^3 c^2}\left[\langle 0| \left(\hat{a} - \hat{a}^\dagger\right)\right]\left(\hat{a} - \hat{a}^\dagger\right)^2\left[\left(\hat{a} - \hat{a}^\dagger\right)|0\rangle\right] \tag{11.123}$$

where the terms in square brackets have been written in a way to facilitate evaluation. We first evaluate the terms in square brackets using Eqs. (7.28) and (7.29) which, for convenience, we reproduce here.

$$\hat{a}|n\rangle = \sqrt{n}|n-1\rangle \quad \text{and} \quad \hat{a}^\dagger|n\rangle = \sqrt{n+1}|n+1\rangle \tag{11.124}$$

Using these raising and lowering operations we have

$$\left[\langle 0| \left(\hat{a} - \hat{a}^\dagger\right)\right] = \langle 1| \quad \text{and} \quad \left[\left(\hat{a} - \hat{a}^\dagger\right)|0\rangle\right] = -|1\rangle \tag{11.125}$$

where we have used the fact that $\hat{a}$ and $\hat{a}^\dagger$ are Hermitian adjoints so that operating on a ket with $\hat{a}$ is equivalent to operating with $\hat{a}^\dagger$ on a bra. To arrive at Eq. (11.125) we have also used the fact that applying the lowering operator to $|0\rangle$ annihilates it, that is $\hat{a}\,|0\rangle \equiv 0$.

Continuing, we have

$$E_0^{(1)} = \frac{\alpha^4 \hbar^4}{32\mu^3 c^2} \left[\langle 1| \left(\hat{a} - \hat{a}^\dagger \right) \right] \left[\left(\hat{a} - \hat{a}^\dagger \right) |1\rangle \right]$$

and

$$\left[\langle 1| \left(\hat{a} - \hat{a}^\dagger \right) \right] = \sqrt{2}\,\langle 2| - \langle 0|$$
$$\left[\left(\hat{a} - \hat{a}^\dagger \right) |1\rangle \right] = |0\rangle - \sqrt{2}\,|2\rangle \qquad (11.126)$$

and

$$\begin{aligned}
E_0^{(1)} &= \frac{\alpha^4 \hbar^4}{32\mu^3 c^2} \left(\sqrt{2}\,\langle 2| - \langle 0| \right) \left(|0\rangle - \sqrt{2}\,|2\rangle \right) \\
&= \frac{\alpha^4 \hbar^4}{32\mu^3 c^2} \left(-2 - 1 \right) \\
&= -\frac{3\alpha^4 \hbar^4}{32\mu^3 c^2} = \frac{3\hbar^4}{32\mu^3 c^2} \left(\sqrt{\frac{m\omega}{\hbar}} \right)^4 \\
&= -\frac{3\hbar^2 \omega^2}{32\mu c^2} \qquad (11.127)
\end{aligned}$$

Note that $E_0^{(1)}$ has units of energy as it must.

(b) The ground state vibrational energy of the hydrogen molecule, the zero-point energy, is $E_0^{(0)} = \frac{1}{2}\hbar\omega \approx 0.27\,\text{eV}$ so the ratio $\left| E_0^{(1)}/E_0^{(0)} \right|$ is

$$\begin{aligned}
\left| \frac{E_0^{(1)}}{E_0^{(0)}} \right| &= \frac{12 \left(\frac{1}{2}\hbar\omega \right)}{32 \left(mc^2 \right)} \\
&= -\frac{3\,(0.27)\,\text{eV}}{8\,(4.7 \times 10^8)\,\text{eV}} \\
&= 2.16 \times 10^{-10} \qquad (11.128)
\end{aligned}$$

This is a very small number so the use of perturbation theory is justified.

8. Find the first-order correction to the energy levels of the H-atom due to relativistic motion of the electron.

Solution

The relativistic energy is

$$E^2 = \hat{p}^2 c^2 + m^2 c^4$$

$$E \approx \frac{\hat{p}^2}{2m} - \frac{1}{8} \frac{\hat{p}^4}{m^3 c^2} \tag{11.129}$$

so the relativistic hydrogen atom Hamiltonian to first order is

$$\hat{H} = \frac{\hat{p}^2}{2m} - \frac{1}{8} \frac{\hat{p}^4}{m^3 c^2} - \left(\frac{1}{4\pi \epsilon_0} \right) \frac{e^2}{r}$$

$$= \hat{H}_0 + \hat{H}_1 \tag{11.130}$$

where

$$\hat{H}_0 = \frac{\hat{p}^2}{2m} - \left(\frac{1}{4\pi \epsilon_0} \right) \frac{e^2}{r} - \frac{1}{8} \frac{\hat{p}^4}{m^3 c^2} \tag{11.131}$$

and

$$\hat{H}_1 = -\frac{1}{8} \frac{\hat{p}^4}{m^3 c^2} = -\frac{1}{2mc^2} \left(\frac{\hat{p}^2}{2m} \right)^2 \tag{11.132}$$

Because $\hat{p}^2/2m$ is the nonrelativistic kinetic energy we may write

$$\hat{H}_1 = -\frac{1}{2mc^2} \left(\frac{\hat{p}^2}{2m} \right)^2$$

$$= -\frac{1}{2mc^2} \left[\hat{H}_0 + \left(\frac{1}{4\pi \epsilon_0} \right) \frac{e^2}{r} \right]^2$$

$$= -\frac{1}{2mc^2} \left\{ \hat{H}_0^2 + 2\hat{H}_0 \left(\frac{1}{4\pi \epsilon_0} \right) \frac{e^2}{r} + \left[\frac{e^2}{r} \left(\frac{1}{4\pi \epsilon_0} \right) \right]^2 \left(\frac{e^2}{r} \right)^2 \right\} \tag{11.133}$$

The first-order correction to the nonrelativistic energy is the expectation value of $\hat{H}_1$, Eq. (11.65) so

$$\langle n\ell m | \hat{H}_1 | n\ell m \rangle = -\frac{1}{2mc^2} \left\{ \left(E_n^{(0)} \right)^2 + 2E_n^{(0)} \left(\frac{e^2}{4\pi \epsilon_0} \right) \left\langle \frac{1}{r} \right\rangle + \left(\frac{e^2}{4\pi \epsilon_0} \right)^2 \left\langle \frac{1}{r^2} \right\rangle \right\} \tag{11.134}$$

Putting Eq. (11.134) in terms of the fine structure constant α and using Table T.3 we have

$$
E_n^{(1)} = \left\langle \hat{H}_1 \right\rangle = -\frac{1}{2mc^2} \left\{ \left(E_n^{(0)} \right)^2 + 2E_n^{(0)} \hbar c \alpha \left\langle \frac{1}{r} \right\rangle + (\hbar c \alpha)^2 \left\langle \frac{1}{r^2} \right\rangle \right\}
$$

$$
= -\frac{1}{2mc^2} \left\{ \left(-\frac{(mc^2\alpha^2)}{2n^2} \right)^2 - 2 \left(\frac{(m^2 c^4) \alpha^4}{2n^2} \right) \left(\frac{1}{n^2} \right) \right.
$$

$$
\left. + (m^2 c^4 \alpha^4) \left[\frac{1}{n^3 \left(\ell + \frac{1}{2} \right)} \right] \right\}
$$

$$
= -\frac{m^2 c^4 \alpha^4}{2mc^2} \left\{ \left(-\frac{1}{2n^2} \right)^2 - \left(\frac{1}{n^4} \right) + \left[\frac{1}{n^3 \left(\ell + \frac{1}{2} \right)} \right] \right\}
$$

$$
= -\frac{mc^2 \alpha^4}{2n^4} \left\{ \left(-\frac{1}{2} \right)^2 - 1 + \left[\frac{n}{\left(\ell + \frac{1}{2} \right)} \right] \right\}
$$

$$
= \frac{\left(mc^2 \alpha^2 / 2n^2 \right) \alpha^2}{2n^2} \left\{ \frac{1}{4} - 1 + \left[\frac{n}{\left(\ell + \frac{1}{2} \right)} \right] \right\}
$$

$$
= -E_n^{(0)} \frac{\alpha^2}{n^2} \left[\frac{3}{4} - \left(\frac{n}{\ell + \frac{1}{2}} \right) \right] \tag{11.135}
$$

The ratio of the energy correction to the unperturbed energy is

$$
\frac{E_n^{(1)}}{E_n^{(0)}} = -\frac{\alpha^2}{n^2} \left[\frac{3}{4} - \left(\frac{n}{\ell + \frac{1}{2}} \right) \right]
$$

$$
\approx -\frac{5 \times 10^{-5}}{n^2} \left[\frac{3}{4} - \left(\frac{n}{\ell + \frac{1}{2}} \right) \right] \tag{11.136}
$$

This relativistic correction to the hydrogen atom energies, termed "fine structure" correction, is proportional to α^2 and is small. There are additional corrections such as that resulting from the coupling between the orbital magnetic moment and the intrinsic magnetic moment of the electron (the spin). These various corrections account for the splitting of spectral lines observed in the spectrum of atomic hydrogen.

9. A harmonic oscillator is perturbed by a potential $H_1 = Cx^4$ where C is a positive constant.

(a) Find the first-order correction to the ground state energy of the harmonic oscillator.

(b) If the quartic perturbation is replaced by a cubic perturbation, $H_1 = Bx^3$ (B is a constant) find the ground state energy correction to first order. [Hint: Think about symmetry and you should be able to find the answer without writing anything down.]

Solution

(a) We write the Hamiltonian as

$$\hat{H} = \frac{\hat{p}^2}{2m} + \frac{1}{2}m\omega^2 x^2 + Cx^4 \tag{11.137}$$

The first-order correction to the ground state $(n = 0)$ is

$$E^{(1)}_{n=0} = \langle 0| \hat{H}_1 |0\rangle$$
$$= C \langle 0| x^4 |0\rangle \tag{11.138}$$

The integral in Eq. (11.138) has already been computed using the harmonic oscillator ladder operators (see Problem 8 of Chap. 7, Eq. (7.84)). For an arbitrary state $|n\rangle$ the expectation value of x^4 is

$$\langle n| \hat{x}^4 |n\rangle = \frac{3}{4\alpha^4} \left(2n^2 + 2n + 1\right) \tag{11.139}$$

so the first-order correction to the energy for a quartic perturbation to a harmonic oscillator is

$$E^{(1)}_{n=0} = \frac{3C}{4\alpha^4} \tag{11.140}$$

(b) The cubic perturbation is an odd function, so $\langle n| \hat{x}^3 |n\rangle$ vanishes for any n, and the first-order correction is zero. This is important because a real diatomic potential function can be expanded around its equilibrium position in a Taylor series, with the first term being a quadratic, the second a cubic, and so forth. The vanishing of the first-order cubic correction enhances the utility of the quadratic (harmonic oscillator) approximation for low vibrational levels. This utility decreases for higher vibrational levels where the second order cubic and first-order quadratic corrections become important.

10. A matrix that has been constructed on an orthonormal basis set and that represents the Hamiltonian of a particular system is given by

$$\hat{H} = \begin{pmatrix} 1 & \epsilon & 0 \\ \epsilon & 2 & 0 \\ 0 & 0 & 3-\epsilon \end{pmatrix} \tag{11.141}$$

(a) Write this matrix as the sum of two matrices $\hat{H} = \hat{H}_0 + \hat{H}_1$ for the purpose of applying perturbation theory to approximate the eigenvalues of $\hat{H}$.

(b) Find the eigenvalues to second order.

(c) Solve the problem exactly and compare with the perturbation theory result.

Solution

(a) Splitting the matrix as suggested we have

$$\hat{H} = \begin{pmatrix} 1 & 0 & 0 \\ 0 & 2 & 0 \\ 0 & 0 & 3 \end{pmatrix} + \begin{pmatrix} 0 & \epsilon & 0 \\ \epsilon & 0 & 0 \\ 0 & 0 & -\epsilon \end{pmatrix} \tag{11.142}$$

where

$$\hat{H}_0 = \begin{pmatrix} 1 & 0 & 0 \\ 0 & 2 & 0 \\ 0 & 0 & 3 \end{pmatrix} \text{ and } \hat{H}_1 = \begin{pmatrix} 0 & \epsilon & 0 \\ \epsilon & 0 & 0 \\ 0 & 0 & -\epsilon \end{pmatrix} \tag{11.143}$$

Notice that we could have written the $\hat{H}_0$ matrix with $3 - \epsilon$ as the entry in the lower right-hand corner in which case the $\hat{H}_1$ matrix would have a zero in the lower right-hand corner and we could have reduced the problem to working with 2×2 matrices. We will, however, continue this solution the hard way.

(b) The unperturbed eigenkets are the eigenkets of $\hat{H}_0$, namely

$$\left|1^{(0)}\right\rangle = \begin{pmatrix} 1 \\ 0 \\ 0 \end{pmatrix} \; ; \; \left|2^{(0)}\right\rangle = \begin{pmatrix} 0 \\ 1 \\ 0 \end{pmatrix} \; ; \; \left|3^{(0)}\right\rangle = \begin{pmatrix} 0 \\ 0 \\ 1 \end{pmatrix} \tag{11.144}$$

and the unperturbed eigenvalues are

$$E_1^{(0)} = 1 \; ; \; E_2^{(0)} = 2 \; ; \; E_3^{(0)} = 3 \tag{11.145}$$

First order:

$$E_i^{(1)} = \left\langle i^{(0)} \middle| \hat{H}_1 \middle| i^{(0)} \right\rangle \quad i = 1, 2, 3 \tag{11.146}$$

so

$$E_1^{(1)} = \begin{pmatrix} 1 & 0 & 0 \end{pmatrix} \begin{pmatrix} 0 & \epsilon & 0 \\ \epsilon & 0 & 0 \\ 0 & 0 & -\epsilon \end{pmatrix} \begin{pmatrix} 1 \\ 0 \\ 0 \end{pmatrix} = 0$$

$$E_2^{(1)} = \begin{pmatrix} 0 & 1 & 0 \end{pmatrix} \begin{pmatrix} 0 & \epsilon & 0 \\ \epsilon & 0 & 0 \\ 0 & 0 & -\epsilon \end{pmatrix} \begin{pmatrix} 0 \\ 1 \\ 0 \end{pmatrix} = 0$$

$$E_3^{(1)} = \begin{pmatrix} 0 & 0 & 1 \end{pmatrix} \begin{pmatrix} 0 & \epsilon & 0 \\ \epsilon & 0 & 0 \\ 0 & 0 & -\epsilon \end{pmatrix} \begin{pmatrix} 0 \\ 0 \\ 1 \end{pmatrix} = -\epsilon \qquad (11.147)$$

Second order:

$$E_i^{(2)} = \sum_{k \neq i} \frac{\left| \langle k^{(0)} | \hat{H}_1 | i^{(0)} \rangle \right|^2}{\left(E_i^{(0)} - E_k^{(0)} \right)} \qquad (11.148)$$

Then

$$E_1^{(2)} = \frac{\left| \langle 2^{(0)} | \hat{H}_1 | 1^{(0)} \rangle \right|^2}{\left(E_1^{(0)} - E_2^{(0)} \right)} + \frac{\left| \langle 3^{(0)} | \hat{H}_1 | 1^{(0)} \rangle \right|^2}{\left(E_1^{(0)} - E_3^{(0)} \right)} \qquad (11.149)$$

Digress to evaluate all three off-diagonal matrix elements. (Only three are needed because $\hat{H}$ is Hermitian.)

$$\langle 2^{(0)} | \hat{H}_1 | 1^{(0)} \rangle = \begin{pmatrix} 0 & 1 & 0 \end{pmatrix} \begin{pmatrix} 0 & \epsilon & 0 \\ \epsilon & 0 & 0 \\ 0 & 0 & -\epsilon \end{pmatrix} \begin{pmatrix} 1 \\ 0 \\ 0 \end{pmatrix} = \epsilon$$

$$\langle 3^{(0)} | \hat{H}_1 | 1^{(0)} \rangle = \begin{pmatrix} 0 & 0 & 1 \end{pmatrix} \begin{pmatrix} 0 & \epsilon & 0 \\ \epsilon & 0 & 0 \\ 0 & 0 & -\epsilon \end{pmatrix} \begin{pmatrix} 1 \\ 0 \\ 0 \end{pmatrix} = 0$$

$$\langle 3^{(0)} | \hat{H}_1 | 2^{(0)} \rangle = \begin{pmatrix} 0 & 0 & 1 \end{pmatrix} \begin{pmatrix} 0 & \epsilon & 0 \\ \epsilon & 0 & 0 \\ 0 & 0 & -\epsilon \end{pmatrix} \begin{pmatrix} 0 \\ 1 \\ 0 \end{pmatrix} = 0 \qquad (11.150)$$

Now back to second order.

$$E_1^{(2)} = \frac{\epsilon^2}{(1-2)} + \frac{0}{(1-3)} = -\epsilon^2 \qquad (11.151)$$

$$E_2^{(2)} = \frac{\left|\langle 1^{(0)}|\hat{H}_1|2^{(0)}\rangle\right|^2}{\left(E_2^{(0)} - E_1^{(0)}\right)} + \frac{\left|\langle 3^{(0)}|\hat{H}_1|2^{(0)}\rangle\right|^2}{\left(E_2^{(0)} - E_3^{(0)}\right)}$$

$$= \frac{\epsilon^2}{(2-1)} + \frac{0}{(2-3)} = \epsilon^2 \tag{11.152}$$

$$E_3^{(2)} = \frac{\left|\langle 1^{(0)}|\hat{H}_1|3^{(0)}\rangle\right|^2}{\left(E_3^{(0)} - E_1^{(0)}\right)} + \frac{\left|\langle 2^{(0)}|\hat{H}_1|3^{(0)}\rangle\right|^2}{\left(E_3^{(0)} - E_2^{(0)}\right)}$$

$$= \frac{0}{(3-1)} + \frac{0}{(3-2)} = 0 \tag{11.153}$$

The energies correct to second order are therefore

$$E_i = E_i^{(0)} + E_i^{(1)} + E_i^{(2)}$$
$$E_1 = 1 + 0 - \epsilon^2 = 1 - \epsilon^2$$
$$E_2 = 2 + 0 + \epsilon^2 = 2 + \epsilon^2$$
$$E_3 = 3 - \epsilon + 0 = 3 - \epsilon \tag{11.154}$$

(c) To solve exactly we solve the secular equation for the whole Hamiltonian.

$$\begin{vmatrix} 1-E & \epsilon & 0 \\ \epsilon & 2-E & 0 \\ 0 & 0 & 3-\epsilon-E \end{vmatrix} = 0 \tag{11.155}$$

Expanding along the bottom row we have

$$(3 - \epsilon - E)\left[(1-E)(2-E) - \epsilon^2\right] = 0$$
$$\text{or } (3 - \epsilon - E)\left(2 - 3E + E^2 - \epsilon^2\right) = 0 \tag{11.156}$$

Clearly one root is

$$E = 3 - \epsilon \tag{11.157}$$

The other two are given by solving the quadratic equation

$$E^2 - 3E + \left(2 - \epsilon^2\right) = 0$$

and are

$$E = \frac{1}{2}\left(3 \pm \sqrt{9 - 4\left(2 - \epsilon^2\right)}\right)$$

$$= \frac{1}{2}\left(3 \pm \sqrt{1 + 4\epsilon^2}\right) \tag{11.158}$$

Now the root $E = 3 - \epsilon$ is equal to E_3 from above. This is expected because it is the only entry in the last row and column of $\hat{H}$. To compare the other two roots of the quadratic equation we expand the radical using the binomial expansion.

$$E = \frac{1}{2}\left(3 \pm \sqrt{1 + 4\epsilon^2}\right)$$

$$\approx \frac{3}{2} \pm \frac{1}{2}\left(1 + \frac{1}{2}4\epsilon^2 + \cdots\right)$$

$$= \left(2 + \epsilon^2\right), \left(1 - \epsilon^2\right) \tag{11.159}$$

so the exact result reduces to the second-order perturbation theory result for small ϵ.

11.4 The Helium Atom

Although it looks deceptively simple, the Schrödinger equation for this two-electron system cannot be solved analytically. The helium atom is, however, amenable to treatment using one or more of the approximation methods discussed in this chapter. It is also a system for which indistinguishability of the two electrons plays an important role. The Hamiltonian in atomic units (see Appendix C) for a two electron system having Z protons in the nucleus is

$$H = \frac{p_1^2}{2m_e} + \frac{p_2^2}{2m_e} - \frac{Z}{r_1} - \frac{Z}{r_2} + \frac{1}{r_{12}} \tag{11.160}$$

where r_1 and r_2 are the distances of electrons 1 and 2 from the nucleus; $r_{12} = |r_1 - r_2|$ is the distance between electrons 1 and 2. If the r_{12} term were absent, the Hamiltonian discussed in chapter would represent two independent hydrogenic atoms with nuclear charge Z ($Z = 2$ for He). The eigenfunctions would be products of H-atom eigenfunctions scaled by Z, and, using the notation of Sect. 11.3 the zeroth-order energies would be

$$E_{n_1 n_2}^{(0)} = -\frac{Z^2}{2n_1} - \frac{Z^2}{2n_2} \tag{11.161}$$

where n_1, n_2 are the principal quantum numbers of electrons 1 and 2 (we have ignored indistinguishability here, but including it would not change the allowed energies). For $Z = 2$, the ground state energy neglecting the r_{12} term would be $E_{11}^{(0)} = -4$ a.u. or about -109 eV. The observed energy for the He ground state is $E_{\text{ex}} = -78.8$ eV. Introducing the r_{12} electron repulsion term obviously produces a significant positive correction to the energy. Utilization of the approximation methods discussed above allows considerable improvement to the calculated energies and wave functions for He, or any other two-electron atom, such as H^- or Li^+.

Problems

1. Evaluate the ground state energy for He using first-order perturbation theory.

Solution

First we must construct the proper zero-order ground state wave function. This function is simply the product of two ground state hydrogenic wave functions as mentioned above. The spatial part of this wave function in ket notation is

$$|n_1\ell_1 m_1, n_2\ell_2 m_2\rangle = |n_1\ell_1 m_1\rangle |n_2\ell_2 m_2\rangle$$

$$|100, 100\rangle = |100\rangle |100\rangle \tag{11.162}$$

where $n = 1$ and $\ell = m = 0$ for electrons 1 and 2. This ket is symmetric when the coordinates for electrons 1 and 2 are interchanged, so the proper spin function must be antisymmetric. As in Problem 2 of Chap. 9, we designate this spin state $|00\rangle$, and from Problem 13 of Chap. 8 the spin zero coupled state $|00\rangle$ in terms of the uncoupled states is

$$|00\rangle = -\frac{1}{\sqrt{2}} \left(\left|-\tfrac{1}{2}, \tfrac{1}{2}\right\rangle - \left|\tfrac{1}{2}, -\tfrac{1}{2}\right\rangle \right) \tag{11.163}$$

The complete ground state ket, symmetric in space and antisymmetric in spin, is simply $|1_1 0_1 0_1 1_2 0_2 0_2\rangle |00\rangle$ and the first-order correction to the zeroth-order ground state energy is

$$E_{11}^{(1)} = \langle 1_1 0_1 0_1 1_2 0_2 0_2 | \frac{1}{r_{12}} | 1_1 0_1 0_1 1_2 0_2 0_2 \rangle \tag{11.164}$$

The nonrelativistic Hamiltonian, Eq. (11.160), does not contain spin variables, so the inner product of the spin bra and ket is unity, and thus does not affect the calculated $E_{11}^{(1)}$ value. The actual evaluation of $E_{11}^{(1)}$ requires evaluation of the integral that is the inner product in Eq. (11.164). This becomes

$$E_{11}^{(1)} = \frac{2^2 Z^6}{(4\pi)^2} \int_{\text{all space}} \exp\left[-2Z\right](r_1 + r_2)\left(\frac{1}{r_{12}}\right)$$

$$\times (4\pi)\left(r_1^2 \sin\theta_1 d\theta_1 dr_1\right)\left(r_2^2 \sin\theta_2 d\theta_2 dr_2\right) \qquad (11.165)$$

This integral can be evaluated by expanding $1/r_{12}$ in terms of Legendre polynomials as shown in Appendix P. The result after some algebra is

$$E_{11}^{(1)} = \frac{5}{8}Z$$

$$= \frac{5}{4} \text{ a.u. for He} \qquad (11.166)$$

so the first-order correction to the total He energy is

$$E_{11}^{(0)} + E_{11}^{(1)} = -4 \text{ a.u.} + 1.25 \text{ a.u.}$$

$$= -2.75 \text{ a.u.} = -74.8\,\text{eV} \qquad (11.167)$$

This is an approximation to the total amount of energy required to liberate *both* electrons from the Coulomb attraction of the nucleus. While $E_{11}^{(1)}$ due to the electron–electron repulsion is a significant fraction of $E_{11}^{(0)}$, in excess of 30 %, the answer is remarkably close to the actual energy, -78.95, when one considers that it was evaluated using perturbation theory.

2. Use the perturbation result from Problem 1 along with the variational method to improve the perturbation result for the ground state energy of a He atom. Use Z as the variational parameter; that is, assume that the He wave function can be improved by using a nuclear charge less than or greater than $Z = 2$.

Solution

We begin by replacing Z by ς in the two-particle unperturbed hydrogenic radial wave function (Table T.2), so that (in a.u.)

$$|100\rangle\,|100\rangle \rightarrow \frac{\varsigma^3}{\pi} \exp\left[-\varsigma\left(r_1 - r_2\right)\right] \qquad (11.168)$$

We must find the value of ς that minimizes the calculated first-order perturbation theory energy as obtained in Problem 1. Although we have introduced a variable ς in the wave function, the Hamiltonian still contains the actual Z, so we simplify the problem by rewriting the Hamiltonian, Eq. (11.160), as

$$\langle H \rangle = \frac{p_1^2}{2m_e} + \frac{p_2^2}{2m_e} - \frac{\varsigma}{r_1} - \frac{\varsigma}{r_2} + \frac{1}{r_{12}} + \left[(\varsigma - Z) \left(\frac{1}{r_1} + \frac{1}{r_2} \right) \right] \quad (11.169)$$

Neglecting the term in brackets, this Hamiltonian is identical to that in Problem 1 with $Z \rightarrow \varsigma$ so, using Eq. (11.161), we can immediately write

$$\langle H \rangle = \varsigma^2 - \frac{5}{8}\varsigma + 2(\varsigma - Z) \left\langle \frac{1}{r} \right\rangle$$

$$= \varsigma^2 - \frac{5}{8}\varsigma + 2(\varsigma - Z)\varsigma \quad (11.170)$$

where we have used the expectation value of $1/r$ (see Table T.3). Differentiating with respect to the adjustable parameter ς and setting the result equal to zero, we obtain $\varsigma = \varsigma_0$ the value of ς that minimizes $\langle H \rangle$.

$$\varsigma_0 = Z - \frac{5}{16} \quad (11.171)$$

Inserting this into Eq. (11.170) for $\langle H \rangle$, the variational energy $E^{(v)}(Z)$ for ground state of He becomes

$$E^{(v)}(Z) = - \left(Z - \frac{5}{16} \right)^2 \quad \text{in a.u.} \quad (11.172)$$

which for $Z = 2$ is

$$E^{(v)}(Z = 2) = -2.85 \quad \text{in a.u.}$$

$$= -77.5\,\text{eV} \quad (11.173)$$

This result is fairly close to the experimental value of -78.8 eV, but remains above it as guaranteed by the variational method. Indeed, it is considerably closer to this value than the -74.8 eV obtained using only perturbation theory as in Problem 1. The fact that the optimum value of Z (ς_0) is less than $Z = 2$ reflects the screening each electron provides the other, reducing the *effective* nuclear charge.

As might be evident from the discussion of the variational method in Sect. 11.2 the total energy may be calculated with progressively increasing accuracy as we add more variational parameters.

11.5 Degenerate Perturbation Theory

Examination of Eqs. (11.66) and (11.67) reveals a serious problem if one or more of the unperturbed levels are degenerate. Using the same notation as that in Sect. 11.3 it is seen that if $\left|\psi_n^{(0)}\right\rangle$ and $\left|\psi_k^{(0)}\right\rangle$ are degenerate then $E_n^{(0)} = E_k^{(0)}$ and the first-order corrections to the unperturbed wave functions blows up. This problem can be solved by constructing linear combinations of the degenerate eigenfuntions that are not only eigenfunctions of $\hat{H}_0$, but also eigenfunctions of the perturbing Hamiltonian $\hat{H}_1$. We call this new set of eigenfunctions the *select set* and designate them $\left|\phi_i^{(0)}\right\rangle$. Thus

$$\left|\phi_i^{(0)}\right\rangle = \sum_{j=1}^{q} c_{ij} \left|\psi_j^{(0)}\right\rangle \tag{11.174}$$

is a select ket such that

$$\hat{H}_1 \left|\phi_i^{(0)}\right\rangle = E_i^{(1)} \left|\phi_i^{(0)}\right\rangle \tag{11.175}$$

where $E_i^{(1)}$ is the first-order correction to the energies of each of the degenerate states. If there are q degenerate eigenstates of $\hat{H}_0$, there will be q select states. The task then is to find the expansion coefficients c_{ij} in Eq. (11.174).

After applying first-order time independent perturbation theory [1] we arrive at the following relationship

$$\sum_{j=1}^{q} c_{ij} \left\langle \psi_k^{(0)} \middle| \hat{H}_1 \middle| \psi_j^{(0)} \right\rangle = E_i^{(1)} c_{ik} \quad k \le q \tag{11.176}$$

where $\left\langle \psi_k^{(0)} \middle|$ and $\left| \psi_j^{(0)} \right\rangle$ represent degenerate (unperturbed) states of $\hat{H}_0$. Equation (11.176) provides the matrix elements for the matrix representation of $\hat{H}_1$. Because the degenerate unperturbed eigenkets $\left| \psi_i^{(0)} \right\rangle$ are not eigenkets of $\hat{H}_1$ the $\hat{H}_1$ matrix is not diagonal on this basis set. It is, however, diagonal on the $\left| \phi_j^{(0)} \right\rangle$ basis set, which is the reason for constructing this select set. Note that the matrix representing the unperturbed Hamiltonian, $\hat{H}_0$, is diagonal on both basis sets. Problem 1 below illustrates this procedure in a step-by-step fashion.

Problems

1. A particle of mass m is trapped in a two-dimensional a-box and a perturbation $\hat{H}_1(x, y) = -Kxy$ is applied.

   (a) Find the eigenfunctions and eigenvalues for the unperturbed ground state and for the doubly degenerate first excited state.
   (b) Find the matrix that represents the complete Hamiltonian $\hat{H}$ for the excited states and find the energies of the ground state and the first excited states to first order.

Solution

(a) The unperturbed Hamiltonian is separable in x and y so

$$\hat{H}_0(x, y) = \hat{H}_x(x) + \hat{H}_y(y) \tag{11.177}$$

When the Hamiltonian can be separated as in Eq. (11.177) [1] the eigenfunctions are products of the eigenfunctions of the individual Hamiltonians and the energies are the sums of the individual energy eigenvalues. The eigenfunctions for a one-dimensional a-box are (see Appendix M):

$$\psi_n(x) = \sqrt{\frac{2}{a}} \cos\left(\frac{n\pi x}{a}\right) \quad ; \quad -\frac{a}{2} \le x \le \frac{a}{2} \quad n = 1, 3, 5 \ldots$$

$$\psi_n(x) = \sqrt{\frac{2}{a}} \sin\left(\frac{n\pi x}{a}\right) \quad ; \quad -\frac{a}{2} \le x \le \frac{a}{2} \quad n = 2, 4, 6 \ldots$$

$$= 0 \quad ; \quad -\infty < x < -a/2 \ ; \ a/2 < x < \infty \text{ for all } n \tag{11.178}$$

Therefore

$$\psi_{11}^{(0)}(x, y) = \psi_1^{(0)}(x)\,\psi_1^{(0)}(y)$$

$$= \frac{2}{a} \cos\left(\frac{\pi x}{a}\right) \cos\left(\frac{\pi y}{a}\right) \tag{11.179}$$

One of the degenerate first excited states is

$$\psi_{12}^{(0)}(x, y) = \psi_1(x)\,\psi_2(y)$$

$$= \frac{2}{a} \cos\left(\frac{\pi x}{a}\right) \sin\left(\frac{2\pi y}{a}\right) \tag{11.180}$$

while the other is

$$\psi_{21}^{(0)}(x, y) = \psi_2(x)\,\psi_1(y)$$

$$= \frac{2}{a}\cos\left(\frac{\pi y}{a}\right)\sin\left(\frac{2\pi x}{a}\right) \tag{11.181}$$

The unperturbed energies of these states are (Appendix M)

$$E_{11}^{(0)} = 2\cdot\frac{1^2\pi^2\hbar^2}{2mL^2}$$

$$E_{12}^{(0)} = \frac{\pi^2\hbar^2}{2mL^2} + \frac{2^2\pi^2\hbar^2}{2mL^2} = \frac{5\pi^2\hbar^2}{2mL^2} = E_{21}^{(0)} \tag{11.182}$$

Therefore, $\psi_{12}^{(0)}(x, y) = |12\rangle$ and $\psi_{21}^{(0)}(x, y) = |21\rangle$ are degenerate states because their (unperturbed) energies are the same.

(b) The first-order correction to the ground state energy is clearly zero because the perturbation is odd in both x and y.

For the excited states the Hamiltonian in matrix form is

$$\hat{H} = \hat{H}_0 + \hat{H}_1$$

$$= \begin{pmatrix} \langle 12|\,\hat{H}_0\,|12\rangle & \langle 21|\,\hat{H}_0\,|12\rangle \\ \langle 12|\,\hat{H}_0\,|21\rangle & \langle 21|\,\hat{H}_0\,|21\rangle \end{pmatrix}$$

$$-K\begin{pmatrix} \langle 12|\,xy\,|12\rangle & \langle 21|\,xy\,|12\rangle \\ \langle 12|\,xy\,|21\rangle & \langle 21|\,xy\,|21\rangle \end{pmatrix} \tag{11.183}$$

The diagonal matrix elements of $\hat{H}_0$ are simply $E_{12}^{(0)}$ while the off-diagonal matrix elements of $\hat{H}_0$ vanish. Therefore

$$\hat{H} = \begin{pmatrix} E_{12}^{(0)} & 0 \\ 0 & E_{12}^{(0)} \end{pmatrix} - K\begin{pmatrix} \langle 12|\,xy\,|12\rangle & \langle 21|\,xy\,|12\rangle \\ \langle 12|\,xy\,|21\rangle & \langle 21|\,xy\,|21\rangle \end{pmatrix} \tag{11.184}$$

Because $\hat{H}_0$ is represented by a diagonal matrix the unperturbed kets must be the column matrices

$$|12\rangle = \begin{pmatrix} 1 \\ 0 \end{pmatrix} \text{ and } |21\rangle = \begin{pmatrix} 0 \\ 1 \end{pmatrix} \tag{11.185}$$

Symmetry considerations dictate that the diagonal matrix elements in the second matrix of Eq. (11.184) vanish because the perturbation $-Kxy$ is an odd function. We must, however, evaluate the off-diagonal elements.

$$\langle 12|\,xy\,|21\rangle = \langle 21|\,xy\,|12\rangle$$

$$= -K\frac{4}{a^2}\int_{-a/2}^{a/2} dx\int_{-a/2}^{a/2} dy\left[\cos\left(\frac{\pi x}{a}\right)\sin\left(\frac{2\pi y}{a}\right)\right]$$

$$\times (xy) \left[\cos\left(\frac{\pi y}{a}\right) \sin\left(\frac{2\pi x}{a}\right) \right] \qquad (11.186)$$

The x-integral is

$$
\begin{aligned}
I_x &= \int_{-a/2}^{a/2} x \cos\left(\frac{\pi x}{a}\right) \sin\left(\frac{2\pi x}{a}\right) dx \\
&= 2 \int_{-a/2}^{a/2} x \cos^2\left(\frac{\pi x}{a}\right) \sin\left(\frac{\pi x}{a}\right) dx \\
&= \frac{8a^2}{9\pi^2} \qquad (11.187)
\end{aligned}
$$

where we have integrated by parts and used Eq. (E.8). Because the y-integral is identical to the x-integral we have

$$
\begin{aligned}
\langle 12| \, xy \, |21\rangle &= -K\frac{4}{a^2}\left(\frac{8a^2}{9\pi^2}\right)^2 \\
&= -K\frac{256a^2}{81\pi^4} \\
&\equiv \kappa \qquad (11.188)
\end{aligned}
$$

The Hamiltonian in matrix form is then

$$\hat{H} = \begin{pmatrix} E_{12}^{(0)} & 0 \\ 0 & E_{12}^{(0)} \end{pmatrix} + \kappa \begin{pmatrix} 0 & 1 \\ 1 & 0 \end{pmatrix} \qquad (11.189)$$

If the second matrix can be diagonalized it will be a simple matter to obtain the perturbed eigenvalues. To do this we solve the eigenvalue problem for that matrix to obtain the eigenkets (in terms of the unperturbed eigenfunctions) that make the matrix diagonal. These eigenkets are the select set.

We have

$$\kappa \begin{pmatrix} 0 & 1 \\ 1 & 0 \end{pmatrix}\begin{pmatrix} a \\ b \end{pmatrix} = E_1^{(1)} \begin{pmatrix} 1 & 0 \\ 0 & 1 \end{pmatrix}\begin{pmatrix} a \\ b \end{pmatrix} \qquad (11.190)$$

where the column matrix represents the select kets. We have inserted the identity matrix on the rhs for clarity. Multiplying the matrices in Eq. (11.190) gives two equations.

$$\kappa b - E_1^{(1)} a = 0$$
$$\kappa a - E_1^{(1)} b = 0 \qquad (11.191)$$

These are homogeneous equations in a and b so the only non-zero solution is obtained by solving the secular equation which is

$$\det \begin{pmatrix} E_1^{(1)} & -\kappa \\ -\kappa & E_1^{(1)} \end{pmatrix} = 0 \qquad (11.192)$$

which yields

$$\left[E_1^{(1)} \right]^2 - \kappa^2 = 0 \Rightarrow E_1^{(1)} = \pm \kappa = \mp K \frac{256a^2}{81\pi^4} \qquad (11.193)$$

Thus, the degenerate levels are split, one higher than the degenerate unperturbed energy $E_{12}^{(0)}$ and the other lower. That is,

$$E_{12}^{(0)} + E_{12}^{(1)} = \frac{5\pi^2\hbar^2}{2mL^2} \mp K \frac{256a^2}{81\pi^4} \qquad (11.194)$$

Additionally, inasmuch as $E_1^{(1)} = \pm \kappa$ we see from Eq. (11.191) that the select set of eigenkets is

$$\frac{1}{\sqrt{2}} \begin{pmatrix} 1 \\ 1 \end{pmatrix} \quad \text{and} \quad \frac{1}{\sqrt{2}} \begin{pmatrix} 1 \\ -1 \end{pmatrix} \qquad (11.195)$$

2. A particle of mass μ is confined to move on a circle of radius R in the xy-plane ($\theta = \pi/2$ in spherical coordinates). Along the circular path there is a δ-function perturbation applied in the form $\hat{H}_1(\phi) = U_0\delta(\phi)$ where ϕ is the azimuthal angle in spherical coordinates.

   (a) Find the energy eigenfunctions and eigenvalues of the unperturbed Hamiltonian $\hat{H}_0(\phi)$ for all states of this system.
   (b) Find the matrix that represents the complete Hamiltonian $\hat{H}(\phi)$ and find the energies of all states to first order.
   (c) Find the select set of kets and show that they are eigenkets of both $\hat{H}_0$ and $\hat{H}_1$.

Solution

(a) There is no potential energy in $\hat{H}_0(\phi)$ so the total unperturbed energy is the kinetic energy of a rigid rotor in a plane. Using Eq. (Q.3), the unperturbed Hamiltonian is

$$\hat{H}_0(\phi) = \frac{L_z^2}{2I} = \frac{1}{2\mu R^2} \left(\frac{\hbar}{i} \frac{d}{d\phi} \right)^2 \qquad (11.196)$$

where I is the moment of inertia of the mass about the origin. This problem is a one-dimensional problem in ϕ. The unperturbed Hamiltonian $\hat{H}_0\,(\phi)$ is simply the square of the angular momentum operator so the eigenvalue equation is

$$\hat{H}_0\,(\phi)\,\psi^{(0)}\,(\phi) = \frac{1}{2\mu R^2}\left(\frac{\hbar}{i}\frac{d}{d\phi}\right)^2\psi^{(0)}\,(\phi) \tag{11.197}$$

Solving this differential equation for $\psi_m^{(0)}\,(\phi)$ and normalizing we have

$$\psi_m^{(0)}\,(\phi) = \sqrt{\frac{1}{2\pi}}\,e^{im\phi}\quad m = \pm 1, \pm 2, \pm 3,\dots \tag{11.198}$$

where m must be an integer to satisfy the criterion that the wave function must be single valued. The $\pm$ m-values correspond to clockwise and counterclockwise motion on the circle. Inserting Eq. (11.198) into Eq. (11.197) leads to $E_m^{(0)}$ the unperturbed energy levels

$$E_m^{(0)} = \frac{m^2\hbar^2}{2\mu R^2}\quad m = \pm 1, \pm 2, \pm 3,\dots \tag{11.199}$$

Because m is squared in Eq. (11.199) the mth energy is twofold degenerate. This is the same as saying that the energy is independent of the direction of rotation.

(b) Designating the eigenkets corresponding to $\psi_{\pm m}^{(0)}\,(\phi)$ by $|\pm m\rangle$ and using the pair of kets $|\pm m\rangle$ as the basis set to split the twofold degeneracy, the complete Hamiltonian in matrix form is

$$\hat{H} = \hat{H}_0 + \hat{H}_1$$

$$= \begin{pmatrix} E_m^{(0)} & 0 \\ 0 & E_m^{(0)} \end{pmatrix} + U_0\begin{pmatrix} \langle m|\,\delta\,|m\rangle & \langle m|\,\delta\,|-m\rangle \\ \langle -m|\,\delta\,|m\rangle & \langle -m|\,\delta\,|-m\rangle \end{pmatrix} \tag{11.200}$$

where $\delta = \delta\,(\phi)$. Using the sifting property of the δ-function (see Table J.1), the diagonal elements in $\hat{H}_1$ are

$$\langle m|\,\delta\,|m\rangle = \langle -m|\,\delta\,|-m\rangle$$

$$= \frac{1}{2\pi}\int_{-\pi}^{\pi} e^{im\phi}\delta\,(\phi)\,e^{im\phi}\,d\phi$$

$$= \frac{1}{2\pi} \tag{11.201}$$

The off-diagonal integrals are

$$\langle m|\,\delta\,|-m\rangle = \langle -m|\,\delta\,|m\rangle$$

$$= \frac{1}{2\pi} \int_{-\pi}^{\pi} e^{im\phi} \delta(\phi) e^{-im\phi} d\phi$$

$$= \frac{1}{2\pi} \tag{11.202}$$

Then

$$\hat{H}_1 = U_0 \begin{pmatrix} \langle m| \delta |m\rangle & \langle m| \delta |-m\rangle \\ \langle -m| \delta |m\rangle & \langle -m| \delta |-m\rangle \end{pmatrix} = \frac{U_0}{2\pi} \begin{pmatrix} 1 & 1 \\ 1 & 1 \end{pmatrix} \tag{11.203}$$

Now we solve the eigenvalue problem for $\hat{H}_1$.

$$\frac{U_0}{2\pi} \begin{pmatrix} 1 & 1 \\ 1 & 1 \end{pmatrix} \begin{pmatrix} a \\ b \end{pmatrix} = E_m^{(1)} \begin{pmatrix} a \\ b \end{pmatrix} \tag{11.204}$$

so

$$U_0 a + U_0 b = 2\pi a E_m^{(1)}$$

$$U_0 a + U_0 b = 2\pi b E_m^{(1)} \tag{11.205}$$

or

$$\left[U_0 - 2\pi E_m^{(1)} \right] a + U_0 b = 0$$

$$U_0 a + \left[U_0 - 2\pi E_m^{(1)} \right] b = 0 \tag{11.206}$$

The only non-zero solutions to these simultaneous equations are found by solving the secular equation which is

$$\left[U_0 - 2\pi E_m^{(1)} \right]^2 - U_0^2 = 0$$

$$-4\pi E_m^{(1)} U_0 + 4\pi^2 E_m^{(1)} = 0$$

$$E_m^{(1)} \left[\pi E_m^{(1)} - U_0 \right] = 0 \tag{11.207}$$

Thus the corrections to the unperturbed energy $E_m^{(0)}$ are

$$E_m^{(1)} = 0 \quad \text{and} \quad E_m^{(1)} = \frac{U_0}{\pi} \tag{11.208}$$

The perturbation separates the m-level energies by U_0/π to first order. The total energies to first order are

$$E_m^{(0)} + E_m^{(1)} = \frac{m^2 \hbar^2}{2\mu R^2} \quad \text{and} \quad E_m^{(0)} + E_m^{(1)} = \frac{m^2 \hbar^2}{2\mu R^2} + \frac{U_0}{\pi} \tag{11.209}$$

(c) Solving Eq. (11.206) for a and b we have

$$a = \frac{U_0}{\left[U_0 - 2\pi E_m^{(1)}\right]} b$$

$$b = -\frac{U_0}{\left[U_0 - 2\pi E_m^{(1)}\right]} a \qquad (11.210)$$

First consider the case for which $E_m^{(1)} = 0$. From Eq. (11.210) we have $a = \pm b$ so the eigenkets corresponding to $E_m^{(1)} = 0$, the select set, are

$$\frac{1}{\sqrt{2}}\begin{pmatrix} 1 \\ 1 \end{pmatrix} \quad \text{and} \quad \frac{1}{\sqrt{2}}\begin{pmatrix} 1 \\ -1 \end{pmatrix} \qquad (11.211)$$

We would get the same result if we had chosen $E_m^{(1)} = U_0/\pi$ because there is only one select set. Now we wish to solve the eigenvalue problem with the Hamiltonian $\hat{H} = \hat{H}_0 + \hat{H}_1$ to verify that the eigenkets in Eq. (11.211) are eigenkets of both $\hat{H}_0$ and $\hat{H}_1$ (the necessary criterion for the select set). The eigenvalues that are obtained should be the same as those in Eq. (11.209).

Using the first of the select kets in Eq. (11.211), call it $(1, 1)$, we have

$$\hat{H}\,|(1, 1)\rangle = E_m^{(0)} \frac{1}{\sqrt{2}} \begin{pmatrix} 1 & 0 \\ 0 & 1 \end{pmatrix} \begin{pmatrix} 1 \\ 1 \end{pmatrix} + \frac{U_0}{2\pi} \frac{1}{\sqrt{2}} \begin{pmatrix} 1 & 1 \\ 1 & 1 \end{pmatrix} \begin{pmatrix} 1 \\ 1 \end{pmatrix}$$

$$= \frac{E_m^{(0)}}{\sqrt{2}} \begin{pmatrix} 1 \\ 1 \end{pmatrix} + \frac{U_0}{\sqrt{2\pi}} \begin{pmatrix} 1 \\ 1 \end{pmatrix}$$

$$= \left(E_m^{(0)} + \frac{U_0}{\pi}\right) \left[\frac{1}{\sqrt{2}}\begin{pmatrix} 1 \\ 1 \end{pmatrix}\right] \qquad (11.212)$$

Using the second of the select kets in Eq. (11.211), $(1, -1)$, we have

$$\hat{H}\,|(1, -1)\rangle = E_m^{(0)} \frac{1}{\sqrt{2}} \begin{pmatrix} 1 & 0 \\ 0 & 1 \end{pmatrix} \begin{pmatrix} 1 \\ \pm 1 \end{pmatrix} + \frac{U_0}{2\pi} \frac{1}{\sqrt{2}} \begin{pmatrix} 1 & 1 \\ 1 & 1 \end{pmatrix} \begin{pmatrix} 1 \\ \pm 1 \end{pmatrix}$$

$$= \frac{E_m^{(0)}}{\sqrt{2}} \begin{pmatrix} 1 \\ -1 \end{pmatrix} + \frac{U_0}{\pi} \frac{1}{\sqrt{2}} \begin{pmatrix} 0 \\ 0 \end{pmatrix}$$

$$= E_m^{(0)} \left[\frac{1}{\sqrt{2}}\begin{pmatrix} 1 \\ -1 \end{pmatrix}\right] \qquad (11.213)$$

The respective energy eigenvalues, $E_m^{(0)} + 0$ and $E_m^{(0)} + U_0/\pi$, are the same as those in Eq. (11.209) as they should be.

3. A perturbation $\hat{H}_1(x, y) = Cxy$ is applied to an isotropic two-dimensional harmonic oscillator ($\omega_x = \omega_y$).

   (a) Find the eigenfunctions and eigenvalues for the unperturbed ground state and for the doubly degenerate first excited state.
   (b) Find the matrix that represents the complete Hamiltonian of the excited states and find the energies of the perturbed first excited states to first order.

Solution

(a) The unperturbed Hamiltonian is separable in x and y so

$$\hat{H}_0(x, y) = \hat{H}_x(x) + \hat{H}_y(y) \tag{11.214}$$

   The Hamiltonian $\hat{H}_0(x, y)$ may be written in terms of the ladder operators (see Chap. 7).

$$\hat{H}_0(x, y) = \hbar\omega \left(\hat{a}_x^\dagger \hat{a}_x + \frac{1}{2}\right) + \hbar\omega \left(\hat{a}_y^\dagger \hat{a}_y + \frac{1}{2}\right)$$

$$= \hbar\omega \left(\hat{a}_x^\dagger \hat{a}_x + \hat{a}_y^\dagger \hat{a}_y + 1\right) \tag{11.215}$$

where the x and y subscripts on the ladder operators indicate that these operators affect only that coordinate. As in Problem 1 of this chapter the eigenfunctions of $\hat{H}_0(x, y)$ are a product of the eigenfunctions of $\hat{H}_x(x)$ and $\hat{H}_y(y)$. In Dirac notation this means that an arbitrary unperturbed eigenket with quantum numbers n_x and n_y is written

$$|n_x\rangle |n_y\rangle = |n_x n_y\rangle \tag{11.216}$$

and the unperturbed eigenvalues are

$$E_{n_x n_y}^{(0)} = \left(n_x + n_y + 1\right) \hbar\omega \tag{11.217}$$

   The eigenvalues of the ground are

$$\langle 00| \hat{H}_0 |00\rangle = \hbar\omega \tag{11.218}$$

   The first excited states $|01\rangle$ and $|10\rangle$ are clearly degenerate with energy

$$\langle 10| \hat{H}_0 |10\rangle = 2\hbar\omega \tag{11.219}$$

(b) The first-order correction to the ground state vanishes because the perturbation is odd in both x and y.

Analogous to the Hamiltonian in Problem 1 of this chapter the matrix that represents the total Hamiltonian for the degenerate excited states $|01\rangle$ and $|10\rangle$ is

$$\hat{H} = 2\hbar\omega \begin{pmatrix} 1 & 0 \\ 0 & 1 \end{pmatrix} + C \begin{pmatrix} 0 & \langle 10|\, xy\, |01\rangle \\ \langle 01|\, xy\, |10\rangle & 0 \end{pmatrix} \qquad (11.220)$$

where the diagonal elements of $\hat{H}_1\,(x, y)$ vanish because of parity. We must diagonalize the $\hat{H}_1\,(x, y)$ matrix. Rather than use integrals of the algebraic eigenfunctions with their Hermite polynomials to evaluate $\langle 10|\, xy\, |01\rangle$ and $\langle 01|\, xy\, |10\rangle$. We will use the ladder operators. Replace x and y with

$$\hat{x} = \frac{1}{\sqrt{2}\alpha}\left(\hat{a}_x + \hat{a}_x^\dagger\right) \text{ and } \hat{y} = \frac{1}{\sqrt{2}\alpha}\left(\hat{a}_y + \hat{a}_y^\dagger\right) \qquad (11.221)$$

so that

$$\hat{x}\hat{y} = \frac{1}{2\alpha^2}\left(\hat{a}_x\hat{a}_y + \hat{a}_x\hat{a}_y^\dagger + \hat{a}_x^\dagger\hat{a}_y + \hat{a}_x^\dagger\hat{a}_y^\dagger\right) \qquad (11.222)$$

and employ the raising and lowering properties (see Appendix O) to evaluate the four terms in $\langle 10|\, xy\, |01\rangle$.

$$\frac{1}{2\alpha^2}\langle 10|\,\hat{a}_x\hat{a}_y\,|01\rangle = \frac{1}{2\alpha^2}\sqrt{2}\,\langle 20|\, 00\rangle = 0$$

$$\frac{1}{2\alpha^2}\langle 10|\,\hat{a}_x\hat{a}_y^\dagger\,|01\rangle = \frac{1}{\alpha^2}\langle 20|\, 02\rangle = 0$$

$$\frac{1}{2\alpha^2}\langle 10|\,\hat{a}_x^\dagger\hat{a}_y\,|01\rangle = \frac{1}{2\alpha^2}\langle 00|\, 00\rangle = \frac{1}{2\alpha^2}$$

$$\frac{1}{2\alpha^2}\langle 10|\,\hat{a}_x^\dagger\hat{a}_y^\dagger\,|01\rangle = \frac{1}{\sqrt{2}\alpha^2}\langle 00|\, 02\rangle = 0 \qquad (11.223)$$

so that the matrix elements of $\hat{H}_1$ are

$$C\,\langle 10|\, xy\, |01\rangle = C\,\langle 01|\, xy\, |10\rangle$$

$$= \frac{C}{2\alpha^2} = C\frac{\hbar^2}{2m^2\omega^2} \qquad (11.224)$$

We must now diagonalize the $\hat{H}_1$ matrix. Because the unperturbed Hamiltonian is symmetric in x and y and the perturbation is the same as that in Problem 1 we know that the $\hat{H}_1$ matrix will be of the same form.

$$\frac{C}{2\alpha^2}\begin{pmatrix} 0 & 1 \\ 1 & 0 \end{pmatrix} \qquad (11.225)$$

Using the result of Problem 1 we have

$$E_1^{(1)} = \pm \frac{C\hbar}{2m\omega} \tag{11.226}$$

Therefore, the energies of the degenerate levels of $\hat{H}_0 (x, y)$ are shifted by $E_1^{(1)}$, one lower and the other higher. That is,

$$E_{12}^{(0)} + E_{12}^{(1)} = 2\hbar\omega \pm \frac{C\hbar}{2m\omega} \tag{11.227}$$

The select set is

$$\frac{1}{\sqrt{2}} \begin{pmatrix} 1 \\ 1 \end{pmatrix} \quad \text{and} \quad \frac{1}{\sqrt{2}} \begin{pmatrix} 1 \\ -1 \end{pmatrix} \tag{11.228}$$

4. Consider a Hamiltonian that is represented by the matrix

$$\hat{H} = \begin{pmatrix} 1 & \epsilon & 0 \\ \epsilon & 1 & 0 \\ 0 & 0 & 6 \end{pmatrix} \tag{11.229}$$

(a) Find the select basis set and show that these eigenvectors diagonalize $\hat{H}$.
(b) Verify that the off-diagonal matrix elements of $\hat{H}$ vanish with the select set and show that the eigenvalues are $(1 \pm \epsilon)$ and 6.
(c) Solve the problem exactly and compare with the perturbation theory result.

Solution

(a) The entry in the lower right is a non-degenerate eigenvalue because there are no other entries in the same column or row. The rest of the Hamiltonian is a 2×2 matrix which may be written

$$\hat{H} = \hat{H}_0 + \hat{H}_1$$
$$= \begin{pmatrix} 1 & 0 \\ 0 & 1 \end{pmatrix} + \begin{pmatrix} 0 & \epsilon \\ \epsilon & 0 \end{pmatrix} \tag{11.230}$$

Thus, the unperturbed eigenkets for this 2×2 system are

$$\left| 1^{(0)} \right\rangle = \begin{pmatrix} 1 \\ 0 \end{pmatrix} \quad \text{and} \quad \left| 2^{(0)} \right\rangle = \begin{pmatrix} 0 \\ 1 \end{pmatrix} \tag{11.231}$$

But these are not the select set. To find the select set we must diagonalize $\hat{H}_1$ which we do by solving the secular equation

$$\det \begin{pmatrix} -E & \epsilon \\ \epsilon & -E \end{pmatrix} = 0 \implies E = \pm \epsilon \qquad (11.232)$$

To find the select set of eigenvectors we must find the eigenkets corresponding to these eigenvalues.

$$\epsilon \begin{pmatrix} 0 & 1 \\ 1 & 0 \end{pmatrix} \begin{pmatrix} a \\ b \end{pmatrix} = \pm \epsilon \begin{pmatrix} a \\ b \end{pmatrix} \qquad (11.233)$$

where the column vectors represent the as yet unknown select kets. We have

$$b\epsilon = \pm \epsilon a \implies b = \pm a \qquad (11.234)$$

Therefore, the (orthonormal) select set which we designate $\left| i^{(s)} \right\rangle$ is

$$\left| 1^{(s)} \right\rangle = \frac{1}{\sqrt{2}} \left(\left| 1^{(0)} \right\rangle + \left| 2^{(0)} \right\rangle \right) = \frac{1}{\sqrt{2}} \begin{pmatrix} 1 \\ 1 \end{pmatrix}$$

$$\left| 2^{(s)} \right\rangle = \frac{1}{\sqrt{2}} \left(\left| 1^{(0)} \right\rangle - \left| 2^{(0)} \right\rangle \right) = \frac{1}{\sqrt{2}} \begin{pmatrix} 1 \\ -1 \end{pmatrix} \qquad (11.235)$$

The unperturbed state $\left| 3^{(0)} \right\rangle$ with eigenvalue 6 is non-degenerate so no manipulation is required.

(b) Although $\left| 1^{(s)} \right\rangle$ and $\left| 2^{(s)} \right\rangle$ were constructed so that the off-diagonal matrix elements of $\hat{H}$ on this select set vanish, let us verify this.

$$\left\langle 1^{(s)} \middle| \hat{H} \middle| 2^{(s)} \right\rangle = \frac{1}{2} (1\ 1) \left[\begin{pmatrix} 1 & 0 \\ 0 & 1 \end{pmatrix} + \begin{pmatrix} 0 & \epsilon \\ \epsilon & 0 \end{pmatrix} \right] \begin{pmatrix} 1 \\ -1 \end{pmatrix}$$

$$= \frac{1}{2} (1\ 1) \begin{pmatrix} 1 \\ -1 \end{pmatrix} + \frac{1}{2} (1\ 1) \begin{pmatrix} -\epsilon \\ \epsilon \end{pmatrix}$$

$$= \frac{1}{2} (1\ 1) \begin{pmatrix} 1 - \epsilon \\ -[1 - \epsilon] \end{pmatrix}$$

$$= \frac{1}{2} [(1 - \epsilon) - (1 - \epsilon)] = 0 \qquad (11.236)$$

Clearly $\left\langle 2^{(s)} \middle| \hat{H} \middle| 1^{(s)} \right\rangle$ also vanishes. Now for the diagonal elements.

$$\left\langle 1^{(s)} \middle| \hat{H} \middle| 1^{(s)} \right\rangle = \frac{1}{2} (1\ 1) \left[\begin{pmatrix} 1 & 0 \\ 0 & 1 \end{pmatrix} + \begin{pmatrix} 0 & \epsilon \\ \epsilon & 0 \end{pmatrix} \right] \begin{pmatrix} 1 \\ 1 \end{pmatrix}$$

$$= \frac{1}{2} (1\ 1) \begin{pmatrix} 1 \\ 1 \end{pmatrix} + \frac{1}{2} (1\ 1) \begin{pmatrix} \epsilon \\ \epsilon \end{pmatrix}$$

$$= \frac{1}{2} (1\ 1) \begin{pmatrix} 1 + \epsilon \\ 1 + \epsilon \end{pmatrix}$$

$$= \frac{1}{2} (2 + 2\epsilon) = 1 + \epsilon \qquad (11.237)$$

and

$$\langle 2^{(s)} | \hat{H} | 2^{(s)} \rangle = \frac{1}{2} (1\ -1) \left[\begin{pmatrix} 1\ 0 \\ 0\ 1 \end{pmatrix} + \begin{pmatrix} 0\ \epsilon \\ \epsilon\ 0 \end{pmatrix} \right] \begin{pmatrix} 1 \\ -1 \end{pmatrix}$$

$$= \frac{1}{2} (1\ -1) \left[\begin{pmatrix} 1 \\ -1 \end{pmatrix} + (1\ -1) \begin{pmatrix} -\epsilon \\ \epsilon \end{pmatrix} \right]$$

$$= \frac{1}{2} (1\ -1) \begin{pmatrix} 1 - \epsilon \\ -[1 - \epsilon] \end{pmatrix}$$

$$= \frac{1}{2} (2 - 2\epsilon) = 1 - \epsilon \qquad (11.238)$$

(c) Ignoring the already diagonalized portion of $\hat{H}$, the part of the matrix representing $\hat{H}$ is

$$\begin{pmatrix} 1 & \epsilon \\ \epsilon & 1 \end{pmatrix} \qquad (11.239)$$

The eigenvalue equation is

$$\begin{pmatrix} 1 & \epsilon \\ \epsilon & 1 \end{pmatrix} \begin{pmatrix} A \\ B \end{pmatrix} = E \begin{pmatrix} A \\ B \end{pmatrix} \qquad (11.240)$$

which becomes

$$\begin{pmatrix} 1 - E & \epsilon \\ \epsilon & 1 - E \end{pmatrix} \begin{pmatrix} A \\ B \end{pmatrix} = 0 \qquad (11.241)$$

Expanding this equation into two simultaneous equations we notice that these will be homogeneous equations so the only non-trivial solution can occur only if the determinant of the coefficients vanishes. We have

$$(1 - E)^2 - \epsilon^2 = 0$$

or

$$E^2 - 2E + \left(1 - \epsilon^2\right) = 0$$

Using the quadratic formula we have

$$E = \left(2 \pm \sqrt{4 - 4\left(1 - \epsilon^2\right)}\right) / 2$$
$$= 1 \pm \epsilon \tag{11.242}$$

Although this problem involved a 3×3 matrix the fact that one of the states in this problem was non-degenerate reduced it to a 2×2 problem.

11.6 Time Dependent Perturbation Theory

In time dependent perturbation theory the Hamiltonian $\hat{H}(r, t)$ is written as the sum of two terms, the unperturbed Hamiltonian $\hat{H}_0$, which is time independent and the time dependent perturbation $\hat{W}(r, t)$.

$$\hat{H}(r, t) = \hat{H}_0(r) + \hat{W}(r, t) \tag{11.243}$$

It is assumed that $\hat{W}(r, t)$ is turned on at $t = t_0$ and that the eigenkets and energy eigenvalues of $\hat{H}_0$ are known. Additionally, each eigenket $|\psi_n\rangle$ has an unperturbed energy eigenvalue E_n.

Time dependent perturbation theory is used to determine the probability of finding a system that was initially in state i to be in some final state f after having turned on the perturbation $\hat{W}(r, t)$ at $t = t_0$. The details of this subject are covered in most quantum mechanics textbooks [1]. The fundamental result is that the probability of the transition $i \to f$ designated by $P_{i \to f}^{(1)}$ is given by

$$P_{i \to f}^{(1)}(t) = \frac{1}{\hbar^2} \left| \int_{t_0}^{t} \hat{W}_{fi}(r, t') e^{i\omega_{fi}t'} dt' \right|^2 \tag{11.244}$$

where

$$\omega_{fi} = \frac{E_f - E_i}{\hbar} \tag{11.245}$$

and

$$\hat{W}_{fi} = \langle \psi_f | \hat{W}(r, t) | \psi_i \rangle \tag{11.246}$$

It is important to note that Eq. (11.244) is valid only when the sum of the probabilities $P_{i \to f}^{(1)}$ ($f \neq i$) is much less than unity.

If the limits of integration in Eq. (11.244) are $-\infty$ to ∞ the integral is simply the Fourier transform of the perturbation evaluated at the transition frequency.

Problems

1. A particle of mass m is in the ground state of an L-box (see Appendix M). At $t = 0$ a perturbation $\hat{W}(x, t)$ is applied;

$$\hat{W}(x, t) = Axe^{-(t/\tau)^2} \tag{11.247}$$

where A and τ are constants. What are the units of A? Use first-order time-dependent perturbation theory to find the probability that the system will undergo a transition to the first excited state after a long time, $t = +\infty$. Do the calculated probabilities have the correct units?

Solution

The perturbation must have units of energy so the units of A are J/m (SI).
     Use the fundamental result of time dependent perturbation theory, Eq. (11.244), to obtain

$$P_{1\to2}^{(1)} = \frac{1}{\hbar^2} \left| \int_0^\infty \hat{W}_{12}(x, t')\, e^{i\omega_{21}t'}\, dt' \right|^2 \tag{11.248}$$

where $\hbar\omega_{21}$ is the difference between the L-box ground state and first excited state energies (see Appendix M). That is

$$\hbar\omega_{21} = \frac{2^2\pi^2\hbar^2}{2mL^2} - \frac{1^2\pi^2\hbar^2}{2mL^2}$$

$$= \frac{3\pi^2\hbar^2}{2mL^2} \tag{11.249}$$

The matrix element $\hat{W}_{12}(x, t')$ is

$$\hat{W}_{12}(x, t') = Ae^{-(t/\tau)^2} \langle 1| x |2\rangle \tag{11.250}$$

where the bra and ket represent the ground and first excited states, respectively. The matrix element $\langle 1| x |2\rangle$ is, using the eigenfunctions given in Appendix M

$$A \langle 1 | x | 2 \rangle = \frac{2A}{L} \int_0^L \sin\left(\frac{\pi x}{L}\right) \cdot x \cdot \sin\left(\frac{2\pi x}{L}\right) dx$$

$$= \frac{4A}{L} \int_0^L x \sin^2\left(\frac{\pi x}{L}\right) \cos\left(\frac{\pi x}{L}\right) dx$$

$$= \frac{16AL}{9\pi^2} \tag{11.251}$$

This result is obtained by integrating Eq. (11.251) by parts and using the integral given in Eq. (G.13).

We have now

$$P_{1\to 2}^{(1)} = \left(\frac{16AL}{9\pi^2\hbar}\right)^2 \left| \int_0^\infty e^{\left(i\omega_{21}t' - t'^2/\tau^2\right)} dt' \right|^2 \tag{11.252}$$

We can solve this integral by completing the square in the exponent. Dropping the primes and subscripts for convenience

$$q = \left(\frac{t}{\tau^2} - \frac{i}{2}\omega\right) \Rightarrow dq = \frac{dt}{\tau^2}$$

$$q^2 = \frac{t^2}{\tau^4} - \frac{i\omega t}{\tau^2} - \frac{\omega^2}{4} \tag{11.253}$$

so that

$$i\omega t - \frac{t^2}{\tau^2} = -\tau^2\left(\frac{\omega^2}{4} + q^2\right) \tag{11.254}$$

Inserting this into Eq. (11.252) we have

$$P_{1\to 2}^{(1)} = \left(\frac{16AL}{9\pi^2\hbar}\right)^2 \left| \int_0^\infty e^{-\tau^2\left(\omega^2/4 + q^2\right)} \left(\tau^2 dq\right) \right|^2$$

$$= \left(\frac{16AL}{9\pi^2\hbar}\right)^2 \tau^4 e^{-\omega^2\tau^2/4} \left| \int_0^\infty e^{-\tau^2 q^2} dq \right|^2 \tag{11.255}$$

Before evaluating this integral, which is finally in a comfortably recognizable form, it is worthwhile to check the units of the constants in this problem. Because exponents must be unitless it is clear from the equations above that τ and q have units s and s^{-1}, respectively.

Now, we simply use the integral given in Eq. (G.3) and find that

$$P_{1\to 2}^{(1)} = \left(\frac{16AL\tau^2}{9\pi^2\hbar}\right)^2 e^{-\omega_{21}^2\tau^2/4} \left| \frac{1}{2}\sqrt{\frac{\pi}{\tau^2}} \right|^2$$

$$= \pi \left(\frac{8AL\tau}{9\pi^2 \hbar} \right)^2 e^{-\omega_{21}^2 \tau^2/4} \tag{11.256}$$

This probability must be unitless. To check we use our deduction above that A has units J/m. We have already shown that the exponential is unitless, so

$$P = \frac{[J^2]}{[m]^2} \cdot \frac{[m]^2 \cdot [s]^2}{[J \cdot s]^2}$$
$$\rightarrow \text{unitless} \tag{11.257}$$

2. An electron is in the $n = 1$ state of an L-box (see Appendix M) for which the potential energy is

$$U(x) = 0 \qquad 0 \leq x \leq L$$
$$= \infty \qquad -\infty < x < 0; L < x < \infty \tag{11.258}$$

At $t = 0$ a uniform electric field of magnitude F is applied in the x-direction and turned off after a short time τ. Obtain an expression for the probability of finding the electron in eigenstate n of the L-box after the field is turned off. Evaluate the probability expression for $n = 2$ and $n = 3$.

Solution

Using the fundamental result of time dependent perturbation theory, Eq. (11.248), we have, in the notation of this problem

$$P_{1 \to f}^{(1)} (\tau) = \frac{1}{\hbar^2} \left| \int_0^\tau \hat{W}_{f1} (t') e^{i\omega_{f1} t'} dt' \right|^2 \tag{11.259}$$

where the final state f will be either the first or second excited state of the L-box, $n_f = 2$ or 3. The difference in energies is

$$\hbar\omega_{f1} = \frac{n_f^2 \pi^2 \hbar^2}{2mL^2} - \frac{1^2 \pi^2 \hbar^2}{2mL^2}$$
$$= (n_f^2 - 1) \frac{\pi^2 \hbar^2}{2mL^2} \tag{11.260}$$

The perturbing Hamiltonian $\hat{W}(r, t)$ is simply the potential energy of a particle of charge e in an electric field during the time interval $0 < t < \tau$. Because the field F is uniform and in the x-direction the spatial part of the perturbing Hamiltonian is

$$U(x) = eFx \tag{11.261}$$

where e is the charge (negative) on the electron. The matrix element $\hat{W}_{f1}(\mathbf{r}, t')$ is then given by

$$\hat{W}_{f1}(\mathbf{r}, t') = eF \langle 1 | x | n \rangle \quad 0 < t < \tau \tag{11.262}$$

where the bra and ket represent L-box eigenfunctions and we have let $n = n_f$. Note that the time dependence of the perturbation is given by the time interval over which the spatial part of the electric field acts.

Using the L-box eigenfunctions (see Appendix M), the trigonometric identity given in Eq. (E.5) and the integral given in Eq. (G.8) we have

$$
\begin{aligned}
\langle 1 | x | n \rangle &= \frac{2}{L} \int_0^L \sin\left(\frac{\pi x}{L}\right) \cdot x \cdot \sin\left(\frac{n \pi x}{L}\right) dx \tag{11.263} \\
&= \frac{1}{L} \left\{ \int_0^L \cos\left[\left((1-n)\frac{\pi x}{L}\right)\right] x dx - \int_0^L \cos\left[\left((1+n)\frac{\pi x}{L}\right)\right] x dx \right\} \\
&= \frac{L}{\pi^2} \left\{ \left(\frac{1}{(1-n)^2}\right) \cos\left[(1-n)\frac{\pi x}{L}\right] - \left(\frac{1}{(1+n)^2}\right) \cos\left[(1+n)\frac{\pi x}{L}\right] \right\}_0^L \\
&= \frac{L}{\pi^2} \left\{ \left(\frac{1}{(1-n)^2}\right) \cos\left[(1-n)\pi\right] - \left(\frac{1}{(1+n)^2}\right) \cos\left[(1+n)\pi\right] \right\} \\
&= \left(\frac{L}{\pi^2(1-n)^2}\right) \left[(-)^{1-n} - 1\right] - \left(\frac{L}{\pi^2(1+n)^2}\right) \left[(-)^{1+n} - 1\right] \tag{11.264}
\end{aligned}
$$

where

$$
\cos\left[(1-n)\pi\right] = \left[(-)^{1-n} - 1\right]
$$

$$
\cos\left[(1+n)\pi\right] = \left[(-)^{1+n} - 1\right] \tag{11.265}
$$

and

$$
\left[(-)^{1-n} - 1\right] = \left[(-)^{1+n} - 1\right] \tag{11.266}
$$

so we have

$$
\begin{aligned}
\langle 1 | x | n \rangle &= \left[\left(\frac{L}{\pi^2(1-n)^2}\right) - \left(\frac{L}{\pi^2(1+n)^2}\right)\right] \left[(-)^{1+n} - 1\right] \\
&= \left(\frac{L}{\pi^2}\right) \frac{4n}{(1-n^2)^2} \left[(-)^{1+n} - 1\right] \tag{11.267}
\end{aligned}
$$

We see immediately that excitation to the $n = 3$ state is forbidden; it cannot occur because the matrix element vanishes for $n = 3$ and for all final states with n odd. The probability of a transition to $n = 2$ is given by Eq. (11.259). Using Eq. (11.260),

$$\omega_{21} = \frac{3\pi^2\hbar}{2mL^2} \tag{11.268}$$

and the time independent matrix element $\hat{W}_{21}(r, t') = \hat{W}_{21}$ is

$$\hat{W}_{21} = eF\langle 1|x|n\rangle$$

$$= (eF)\left(\frac{L}{\pi^2}\right)\frac{8}{9}(-2)$$

$$= \frac{16}{9}\frac{eLF}{\pi^2} \tag{11.269}$$

so we obtain

$$P^{(1)}_{1\to 2}(\tau) = \frac{256}{81}\frac{(eLF)^2}{\hbar^2\pi^4}\left|\int_0^\tau e^{i\omega_{21}t'}\,dt'\right|^2$$

$$= \frac{256}{81}(eLF)^2\left(\frac{1}{\hbar^2\pi^4}\right)\left[\frac{1}{\omega_{21}^2}\left|1 - e^{i\omega_{21}\tau}\right|^2\right]$$

$$= \frac{256}{81}(eLF)^2\left(\frac{1}{\hbar^2\pi^4}\right)\left(\frac{1}{\omega_{21}^2}\right)[2 - 2\cos(\omega_{21}\tau)]$$

$$= \frac{4^4}{3^4}(eLF)^2\left(\frac{1}{\hbar^2\pi^4}\right)\left(\frac{1}{\omega_{21}^2}\right)\left[4\sin^2(\omega_{21}\tau/2)\right] \tag{11.270}$$

where we have used Eq. (E.9).

For short time intervals $\omega_{21}\tau \ll 1$ we may use $\sin(\omega_{21}\tau/2) \approx (\omega_{21}\tau/2)$ so that

$$P^{(1)}_{1\to 2}(\tau) \approx \frac{4^5}{3^4}(eLF)^2\left(\frac{1}{\hbar^2\pi^4}\right)\left(\frac{1}{\omega_{21}^2}\right)\left(\frac{\omega_{21}\tau}{2}\right)^2$$

$$= \frac{4^4}{3^4}\left[\frac{(eLF)^2}{\hbar^2\pi^4}\right]\tau^2 \tag{11.271}$$

It is interesting that under the condition that $\sin(\omega_{21}\tau/2) \approx (\omega_{21}\tau/2)$ the probability of a transition is proportional to τ^2 and is independent of the energy separation between states, $\hbar\omega_{fi}$.

3. A particle in a one-dimensional harmonic oscillator potential is in the ground state. At $t = 0$ a perturbation is applied

$$\hat{W}(x, t) = Axe^{-t/\tau} \tag{11.272}$$

where A and τ are constants. What are the units of A? Use first-order time-dependent perturbation theory to find the probability that the system will undergo a transition to *any* excited state after a long time, $t = +\infty$. Do the calculated probabilities have the correct units? It will be helpful to use the matrix element for the harmonic oscillator $\langle m| \hat{x} |n \rangle$ that was calculated in Problem 6 of Chap. 7, Eq. (7.73). This matrix element is

$$\langle m| \hat{x} |n \rangle = \frac{1}{\sqrt{2}\alpha} \left(\sqrt{n}\delta_{m,n-1} + \sqrt{n+1}\delta_{m,n+1} \right) \tag{11.273}$$

Solution

The perturbation must have units of energy so, as in Problem 1 of this chapter, the units of A are J/m.

Again as in Problem 1 of this chapter we apply the fundamental result of time-dependent perturbation theory, Eq. (11.244).

$$P_{0\to f}^{(1)} = \frac{1}{\hbar^2} \left| \int_0^\infty \hat{W}_{f0}(x, t') e^{i\omega_{f0}t'} dt' \right|^2 \tag{11.274}$$

where ω_{f0} is the difference between the ground state and final state energies. In the present case the perturbation is given by Eq. (11.272) so the matrix element is

$$\hat{W}_{f0}(x, t') = Ae^{-t'/\tau} \langle 0| x |f \rangle \tag{11.275}$$

where the ket $|f\rangle$ represents any of the final states.

The matrix element we seek is $\langle 0| \hat{x} |f \rangle$ which from Eq. (11.273) is

$$\langle 0| \hat{x} |f \rangle = \frac{1}{\sqrt{2}\alpha} \left(\sqrt{f}\delta_{0,f-1} + \sqrt{f+1}\delta_{0,f+1} \right) \tag{11.276}$$

The only way the first term will not vanish is if the final state is the first excited state $f = 1$. The second term will vanish for *any* final state inasmuch as there is no state for which $f = -1$. Therefore, the only transition possible under this perturbation is the $0 \to 1$ transition so

$$P_{0\to f}^{(1)} \equiv 0 \quad f > 1 \tag{11.277}$$

For $f = 1$ we have

$$\langle 0| \hat{x} |1 \rangle = \frac{1}{\sqrt{2}\alpha} \tag{11.278}$$

To obtain the remaining part of the transition probability we require the integral

$$\int_0^\infty \hat{W}_{10}\left(x, t'\right) e^{i\omega_{f0} t'} dt' = A \langle 0| x |1 \rangle \int_0^\infty e^{-t'/\tau} e^{i\omega_{10} t'} dt'$$

$$= \frac{A}{\sqrt{2}\alpha} \int_0^\infty e^{(i\omega_{10} - 1/\tau) t'} dt'$$

$$= \frac{A}{\sqrt{2}\alpha} \cdot \frac{1}{(i\omega_{10} - 1/\tau)} \tag{11.279}$$

where $\omega_{10} = \omega$ the oscillator frequency. The transition probability is then

$$P_{0\to1}^{(1)} = \frac{|A|^2}{\hbar^2} \frac{1}{2\alpha^2} \frac{1}{(\omega^2 + 1/\tau^2)} \tag{11.280}$$

Using $\alpha^2 = m\omega/\hbar$ we have

$$P_{0\to1}^{(1)} = \frac{1}{2} \frac{|A|^2}{\hbar^2} \cdot \frac{\hbar}{m\omega} \frac{1}{(\omega^2 + 1/\tau^2)} \tag{11.281}$$

This probability must be unitless. To check we use SI units and our deduction above that A has units J/m.

$$P = \frac{1}{2} \cdot \frac{[J^2]}{[m]^2} \cdot \frac{1}{[J \cdot s]^2} \cdot \frac{[J \cdot s]}{[kg] \cdot \left[\frac{1}{s}\right]} \cdot \frac{1}{\left[\frac{1}{s^2}\right]}$$

$$= \frac{[J]}{[(kg \cdot m^2/s^2)]} = \frac{[J]}{[J]} \quad \text{(unitless)} \tag{11.282}$$

We can calculate the average oscillator energy change $\Delta E_{0\to1}$ corresponding to this probability $0 \to 1$. This is simply the probability multiplied by $\hbar\omega$.

$$\Delta E_{0\to1} = \hbar\omega \cdot P_{0\to1}^{(1)}$$

$$= \frac{|A|^2}{2m} \frac{1}{(\omega^2 + 1/\tau^2)} \tag{11.283}$$

In Problem 5 of Sect. 11.3 we calculated the energy shift of a harmonic oscillator due to the application of a *constant* electric field F. The perturbation was given by

$$\hat{H}_1 = eFx \tag{11.284}$$

In this problem we have an electric field that is decaying with time constant τ. If, however, $\tau \to \infty$ the "decaying" field would be constant in time so we should arrive at the same result for the average energy change of the oscillator as that obtained in Problem 5 of Sect. 11.3 which was

$$\Delta E = \frac{e^2 F^2}{2m\omega^2} \tag{11.285}$$

Comparing the perturbation in this problem, Eq. (11.272) with that of the previous problem, Eq. (11.284), we see that A in this problem corresponds to eF in the previous problem. Taking the limit as $\tau \to \infty$ of $\Delta E_{0 \to 1}$ Eq. (11.283) we have

$$\lim_{\tau \to \infty} \Delta E_{0 \to 1} = \lim_{\tau \to \infty} \frac{|A|^2}{2m} \frac{1}{(\omega^2 + 1/\tau^2)}$$

$$= \frac{|A|^2}{2m\omega^2} \tag{11.286}$$

so that replacing $|A|^2$ with $e^2 F^2$ does indeed produce the same energy change as that obtained for a constant field.

4. A particle of mass m is in the ground state of a one-dimensional harmonic oscillator potential. The oscillator frequency is ω. At $t = 0$ a weak constant force $\mathcal{F}$ is applied and acts until time $t = \tau$ so the perturbing potential is

$$\hat{W}(x, t) = -\mathcal{F}x \quad 0 < t < \tau \tag{11.287}$$

Use first-order time dependent perturbation theory to find the value (or values) of τ, call them τ_{max}, that maximize the probability of a transition to $n = 1$.

Solution

Using Eq. (11.244), the probability of the $n = 0 \to 1$ transition is

$$P^{(1)}_{0 \to 1}(\tau) = \frac{\mathcal{F}^2}{\hbar^2} \left| \langle 0| \hat{x} |1 \rangle \int_0^\tau e^{i\omega t'} dt' \right|^2$$

$$= \frac{\mathcal{F}^2}{\hbar^2} \left| \langle 0| \hat{x} |1\rangle \left[\frac{1}{i\omega} e^{i\omega t'} \right]_{t'=0}^{t'=\tau} \right|^2$$

$$= \frac{\mathcal{F}^2}{\hbar^2 \omega^2} |\langle 0| \hat{x} |1\rangle|^2 \left| \left(e^{i\omega \tau} - 1 \right) \right|^2 \tag{11.288}$$

The matrix element $\langle 0| x |1\rangle$, using Eq. (11.273), is

$$\langle 0| \hat{x} |f\rangle = \frac{1}{\sqrt{2}\alpha} \left(\sqrt{f} \delta_{0,f-1} + \sqrt{f+1} \delta_{0,f+1} \right) \tag{11.289}$$

so

$$\langle 0| \hat{x} |1\rangle = \frac{1}{\sqrt{2}\alpha} \left(\delta_{0,0} + \sqrt{2} \delta_{0,2} \right)$$

$$= \frac{1}{\sqrt{2}\alpha} = \sqrt{\frac{\hbar}{2m\omega}} \tag{11.290}$$

The probability that the harmonic oscillator will be found in the first excited state after a time τ is

$$P_{0 \to 1}^{(1)} = \frac{\mathcal{F}^2}{\hbar^2 \omega^2} \frac{\hbar}{2m\omega} \left| \left(e^{i\omega \tau} - 1 \right) \right|^2$$

$$= \frac{\mathcal{F}^2}{2m\hbar\omega^3} \left| e^{i\omega\tau/2} \left(e^{i\omega\tau/2} - e^{-i\omega\tau/2} \right) \right|^2$$

$$= \frac{2\mathcal{F}^2}{m\hbar\omega^3} \left| \frac{\left(e^{i\omega\tau/2} - e^{-i\omega\tau/2} \right)}{2i} \right|^2$$

$$= \frac{2\mathcal{F}^2}{m\hbar\omega^3} \sin^2 \left(\frac{\omega\tau}{2} \right) \tag{11.291}$$

The $\sin^2 (\omega\tau/2)$ term is the only one that depends upon τ. It has a maximum of 1 when $\omega\tau = \pi, 3\pi, \ldots$. In general the values of $\tau_{\max}$ are

$$\tau_{\max} = (2j + 1) \frac{\pi}{\omega} \quad \text{where } j = 0, 1, \ldots \tag{11.292}$$

While Eq. (11.292) is the answer we sought, it is worthwhile to make sure the units are correct. The $\tau_{\max}$ result has units of time as it should. How about the coefficient of the $\sin^2 (\omega\tau/2)$ term in Eq. (11.291)? It must be unitless because probabilities do not have units. We have

$$\frac{2\mathcal{F}^2}{m\hbar\omega^3} \rightarrow 2\frac{N^2}{(\,kg\,)\,(\,J\cdot s\,)\,(\,s^{-3}\,)}$$

$$\rightarrow 2\frac{N^2}{(\,kg\,)\,(\,N\cdot m\,)\cdot s^{-2}}$$

$$\rightarrow 2\frac{N}{(\,kg\cdot m\cdot s^{-2}\,)} \tag{11.293}$$

Inasmuch as $N = kg\cdot m\cdot s^{-2}$ the coefficient is unitless giving us some confidence that the answer is correct.

Because we have used first order perturbation theory we must assure ourselves that $\mathcal{F}$ is weak enough so that the probabilities are much less than unity at τ_{max}. A reasonable limit is $P^{(1)}_{0\rightarrow 1} \lesssim 0.2$ so that, from Eq. (11.291), $\mathcal{F}^2 \lesssim 0.1 m\hbar\omega^3$.

5. A H-atom in the ground state is immersed in an electric field F that is constant in the z-direction, but varies in time as

$$F = 0 \quad \text{for } t < 0$$

$$F = F_0 e^{-t/\tau} \quad \text{for } t > 0 \tag{11.294}$$

Use time dependent perturbation theory to determine the probability that after a long time, i.e. $t = \infty$, the atom is in the $n = 2, \ell = 1, m = 0$ state.

Solution

Again we require the fundamental result of time dependent perturbation theory, Eq. (11.244).

$$P^{(1)}_{i\rightarrow f} = \frac{1}{\hbar^2}\left|\int_0^\infty \hat{W}_{fi}\left(z, t'\right) e^{i\omega_{ki}t'} dt'\right|^2 \tag{11.295}$$

where ω is the difference in the Bohr energies of the $n = 2$ and $n = 1$ levels of atomic hydrogen.

$$\omega = \frac{(E_2 - E_1)}{\hbar} \tag{11.296}$$

The time dependent perturbation term in the Hamiltonian is the potential energy associated with the electric field

$$\hat{W}\left(z, t\right) = -e\left(-F_0 z e^{-t/\tau}\right)$$

$$= eF_0 r\cos\theta e^{-t/\tau} \tag{11.297}$$

In the present case the matrix element $\hat{W}_{fi}$ is

$$\hat{W}_{fi} = \langle 210| \hat{W}(z,t) |100\rangle \tag{11.298}$$

To perform the integral in Eq. (11.298) we require the eigenfunctions given in Appendix T. The bra and ket in Eq. (11.298) correspond to the H-atom eigenfunctions

$$\psi_{210}(r,\theta,\phi) = \frac{1}{2\sqrt{2\pi}} \left(\frac{1}{a_0}\right)^{3/2} \left(\frac{r}{2a_0}\right) e^{-r/2a_0} \cos\theta$$

$$\psi_{100}(r,\theta,\phi) = \frac{1}{\sqrt{\pi}} \left(\frac{1}{a_0}\right)^{3/2} e^{-r/a_0} \tag{11.299}$$

Noting that these functions contain no ϕ-dependence we have

$$\hat{W}_{fi} = \langle 210| \hat{W}(z,t) |100\rangle$$

$$= \frac{eF_0 e^{-t/\tau}}{2\sqrt{2}} \left(\frac{1}{a_0}\right)^4 \int_0^\infty r^4 e^{-3r/2a_0} dr \int_0^\pi \cos^2\theta \sin\theta d\theta$$

$$= \left[\frac{eF_0 e^{-t/\tau}}{2\sqrt{2}} \left(\frac{1}{a_0}\right)^4\right] \left[\frac{\cos^3\theta}{3}\right]_\pi^0 \int_0^\infty r^4 e^{-3r/2a_0} dr$$

$$= \left[\frac{eF_0 e^{-t/\tau}}{3\sqrt{2}} \left(\frac{1}{a_0}\right)^4\right] \int_0^\infty r^4 e^{-3r/2a_0} dr \tag{11.300}$$

Let

$$u = \frac{3r}{2a_0} \tag{11.301}$$

so that

$$r = \frac{2a_0}{3} u \quad \text{and} \quad dr = \frac{2a_0}{3} du \tag{11.302}$$

The last line of Eq. (11.300) becomes

$$\hat{W}_{fi} = \left[\frac{eF_0 e^{-t/\tau}}{3\sqrt{2}} \left(\frac{1}{a_0}\right)^4\right] \left(\frac{2a_0}{3}\right)^5 \int_0^\infty u^4 e^{-u} du \tag{11.303}$$

Using Eq. (G.2) we have

$$
\hat{W}_{fi} = \left[\frac{eF_0 e^{-t/\tau}}{3\sqrt{2}} \left(\frac{1}{a_0}\right)^4 \right] \left(\frac{2a_0}{3}\right)^5 \cdot 4!
$$

$$
= \frac{2^8 a_0}{3^5 \sqrt{2}} \left(eF_0 e^{-t/\tau} \right)
$$

$$
= C e^{-t/\tau} \tag{11.304}
$$

where C is

$$
C = \frac{2^8 a_0}{3^5 \sqrt{2}} (eF_0) \tag{11.305}
$$

Inserting this into Eq. (11.295) we have

$$
P^{(1)}_{i \to f} = \left(\frac{C}{\hbar}\right)^2 \left| \int_0^\infty e^{-t'/\tau} e^{i\omega_{ki} t'} dt' \right|^2
$$

$$
= \left(\frac{C}{\hbar}\right)^2 \left| \int_0^\infty e^{-t'/\tau + i\omega_{ki} t'} dt' \right|^2
$$

$$
= \left(\frac{C}{\hbar}\right)^2 \lim_{t \to \infty} \left| \frac{(e^{-t/\tau + i\omega_{ki} t} - 1)}{i\omega - 1/\tau} \right|^2
$$

$$
= \left(\frac{C}{\hbar}\right)^2 \left| \frac{1}{i\omega - 1/\tau} \right|^2 = \left(\frac{C}{\hbar}\right)^2 \left(\frac{\tau^2}{\omega^2 \tau^2 + 1} \right)
$$

$$
= \frac{2^{15} (eF_0 a_0)^2}{3^{10}} \cdot \left[\frac{1}{(E_2 - E_1)^2 + (\hbar/\tau)^2} \right] \tag{11.306}
$$

Suppose the final state had been chosen to be $|200\rangle$ rather than $|210\rangle$, in which case $\hat{W}_{fi}$ would be

$$
\hat{W}_{fi} = \langle 200| \hat{W}(z, t) |100\rangle \tag{11.307}
$$

This integral vanishes because $\hat{W}(z, t) \propto z$ so the integrand is necessarily odd and is taken over all space. Recall that the spherical harmonics have definite parity determined by the value of ℓ (see Appendix R). The fact that an $\ell = 0 \to \ell = 0$ transition is forbidden (to first order) is an example of the *selection rules* that govern electric dipole (first order) electromagnetic transitions because the electric dipole operator is proportional to the vector r [1, 1, 2].

6. Consider a system prepared in initial state $|i\rangle$ and perturbed by a periodic potential

$$U(x, t) = Axe^{-i\omega t} \qquad (11.308)$$

where A is a constant. The potential is switched on at time $t = 0$. What is the probability that, at some later time t, the system is in state $|f\rangle$? Assume that first-order perturbation theory is valid.

Solution

Use Eqs. (11.244)–(11.246), and write for the probability of an $i \to f$ transition

$$P^{(1)}_{i \to f}(t) = \left(\frac{A}{\hbar}\right)^2 \langle f|x|i\rangle^2 \left| \int_0^t dt'\, \exp\left[i\left(\omega_{\mathrm{fi}} - \omega\right)t'\right]\right|^2 \qquad (11.309)$$

Integration of the exponential and its complex conjugate are straightforward, and $P^{(1)}_{i \to f}(t)$ becomes

$$P^{(1)}_{i \to f}(t) = \left(\frac{A}{\hbar}\right)^2 \langle f|x|i\rangle^2 \left\{\frac{\sin\left[\left(\omega_{\mathrm{fi}} - \omega\right)t/2\right]}{\left(\omega_{\mathrm{fi}} - \omega\right)/2}\right\}^2 \qquad (11.310)$$

When $(\omega_{\mathrm{fi}} - \omega)t/2 << \pi/2$, the transition probability is

$$P^{(1)}_{i \to f}(t) \approx \left(\frac{A}{\hbar}\right)^2 \langle f|x|i\rangle^2 t^2 \qquad (11.311)$$

and is independent of the both the transition frequency ω_{fi} and the applied frequency ω. The transition probability reaches its first maximum when $(\omega_{\mathrm{fi}} - \omega)t/2 = \pi/2$ so that

$$P^{(1)}_{i \to f}(t) \approx \left(\frac{A}{\hbar}\right)^2 \langle f|x|i\rangle^2 \left[\frac{1}{\left(\omega_{\mathrm{fi}} - \omega\right)/2}\right]^2 \qquad (11.312)$$

The subsequent time variation is such that $P^{(1)}_{i \to f}(t)$ oscillates between 0 and $P^{(1)}_{i \to f}(\max)$. The value of $P^{(1)}_{i \to f}(\max)$ must be much less than unity for perturbation theory to be valid; this requirement sets a limit on the magnitude of the potential parameter A.

The "system" in this problem could be an atom or a molecule irradiated by an electric field oscillating at visible or infrared frequencies. Rather than the transition probability at a particular time t, spectroscopists are more interested in the *rate of transition* $R_{i \to f}(t)$ defined by

$$R_{i \to f}(t) \equiv \lim_{t \to \infty} P^{(1)}_{i \to f}(t)/t \qquad (11.313)$$

If we set $a = (\omega_{\text{fi}} - \omega)/2$, the bracketed term in Eq. (11.310) becomes $[\sin(at)/a]^2$, and we have for large t (Appendix J and references therein).

$$\lim_{t\to\infty} [\sin(at)/a]^2/t = \pi\delta(a) = 2\pi\delta(2a) \tag{11.314}$$

Using Eqs. (11.310) and (11.314) in Eq. (11.313), we obtain

$$R_{i\to f}(t) = 2\pi \left(\frac{A}{\hbar}\right)^2 \langle f|x|i\rangle^2 \,\delta(\omega_{\text{fi}} - \omega) \tag{11.315}$$

Equation (11.315) is called "Fermi's Golden Rule" and it emphasizes the importance of resonance in radiative transitions.

A final remark: The units of the delta function are always the inverse of the argument units. In this case, the argument has units of s^{-1}, so the delta function has units of seconds. Using this fact, one sees that $R_{i\to f}(t)$ has units of s^{-1} (probability per unit time) as it should.

References

1. Burkhardt CE, Leventhal JJ (2008) Foundations of quantum physics. Springer, New York
2. Townsend JS (2000) A modern approach to quantum mechanics. University Science Books, Sausalite
3. Wangsness RK (1986) Electromagnetic fields, 2nd edn. Wiley, New York

Appendix A
Greek Alphabet

Table A.1 The letters of the Greek alphabet

LC	UC	Name	Usage in this book
α	A	Alpha	Fine structure constant (LC), harmonic oscillator (LC), spin up (LC)
β	B	Beta	General parameter, spin down (LC)
γ	Γ	Gamma	Γ function (UC), square root transmission of coefficient T (UC)
δ	Δ	Delta	Dirac δ-function (LC), small increment
ϵ, ε	E	Epsilon	Unitless energy parameter (LC), small quantity (LC)
ζ	Z	Zeta	General parameter
η	H	Eta	General parameter
θ, ϑ	Θ	Theta	Polar angle (LC), function (UC)
ι	I	Iota	–
κ	K	Kappa	Real exponent (LC), hyperfine energy (LC)
λ	Λ	Lambda	Wavelength (LC)
μ	M	Mu	General parameter, reduced mass (LC)
ν	N	Nu	Frequency (radians/ s) (LC)
ξ	Ξ	Xi	Unitless length harmonic oscillator (LC)
o	O	Omicron	–
π	Π	Pi	3.14159…(LC)
ρ	P	Rho	Parameter, unitless length hydrogen
σ	Σ	Sigma	Pauli matrices (LC), summation (U. C.)
τ	T	Tau	Increment of time
υ	Υ	Upsilon	–
ϕ	Φ	Phi	Azimuthal angle (LC), function (UC)
χ	X	Chi	Spin state (LC)
ψ	Ψ	Psi	Wave function (LC) and (UC)
ω	Ω	Omega	Frequency (radians/ s) (LC), Bohr frequency

Where appropriate, their primary usage in this book is indicated. LC and UC refer to lower and uppercase

© Springer International Publishing AG 2017
J.D. Kelley, J.J. Leventhal, *Problems in Classical
and Quantum Mechanics*, DOI 10.1007/978-3-319-46664-4

Appendix B
Acronyms, Descriptors and Coordinates

B.1 Acronyms and Descriptors

Table B.1 Acronyms and Descriptors

Term/Abbreviation	Meaning
a-box	Infinite square well with $-a/2 \leq x \leq a/2$
L-box	Infinite square well with $0 \leq x \leq L$
TDSE	Time dependent Schrödinger equation
TISE	Time independent Schrö dinger equation
TME	Total mechanical energy (the "energy")
WKB	Wentzel, Kramers, Brillouin approximation
SHO	Simple harmonic oscillator
H-atom	Hydrogen atom

B.2 Coordinate Systems

Table B.2 Coordinate systems

System	Coordinates	Unit vectors
Cartesian	x, y, z	$\hat{\imath}, \hat{\jmath}, \hat{k}$
Spherical	r (radial), θ (polar), ϕ (azimuthal)	$\hat{a}_r, \hat{a}_\theta, \hat{a}_\phi$
Cylindrical	ρ (radial), ϕ (azimuthal), z	$\hat{a}_\rho, \hat{a}_\phi, \hat{k}$
Plane polar[a]	r, θ	$\hat{a}_r, \hat{a}_\theta$ or $\frac{r}{r}, \hat{a}_\theta$

[a]Plane polar = cylindrical coordinates with $z = 0$, $\rho \rightarrow r$, and $\phi \rightarrow \theta$

© Springer International Publishing AG 2017
J.D. Kelley, J.J. Leventhal, *Problems in Classical
and Quantum Mechanics*, DOI 10.1007/978-3-319-46664-4

Appendix C
Units

In keeping with modern usage SI units are used throughout except in problems where it is inconvenient and other units are appropriate. Useful units for calculations involving atomic dimensions are "atomic units" abbreviated a.u. In a.u. the unit of length is the Bohr radius $a_0 = \left[(4\pi\epsilon_0)\hbar^2\right]/(m_e e^2)$, the electronic charge e, the mass of the electron m_e, and $\hbar$ are all set equal to unity. The conversion between a.u. and SI units is given in Table C.1.

A few points about a.u. are worth noting:

1. The unit of velocity is the velocity of the electron in the first Bohr orbit.
2. The unit of time is the period of the electron in the first Bohr orbit divided by 2π.
3. The unit of energy is twice the ground-state Bohr energy (27.2 eV).

Table C.1 Atomic units (a.u.)

Quantity	a.u.	SI
Mass	$m_e = 1$	9.10×10^{-31} kg
Charge	$e = 1$	1.60×10^{-19} C
Angular momentum	$\hbar = 1$	1.06×10^{-34} J s
Length	$a_0 = 1$	5.29×10^{-11} m
Velocity	$v_0 = \alpha c$	2.20×10^6 m/s
Time	$a_0/v_0 = 1/\alpha c$	2.42×10^{-17} s
Energy	$e^2/(4\pi\epsilon_0 a_0) = 1$	4.36×10^{-18} J
Electric field	$e/(4\pi\epsilon_0 a_0^2) = 1$	5.14×10^{11} V/m
Bohr magneton	$e\hbar/(2m_e) = 1/2$	9.274×10^{-24} J/T
Fine structure constant	$\alpha = e^2/[(4\pi\epsilon_0)\hbar c] = 1/137$	$1/137$

© Springer International Publishing AG 2017
J.D. Kelley, J.J. Leventhal, *Problems in Classical and Quantum Mechanics*, DOI 10.1007/978-3-319-46664-4

4. The speed of light $c = 137$ a.u.$= $ [(length in a.u.)/(time in a.u.)].
5. The fine structure constant α is a dimensionless number (the same in any system of units). It is easily evaluated using a.u.

$$\alpha = \left[\frac{1}{(4\pi\epsilon_0)}\right]\left[\frac{1}{\hbar c}\right] = \frac{1}{137} \tag{C.1}$$

Appendix D
Conic Sections in Polar Coordinates

Most students are familiar with the equations of conic sections , ellipses, hyperbolas, and parabolas in Cartesian coordinates. For the Kepler problem in which the gravitational force law is an attractive inverse square law the orbits of the bodies are conic sections which are most conveniently written in polar coordinates (r, θ). Although a conic section may be defined as the curve that results from the intersection of a plane and a right circular cone, we require an equivalent, but more quantitative definition.

Discussion of the details is facilitated using Fig. D.1 in which the origin of polar coordinates O is the focus and the vertical line a distance p away from O is the directrix. A conic section is the locus of all points having a constant ratio ϵ, the eccentricity.

$$\epsilon = \frac{|OP|}{|MP|} \tag{D.1}$$

The eccentricity is the key parameter in ascertaining the path of the particle. The point P lies on the curve if and only if

Fig. D.1 Parameters for the derivation of the equation of a conic section

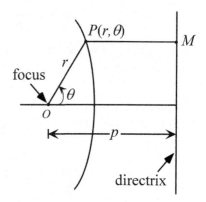

$$|OP| = \epsilon \, |MP| \tag{D.2}$$

Equation (D.2) may be cast in terms of the polar coordinates by noting that

$$|OP| = r \tag{D.3}$$

and

$$|MP| = p - r \cos \theta \tag{D.4}$$

Therefore,

$$r = \epsilon \, (p - r \cos \theta) \tag{D.5}$$

Solving for r we have

$$r = \frac{\epsilon p}{1 + \epsilon \cos \theta} \tag{D.6}$$

This is the equation of a conic section in polar coordinates with the origin at one focus. Different values of the ratio ϵ yield different shapes.

$$\epsilon = 1 \quad \text{parabola}$$
$$\epsilon < 1 \quad \text{ellipse}$$
$$> 1 \quad \text{hyperbola} \tag{D.7}$$

It is known that a circle is an ellipse with $\epsilon = 0$ which, from Eq. (D.1) means that the distance $|MP| \to \infty$.

Equation (D.6) is often written in the form

$$\frac{\alpha}{r} = 1 + \epsilon \cos \theta \tag{D.8}$$

For Keplerian orbits, where the Keplerian potential is given by

$$U(r) = -\frac{k}{r} \tag{D.9}$$

the constants in Eq. (D.8) are given by

$$\alpha = \frac{\ell^2}{\mu k} \quad \text{and} \quad \epsilon = \sqrt{1 + \frac{2E\ell^2}{\mu k^2}} \tag{D.10}$$

where E is the TME of the particle in its orbit and 2α is known as the latus rectum. Notice that if the orbiting object is bound to the force center, an elliptical orbit, then E is negative. Further, if

$$E = -\frac{\mu k^2}{2\ell^2} \qquad\qquad (\text{D.11})$$

then, from Eq. (D.10), $\epsilon = 0$ and the orbit is a circle (see Eq. (D.8)).

Appendix E
Useful Trigonometric Identities

$$\sin(A \pm B) = \sin A \cos B \pm \cos A \sin B \tag{E.1}$$

$$\cos(A \pm B) = \cos A \cos B \mp \sin A \sin B \tag{E.2}$$

$$\tan(A \pm B) = \frac{\tan A \pm \tan B}{1 \mp \tan A \tan B} \tag{E.3}$$

$$\cot(A \pm B) = \frac{\cot A \cot B \mp 1}{\cot B \pm \cot A} \tag{E.4}$$

$$\sin A \sin B = \frac{1}{2}[\cos(A - B) - \cos(A + B)] \tag{E.5}$$

$$\cos A \cos B = \frac{1}{2}[\cos(A - B) + \cos(A + B)] \tag{E.6}$$

$$\sin A \cos B = \frac{1}{2}[\sin(A - B) + \sin(A + B)] \tag{E.7}$$

$$\sin 2A = 2\sin A \cos A \tag{E.8}$$

$$\cos 2A = 1 - 2\sin^2 A = 2\cos^2 A - 1 \tag{E.9}$$

$$\sin 3A = 3\sin A - 4\sin^3 A \tag{E.10}$$

© Springer International Publishing AG 2017
J.D. Kelley, J.J. Leventhal, *Problems in Classical and Quantum Mechanics*, DOI 10.1007/978-3-319-46664-4

Appendix F
Useful Vector Relations

$$A \bullet (B \times C) = C \bullet (A \times B) = B \bullet (C \times A) \qquad \text{(F.1)}$$

$$A \times (B \times C) = B\,(A \bullet C) - C\,(A \bullet B) \qquad \text{(F.2)}$$

$$\frac{d}{dt}\,(A \bullet B) = A \bullet \frac{dB}{dt} + B \bullet \frac{dA}{dt} \qquad \text{(F.3)}$$

$$\frac{d}{dt}\,(A \times B) = A \times \frac{dB}{dt} + \frac{dA}{dt} \times B \qquad \text{(F.4)}$$

$$\int (\nabla \bullet A)\,dV = \oint A \bullet da \quad \text{Gauss' Divergence theorem} \qquad \text{(F.5)}$$

$$\int (\nabla \times A) \bullet da = \oint A \bullet d\ell \quad \text{Stokes' theorem} \qquad \text{(F.6)}$$

© Springer International Publishing AG 2017
J.D. Kelley, J.J. Leventhal, *Problems in Classical
and Quantum Mechanics*, DOI 10.1007/978-3-319-46664-4

Appendix G
Useful Integrals

$$\int x^2 e^{-ax} dx = \frac{e^{-ax}}{-a}\left(x^2 + \frac{2x}{a} + \frac{2}{a^2}\right) \tag{G.1}$$

$$\int_0^\infty x^m e^{-ax} dx = \frac{\Gamma\left[(m+1)\right]}{a^{m+1}} = \frac{m!}{a^{m+1}} \tag{G.2}$$

$$\int_{-\infty}^\infty e^{-ax^2} dx = \sqrt{\frac{\pi}{a}} \tag{G.3}$$

$$\int_0^\infty x^m e^{-ax^2} dx = \frac{\Gamma\left[(m+1)/2\right]}{2a^{(m+1)/2}} \tag{G.4}$$

$$\int \sqrt{a^2 - x^2}\, dx = \frac{x\sqrt{a^2 - x^2}}{2} + \frac{a^2}{2}\sin^{-1}\frac{x}{a} \tag{G.5}$$

$$\int_{-1}^1 \sqrt{1 - u^2}\, du = \frac{\pi}{2} \tag{G.6}$$

$$\int \frac{dx}{\sqrt{a^2 - x^2}} = \sin^{-1}\frac{x}{a} \tag{G.7}$$

$$\int x\cos(\alpha x)\, dx = \frac{\cos(\alpha x)}{\alpha^2} + x\frac{\sin(\alpha x)}{\alpha} \tag{G.8}$$

$$\int x\sin^2(\alpha x)\, dx = \frac{x^2}{4} - \frac{x\sin(2\alpha x)}{4\alpha} - \frac{\cos(2\alpha x)}{8\alpha^2} \tag{G.9}$$

$$\int x^2\sin^2 x\, dx = \frac{x^3}{6} - \left(\frac{x^2}{4} - \frac{1}{8}\right)\sin(2x) - \frac{x\cos(2x)}{4} \tag{G.10}$$

$$\int \cos^3(ax)\, dx = \frac{\sin(ax)}{a} - \frac{\sin^3(ax)}{3a} \tag{G.11}$$

© Springer International Publishing AG 2017
J.D. Kelley, J.J. Leventhal, *Problems in Classical
and Quantum Mechanics*, DOI 10.1007/978-3-319-46664-4

$$\int \cos^4{(ax)}\, dx = \frac{3x}{8} + \frac{\sin{(2ax)}}{4a} + \frac{\sin{(4ax)}}{32a} \tag{G.12}$$

$$\int \sin^3{(ax)}\, dx = -\frac{\cos{ax}}{a} + \frac{\cos^3{ax}}{3a} \tag{G.13}$$

$$\int x\sin{(\alpha x)}\, dx = \frac{\sin{ax}}{a^2} - \frac{x\cos{ax}}{a} \tag{G.14}$$

$$\int x\sin^2{(\alpha x)}\, dx = \frac{x^2}{4} - \frac{x\sin{(2\alpha x)}}{4\alpha} - \frac{\cos{(2\alpha x)}}{8\alpha^2} \tag{G.15}$$

Appendix H
Useful Series

H.1 Taylor Series

The Taylor series expansion of a function $f(x)$ about a point $x = a$ is

$$f(x) = f(a) + \frac{(x-a)}{1!} f'(a) + \frac{(x-a)^2}{2!} f''(a) + \cdots$$

$$= \sum_{n=0}^{\infty} \frac{(x-a)^n}{n!} f^{(n)}(a) \tag{H.1}$$

where the primes signify differentiation with respect to x. For example, $f''(a)$ is the second derivative of the function $f(x)$ with respect to x evaluated at $x = a$.

There are at least four Taylor series that every physics student should have at their command. These series, Eqs. (H.2) through (H.5) below, are easy to remember and are frequently used in problems.

$$e^x = 1 + \frac{x}{1!} + \frac{x^2}{2!} + \frac{x^3}{3!} + \cdots \tag{H.2}$$

$$\sin x = \frac{x}{1!} - \frac{x^3}{3!} + \frac{x^5}{5!} - \frac{x^7}{7!} + \cdots \tag{H.3}$$

$$\cos x = 1 - \frac{x^2}{2!} + \frac{x^4}{4!} - \frac{x^6}{6!} + \cdots \tag{H.4}$$

$$\ln(1+x) = x - \frac{x^2}{2} + \frac{x^3}{3} - \frac{x^4}{4} + \cdots \tag{H.5}$$

Two other useful series are those for the hyperbolic sine and cosine, which are identical to those for the circular sine and cosine.

© Springer International Publishing AG 2017
J.D. Kelley, J.J. Leventhal, *Problems in Classical and Quantum Mechanics*, DOI 10.1007/978-3-319-46664-4

H.2   Binomial Expansion

Binomial series are special cases of Taylor series for $f(x) = (1 + x)^m$ and $a = 0$. The exponent m may be positive or negative and is not restricted to integer values. The binomial expansion is, in three equivalent forms,

$$(1 + x)^m = 1 + mx + \frac{m(m-1)}{2!}x^2 + \frac{m(m-1)(m-2)}{3!}x^3 + \cdots$$

$$= \sum_{n=0}^{\infty} \frac{m!}{n!(m-n)!} x^n$$

$$= \sum_{n=0}^{\infty} \binom{m}{n} x^n \tag{H.6}$$

where

$$\binom{m}{n} \equiv \frac{m!}{n!(m-n)!} \tag{H.7}$$

is called the binomial coefficient. A few of the most common binomial expansions are listed below:

$$(1 + x)^{-1} = 1 - x + x^2 - x^3 + \cdots \tag{H.8}$$

$$(1 + x)^{-2} = 1 - 2x + 3x^2 - 4x^3 + \cdots \tag{H.9}$$

$$(1 + x)^{1/2} = 1 + \frac{1}{2}x - \frac{1}{2 \cdot 4}x^2 + \frac{1}{2 \cdot 4 \cdot 6}x^3 + \cdots \tag{H.10}$$

$$(1 + x)^{-1/2} = 1 - \frac{1}{2}x + \frac{1 \cdot 3}{2 \cdot 4}x^2 - \frac{1 \cdot 3 \cdot 5}{2 \cdot 4 \cdot 6}x^3 + \cdots \tag{H.11}$$

Appendix I
Γ-Functions

I.1   Integral Γ-Functions

$$\Gamma(n) = (n-1)! \quad n = 1, 2, 3, \dots \tag{I.1}$$

$$\Gamma(n+1) = n\Gamma(n) = n! \quad n = 1, 2, 3, \dots \tag{I.2}$$

$$\Gamma(1) = 1 \tag{I.3}$$

$$\Gamma(2) = 1 \tag{I.4}$$

$$\Gamma(3) = 2 \tag{I.5}$$

$$\Gamma(4) = 3! = 6 \tag{I.6}$$

$$\Gamma(5) = 4! = 24 \tag{I.7}$$

I.2   Half-Integral Γ-Functions

$$\Gamma\left(m + \frac{1}{2}\right) = \frac{1 \cdot 3 \cdot 5 \cdots (2m-1)}{2^m} \sqrt{\pi} \quad m = 1, 2, 3, \dots \tag{I.8}$$

$$\Gamma(1/2) = \sqrt{\pi} \tag{I.9}$$

$$\Gamma(3/2) = \frac{\sqrt{\pi}}{2} \tag{I.10}$$

$$\Gamma(5/2) = \frac{3\sqrt{\pi}}{4} \tag{I.11}$$

$$\Gamma(7/2) = \frac{15\sqrt{\pi}}{8} \tag{I.12}$$

© Springer International Publishing AG 2017
J.D. Kelley, J.J. Leventhal, *Problems in Classical and Quantum Mechanics*, DOI 10.1007/978-3-319-46664-4

Appendix J
The Dirac Delta-Function

The Dirac δ-function is defined as follows:

$$\delta\left(x - x'\right) \equiv \frac{1}{2\pi} \int_{-\infty}^{\infty} e^{ik(x-x')} dk \qquad (J.1)$$

The dimension of the δ-function is the reciprocal of the variable of integration. For example, the units of $\delta(x)$ are 1/length.

See Ref. [3], page 665 for more information on the Dirac δ-function.

Table J.1 Some properties of the Dirac delta-functions

Mathematical operation	Property
$f(x_0) = \int_{-\infty}^{\infty} \delta(x - x_0) f(x) \, dx$	Sifting property
$\delta(-x) = \delta(x)$	Parity: even
$\int_{-\infty}^{\infty} \delta(x - x_0) \, dx = 1$	Normalization
$\delta(ax) = (1/\lvert a \rvert)\, \delta(x)$	Scaling

© Springer International Publishing AG 2017
J.D. Kelley, J.J. Leventhal, *Problems in Classical and Quantum Mechanics*, DOI 10.1007/978-3-319-46664-4

Appendix K
Hyperbolic Functions

K.1 Manipulations of Hyperbolic Functions

Some proofs of relations involving hyperbolic functions.
Show that:

$$\cosh^{-1}x = \ln\left(x + \sqrt{x^2 - 1}\right) \qquad\qquad \text{(K.1)}$$

Proof. Let

$$x = \cosh y \qquad y > 0$$
$$= \frac{e^y + e^{-y}}{2}$$
$$= \frac{1}{2e^y}\left(e^{2y} + 1\right) \qquad\qquad \text{(K.2)}$$

or

$$e^{2y} - 2xe^y + 1 = 0 \qquad\qquad \text{(K.3)}$$

Using the quadratic formula

$$e^y = x \pm \sqrt{x^2 - 1} \qquad\qquad \text{(K.4)}$$

or

$$y = \cosh^{-1}x = \ln\left(x + \sqrt{x^2 - 1}\right) \qquad\qquad \text{(K.5)}$$

where the minus sign has been dropped because the principal values of the $\cosh^{-1}x$ are positive.

© Springer International Publishing AG 2017
J.D. Kelley, J.J. Leventhal, *Problems in Classical
and Quantum Mechanics*, DOI 10.1007/978-3-319-46664-4

Show that:

$$\tanh^{-1}x = \frac{1}{2}\ln\left(\frac{1+x}{1-x}\right) \tag{K.6}$$

Proof. Let

$$x = \tanh y$$
$$= \frac{e^y - e^{-y}}{e^y + e^{-y}}$$
$$= \frac{e^{2y} - 1}{e^{2y} + 1} \tag{K.7}$$

Then

$$xe^{2y} + x = e^{2y} - 1$$
$$xe^{2y} - e^{2y} = -(1+x)$$
$$e^{2y}(x-1) = -(1+x) \tag{K.8}$$

so

$$e^{2y} = \left(\frac{1+x}{1-x}\right) \tag{K.9}$$

or

$$y = \tanh^{-1}x = \frac{1}{2}\ln\left(\frac{1+x}{1-x}\right) \tag{K.10}$$

Show that:

$$\tanh^{-1}z = \cosh^{-1}\frac{1}{\sqrt{1-z^2}} \tag{K.11}$$

Proof. Let $x = 1/\sqrt{1-z^2}$ in Eq. (K.5).

$$\cosh^{-1}\frac{1}{\sqrt{1-z^2}} = \ln\left(\frac{1}{\sqrt{1-z^2}} + \sqrt{\frac{1}{1-z^2} - 1}\right)$$
$$= \ln\frac{1+z}{\sqrt{1-z^2}} = \ln\sqrt{\frac{1+z}{1-z}}$$
$$= \frac{1}{2}\ln\left(\frac{1+z}{1-z}\right)$$
$$= \tanh^{-1}z \tag{K.12}$$

where the last step utilized Equation (K.10).

K.2   Relationships Between Hyperbolic and Circular Functions

$$\sin(ix) = i \sinh x \qquad\qquad (K.13)$$

$$\sinh(ix) = i \sin x \qquad\qquad (K.14)$$

$$\cos(ix) = \cosh x \qquad\qquad (K.15)$$

$$\cosh(ix) = \cos x \qquad\qquad (K.16)$$

Appendix L
Useful Formulas

L.1 Classical Mechanics

The Lagrangian

$$\mathcal{L} = T - U \tag{L.1}$$

Lagrange's Equation

$$\frac{\partial \mathcal{L}}{\partial q_i} - \frac{d}{dt}\left(\frac{\partial \mathcal{L}}{\partial \dot{q}_i}\right) = 0 \tag{L.2}$$

Lagrange's Equation with undetermined multiplier

$$\frac{\partial \mathcal{L}}{\partial q_i} - \frac{d}{dt}\frac{\partial \mathcal{L}}{\partial \dot{q}_i} + \lambda \frac{\partial f(q_1, q_2)}{\partial q_i} = 0 \tag{L.3}$$

The Hamiltonian

$$H(q_i, p_i, t) = \sum_j \dot{q}_j p_j - \mathcal{L}(q_i, p_i, t) \tag{L.4}$$

The Hamiltonian is the TME under the following conditions:

(a) The transformation equations from Cartesian to generalized coordinates are time independent.
(b) The potential energy contains only the coordinates and not the velocity.

© Springer International Publishing AG 2017

J.D. Kelley, J.J. Leventhal, *Problems in Classical and Quantum Mechanics*, DOI 10.1007/978-3-319-46664-4

Under these conditions:

$$H = \frac{\hat{p}_x^2}{2m} + U(x) \quad \text{(the total energy)} \tag{L.5}$$

L.2   Quantum Mechanics

The de Broglie wavelength

$$\lambda_{\text{de Broglie}} = \frac{h}{p}$$

$$= \frac{h}{\sqrt{2mE}}; \quad E = \text{ kinetic energy} \tag{L.6}$$

L.2.1   One Dimension

Time dependent Schrödinger Equation (TDSE)

$$\left[-\frac{\hbar^2}{2m} \frac{\partial^2}{\partial x^2} + U(x, t) \right] \Psi(x, t) = -\frac{\hbar}{i} \frac{\partial \Psi(x, t)}{\partial t} \tag{L.7}$$

Time independent Schrödinger Equation (TISE)

$$\left[-\frac{\hbar^2}{2m} \frac{d^2}{dx^2} + U(x) \right] \psi(x) = E\psi(x) \tag{L.8}$$

TISE in terms of the Hamiltonian

$$\hat{H}\psi(x) = E\psi(x)$$

$$= \left[\frac{\hat{p}_x^2}{2m} + U(x) \right] \psi(x) \tag{L.9}$$

Definition of the probability current

$$j(x, t) = \frac{\hbar}{2im} \left[\Psi^*(x, t) \frac{\partial \Psi(x, t)}{\partial x} - \Psi(x, t) \frac{\partial \Psi^*(x, t)}{\partial x} \right] \tag{L.10}$$

The equation of continuity for probability current $j(x,t)$

$$\frac{\partial |\Psi(x,t)|^2}{\partial t} + \frac{\partial}{\partial x} j(x,t) = 0 \tag{L.11}$$

Probability current for an energy eigenfunction (time independent)

$$j(x) = \frac{\hbar}{2im}\left[\psi^*(x)\frac{\partial \psi(x)}{\partial x} - \psi(x)\frac{\partial \psi^*(x)}{\partial x}\right] \tag{L.12}$$

The equation of continuity for time independent probability current $j(x)$

$$\frac{\partial}{\partial x} j(x) = 0 \tag{L.13}$$

L.2.2   Three Dimensions (Central Potentials)

The radial TISE where $R_{n\ell}$ is the radial part of the eigenfunctions and the angular parts are the spherical harmonics $Y_{\ell m}(\theta,\phi)$

$$\left[-\frac{\hbar^2}{2mr^2}\frac{d}{dr}\left(r^2\frac{d}{dr}\right) + \frac{\ell(\ell+1)\hbar^2}{2mr^2} + U(r)\right]R_{n\ell}(r) = E_{n\ell}R_{n\ell}(r) \tag{L.14}$$

The radial TISE with $u(r) = rR(r)$

$$-\frac{\hbar^2}{2m}\frac{d^2u(r)}{dr^2} + \left\{\frac{\ell(\ell+1)\hbar^2}{2mr^2} + U(r)\right\}u(r) = Eu(r) \tag{L.15}$$

The effective potential

$$U_{eff}(r,\ell) = \frac{\ell(\ell+1)\hbar^2}{2mr^2} + U(r) \tag{L.16}$$

Interparticle distance r_{12} in terms of spherical harmonics

$$\frac{1}{r_{12}} = \frac{1}{r_>}\sum_{\ell=0}^{\infty} P_\ell(\cos\gamma)\left(\frac{r_<}{r_>}\right)^\ell$$

$$= \frac{4\pi}{2\ell+1}\left(\frac{1}{r_>}\right)\sum_{\ell=0}^{\infty}\sum_{m=-\ell}^{\ell} Y_{\ell m}(\theta_1,\phi_1)Y_{\ell m}^*(\theta_2,\phi_2)\left(\frac{r_<}{r_>}\right)^\ell \tag{L.17}$$

Appendix M
The Infinite Square Well

The time honored one-dimensional problem of a particle trapped in an infinite square well potential (also known as a particle-in-a-box) is frequently formulated in one of two different configurations which we refer to as an L-box and an a-box. Often one of these configurations is more advantageous for problem solving than the other.

M.1 The L-Box

In this configuration the potential energy is given by

$$U(x) = 0 \qquad 0 \le x \le L$$
$$= \infty \qquad -\infty < x < 0 \,;\, L < x < \infty \qquad (M.1)$$

which is shown in Fig. M.1

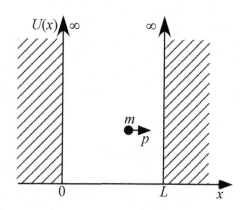

Fig. M.1 The L-box configuration

© Springer International Publishing AG 2017
J.D. Kelley, J.J. Leventhal, *Problems in Classical
and Quantum Mechanics*, DOI 10.1007/978-3-319-46664-4

The eigenfunctions and energy eigenvalues are

$$\psi_n(x) = A \sin\left(\frac{n\pi x}{L}\right) \quad 0 \le x \le L \quad n = 1, 2, 3\ldots$$
$$= 0 \qquad -\infty < x < 0 \,;\, L < x < \infty \qquad \text{(M.2)}$$

and

$$E_n = \frac{n^2 \pi^2 \hbar^2}{2mL^2} \qquad \text{(M.3)}$$

Advantage: The eigenfunctions are the same function for each level, sines.

Disadvantage: The symmetry of the potential well about $x = L/2$ is not helpful in evaluating integrals in problems involving this configuration.

M.2   The a-Box

In this configuration the potential energy is given by

$$U(x) = 0 \quad -a \le x \le a$$
$$= \infty \quad -\infty < x < -a/2 \,;\, a/2 < x < \infty \qquad \text{(M.4)}$$

which is shown in Fig. M.2.

Fig. M.2 The a-box configuration

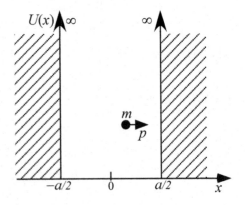

The eigenfunctions are

$$\psi_n(x) = \sqrt{\frac{2}{a}} \cos\left(\frac{n\pi x}{a}\right) \quad -\frac{a}{2} \le x \le \frac{a}{2} \quad n = 1, 3, 5 \ldots \text{ (even parity)}$$

$$\psi_n(x) = \sqrt{\frac{2}{a}} \sin\left(\frac{n\pi x}{a}\right) \quad -\frac{a}{2} \le x \le \frac{a}{2} \quad n = 2, 4, 6 \ldots \text{ even (odd parity)}$$

$$= 0 \quad -\infty < x < -a/2; a/2 < x < \infty \text{ for all } n \quad \text{(M.5)}$$

and the energy eigenvalues are the same as they are for the L-box with $L \to a$.

Advantage: The symmetry of the well and the accompanying definite parity of the eigenfunctions can be quite useful in problems involving this configuration.

Disadvantage: The eigenfunctions alternate between cosine (odd n) and sine (even n) making evaluation of integrals slightly more challenging than it is for the L-box.

Appendix N
Operators, Eigenfunctions, and Commutators

N.1   Eigenfunctions and Eigenvalues of Operators

An operator is a mathematical object that when applied to a function gives a new function. If we have a function $f(x)$ and an operator $\hat{A}$, then $\hat{A}f(x)$ is a some new function $\phi(x)$. In some cases $\phi(x)$ is proportional to $f(x)$; this means that $\hat{A}f(x) = af(x)$ where a is a constant. In this case $f(x)$ is called an eigenfunction of $\hat{A}$ and a is the corresponding eigenvalue.

An operator $\hat{A}$ is linear if $\hat{A}[f(x) + g(x)] = \hat{A}f(x) + \hat{A}g(x)$ where $f(x)$ and $g(x)$ are any two appropriate functions. The operators of importance in quantum mechanics, such as position $\hat{x}$, linear momentum $\hat{p}$, and the Hamiltonian $\hat{H}$, are all linear operators. In quantum mechanics observable quantities such as $\hat{x}, \hat{p}$, and $\hat{H}$ are represented by Hermitian operators . An operator $\hat{A}$ is Hermitian if

$$\int \left[\hat{A}\phi(x)\right]^* \phi(x)\,dx = \int \phi^*(x)\left[\hat{A}\phi(x)\right]dx \qquad \text{(N.1)}$$

Hermitian operators have real eigenvalues, and conversely real eigenvalues result from Hermitian operators . The operators $\hat{x}, \hat{p}$, and $\hat{H}$ are Hermitian.

N.2   Operator Algebra; Commutators

The product of two operators is defined by operating with them on a function. Let the operators be $\hat{A}$ and $\hat{B}$ operate on $f(x)$. The expression $\hat{A}\hat{B}f(x)$ is a new function, and $\hat{A}\hat{B}$ is an operator which we call $\hat{C}$, that is defined as the product of $\hat{A}$ and $\hat{B}$. The meaning of $\hat{A}\hat{B}f(x)$ is that $\hat{B}$ operates first on $f(x)$, giving a new function, and then

© Springer International Publishing AG 2017
J.D. Kelley, J.J. Leventhal, *Problems in Classical and Quantum Mechanics*, DOI 10.1007/978-3-319-46664-4

$\hat{A}$ operates on that new function. Combinations of operators of the form $\hat{A}\hat{B} - \hat{B}\hat{A}$ frequently arise in quantum mechanical calculations. An abbreviated notation that is customarily used is

$$\hat{A}\hat{B} - \hat{B}\hat{A} \equiv \left[\hat{A}, \hat{B}\right] \tag{N.2}$$

The symbol $\left[\hat{A}, \hat{B}\right]$ is referred to as the commutator of $\hat{A}$ and $\hat{B}$. If $\left[\hat{A}, \hat{B}\right] \neq 0$, then $\hat{A}$ and $\hat{B}$ do not commute; if $\left[\hat{A}, \hat{B}\right] = 0$, then $\hat{A}$ and $\hat{B}$ commute. If two operators $\hat{A}$ and $\hat{B}$ commute, then they have common eigenfunctions. This means that the physical quantities associated with $\hat{A}$ and $\hat{B}$ can (in principle) be measured simultaneously and exactly. If $\hat{A}$ and $\hat{B}$ do not commute, the associated physical quantities cannot be accurately measured together. The classic examples of non-commuting operators are position and momentum; the operators, $\hat{x} = i\hbar d/dp$ and $p = -i\hbar d/dx$ do not commute, and therefore the position and momentum of a particle cannot be simultaneously obtained—the uncertainty principle.

N.3   Commutator Identities

$$\left[\hat{A}, \hat{B}\right] \equiv \hat{A}\hat{B} - \hat{B}\hat{A} \tag{N.3}$$

$$\left[\hat{A}, \hat{A}\right] = 0 \tag{N.4}$$

$$\left[\hat{A}, \hat{B}\hat{C}\right] \equiv \left[\hat{A}, \hat{B}\right]\hat{C} + \hat{B}\left[\hat{A}, \hat{C}\right] \tag{N.5}$$

$$\left[\hat{A}\hat{B}, \hat{C}\right] \equiv \left[\hat{A}, \hat{C}\right]\hat{B} + \hat{A}\left[\hat{B}, \hat{C}\right] \tag{N.6}$$

$$\left[\hat{A}\hat{B}, \hat{C}\hat{D}\right] \equiv \left[\hat{A}, \hat{C}\right]\hat{B}\hat{D} + \hat{A}\left[\hat{B}, \hat{C}\right]\hat{D}$$
$$+ \hat{C}\left[\hat{A}, \hat{D}\right]\hat{B} + \hat{C}\hat{A}\left[\hat{B}, \hat{D}\right] \tag{N.7}$$

N.4   Some Quantum Mechanical Commutators

Position and momentum: $[x, \hat{p}] = i\hbar$ $\qquad$ (N.8)

Position and powers of momentum: $\left[x, \hat{p}_x^n\right] = i\hbar n \hat{p}_x^{n-1}$ $\qquad$ (N.9)

Components of angular momentum: $\left[\hat{J}_i, \hat{J}_j\right] = i\hbar \hat{J}_k \epsilon_{ijk}$ $\qquad$ (N.10)

Appendix O
The Quantum Mechanical Harmonic Oscillator

Together with the particle-in-a-box the most often studied problem in one-dimensional bound state quantum mechanics is that of a particle of mass m under the influence of a quadratic potential, that is

$$U(x) = \frac{1}{2}kx^2$$

$$= \frac{1}{2}m\omega^2 \quad (O.1)$$

where k is the force constant and $\omega = \sqrt{k/m}$, the harmonic frequency. One of the most powerful methods for attacking this problem is the use of ladder operators. This subject is summarized in Sect. 7 of this book. In this appendix we summarize the eigenfunctions of the harmonic oscillator which, because the potential is an even function of x, alternate between even and odd functions. The oscillator wave functions are always the product of a Hermite polynomial and a Gaussian exponential. The first four normalized eigenfunctions are listed in Table O.1.
   where

$$\alpha = \sqrt{\frac{m\omega}{\hbar}} = \sqrt{\frac{m}{\hbar}\left(\frac{k^{1/2}}{m^{1/2}}\right)} = m^{1/4}k^{1/4}\sqrt{\frac{1}{\hbar}} \quad (O.2)$$

Note that the units of α are m^{-1}.
   Table O.2 summarizes the properties of the harmonic oscillator ladder operators discussed in Sect. 7.

© Springer International Publishing AG 2017
J.D. Kelley, J.J. Leventhal, *Problems in Classical
and Quantum Mechanics*, DOI 10.1007/978-3-319-46664-4

Table O.1 The first four
normalized harmonic
oscillator eigenfunctions

$$\psi_0(x) = \sqrt{\frac{\alpha}{\sqrt{\pi}}} e^{-\alpha^2 x^2/2}$$

$$\psi_1(x) = \sqrt{\frac{\alpha}{2\sqrt{\pi}}} \left[2(\alpha x) \right] e^{-\alpha^2 x^2/2}$$

$$\psi_2(x) = \sqrt{\frac{\alpha}{2\sqrt{\pi}}} \left[2(\alpha x)^2 - 1 \right] e^{-\alpha^2 x^2/2}$$

$$\psi_3(x) = \sqrt{\frac{\alpha}{3\sqrt{\pi}}} \left[2(\alpha x)^3 - 3(\alpha x) \right] e^{-\alpha^2 x^2/2}$$

Table O.2 Relations
involving the ladder operators

$$\hat{a} = \frac{1}{\sqrt{2}} \left(\alpha \hat{x} + i \frac{1}{\alpha \hbar} \hat{p} \right)$$

$$\hat{a}^\dagger = \frac{1}{\sqrt{2}} \left(\alpha \hat{x} - i \frac{1}{\alpha \hbar} \hat{p} \right)$$

$$\hat{x} = \frac{1}{\sqrt{2}\alpha} \left(\hat{a} + \hat{a}^\dagger \right)$$

$$\hat{N} = \hat{a}^\dagger \hat{a}$$

$$\hat{H} = \hbar\omega \left(\hat{a}^\dagger \hat{a} + \frac{1}{2} \right) = \hbar\omega \left(\hat{N} + \frac{1}{2} \right)$$

$$\left[\hat{a}, \hat{a}^\dagger \right] = 1$$

$$\left[\hat{N}, \hat{a} \right] = -\hat{a} \implies \left[\hat{H}, \hat{a} \right] = -\hbar\omega\hat{a}$$

$$\left[\hat{N}, \hat{a}^\dagger \right] = \hat{a}^\dagger \implies \left[\hat{H}, \hat{a}^\dagger \right] = \hbar\omega\hat{a}^\dagger$$

$$\hat{a} |n\rangle = \sqrt{n} |n-1\rangle \qquad \text{(lowering)}$$

$$\hat{a}^\dagger |n\rangle = \sqrt{n+1} |n+1\rangle \qquad \text{(raising)}$$

Appendix P
Legendre Polynomials

P.1 Properties

The Legendre polynomials $P_\ell (\cos \theta) = P_\ell (x)$ are used in a variety of applications in which there is azimuthal symmetry, that is, no ϕ dependence in spherical coordinates. They are the solutions of Legendre's differential equation

$$\left(1 - x^2\right) \frac{d^2 y}{dx^2} - 2x \frac{dy}{dx} + \ell \left(\ell + 1\right) y = 0 \qquad \text{(P.1)}$$

We present here a few properties of Legendre polynomials as well as the first few of these polynomials that are useful for the problems in this book. The substitution $x = \cos \theta$ is conventional and made for convenience. The variable x is in no way related to the Cartesian coordinate. More detailed information can be found in [1, 3] and [4].

P.2 Legendre Series

A function $f (x)$ that is defined on the interval $(-1, 1)$ may be expanded in a series of Legendre polynomials because they form a complete set. Thus, using the properties in Table P.1

$$f (x) = \sum_{\ell=0}^{\infty} a_\ell P_\ell (x)$$

$$\text{where} \quad a_\ell = \frac{2\ell + 1}{2} \int_{-1}^{1} f (x) P_\ell (x) \, dx \quad (\ell = 0, 1, 2, \cdots) \qquad \text{(P.2)}$$

© Springer International Publishing AG 2017
J.D. Kelley, J.J. Leventhal, *Problems in Classical and Quantum Mechanics*, DOI 10.1007/978-3-319-46664-4

Table P.1 Some properties of Legendre polynomials

Generating function	$\dfrac{1}{\sqrt{1-2tx+t^2}} = \sum\limits_{\ell=0}^{\infty} P_\ell(x)\, t^\ell$
Rodrigues' formula	$P_\ell(x) = \dfrac{1}{2^\ell \ell!} \left(\dfrac{d}{dx}\right)^\ell (x^2-1)^\ell$
Orthogonality	$\int_{-1}^{1} P_\ell(x)\, P_{\ell'}(x)\, dx = \dfrac{2}{2\ell+1} \delta_{\ell\ell'}$
Parity	$P_\ell(-x) = (-)^\ell P_\ell(x)$
Value at $x = 0$	$P_\ell(0) = 0 \quad \ell \text{ odd}$
	$\quad = (-)^{\ell/2} \dfrac{1\cdot 3 \cdot 5 \cdots (\ell-1)}{2 \cdot 4 \cdot 4 \cdots \ell} \quad \ell \text{ even}$
Value at $x = \pm 1$	$P_\ell(1) = 1$
	$P_\ell(-1) = (-1)^\ell$

P.3   The Function $1/|r_1 - r_2|$

There is a useful relationship between the quantity $1/|r_1 - r_2|$ and the Legendre polynomials. This quantity arises in many physical applications. The geometry is shown in Fig. P.1.

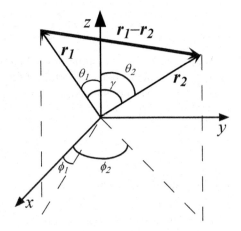

Fig. P.1 Coordinates for deriving the expression for $1/|r_1 - r_2|$ in terms of Legendre polynomials. The angle between the two directions r_1 and r_2 is γ

Applying the law of cosines we write $1/|r_1 - r_2|$ in terms of the angle γ.

$$\frac{1}{|r_1 - r_2|} = \left(r_1^2 + r_2^2 - 2r_1 r_2 \cos\gamma\right)^{-1/2} \tag{P.3}$$

If r_1 is placed along the z-axis, then γ is the spherical coordinate θ. Now assume that $r_2 > r_1$. We may write Eq. (P.3) as

$$\frac{1}{|r_1 - r_2|} = \frac{1}{r_2}\left[1 + \left(\frac{r_1}{r_2}\right)^2 - 2\left(\frac{r_1}{r_2}\right)\cos\gamma\right]^{-1/2} \tag{P.4}$$

Comparing Eq. (P.4) with the first entry in Table P.1 we recognize the generating function for $P_\ell(\mu)$. Letting $\cos\gamma = \mu$ and $t = r_1/r_2$ we write

$$\frac{1}{|r_1 - r_2|} = \frac{1}{r_2}\sum_{\ell=0}^{\infty}P_\ell(\mu)\left(\frac{r_1}{r_2}\right)^\ell \tag{P.5}$$

If we had $r_1 > r_2$ rather than $r_2 > r_1$, then the subscripts in Eq. (P.5) would be reversed. The combined equation for both cases is usually written as

$$\frac{1}{|r_1 - r_2|} = \frac{1}{r_>}\sum_{\ell=0}^{\infty}P_\ell(\mu)\left(\frac{r_<}{r_>}\right)^\ell \tag{P.6}$$

where $r_>$ and $r_<$ designate the larger and smaller of r_1 and r_2.

P.4   Polynomials

Table P.2 Some Legendre polynomials

$P_0(x) = 1$
$P_1(x) = x$
$P_2(x) = (1/2)\left(3x^2 - 1\right)$
$P_3(x) = (1/2)\left(5x^3 - 3x\right)$
$P_4(x) = (1/8)\left(35x^4 - 30x^2 + 3\right)$
$P_5(x) = (1/8)\left(63x^5 - 70x^3 + 15x\right)$

Appendix Q
Orbital Angular Momentum Operators in Spherical Coordinates

$$\hat{L}_x = -\frac{\hbar}{i}\left(\sin\phi\frac{\partial}{\partial\theta} + \cot\theta\cos\phi\frac{\partial}{\partial\phi}\right) \tag{Q.1}$$

$$\hat{L}_y = \frac{\hbar}{i}\left(\cos\phi\frac{\partial}{\partial\theta} - \cot\theta\sin\phi\frac{\partial}{\partial\phi}\right) \tag{Q.2}$$

$$\hat{L}_z = \frac{\hbar}{i}\frac{\partial}{\partial\phi} \tag{Q.3}$$

$$\hat{L}_+ = \hbar e^{i\phi}\left(\frac{\partial}{\partial\theta} + i\cot\theta\frac{\partial}{\partial\phi}\right) \tag{Q.4}$$

$$\hat{L}_- = -\hbar e^{-i\phi}\left(\frac{\partial}{\partial\theta} - i\cot\theta\frac{\partial}{\partial\phi}\right) \tag{Q.5}$$

© Springer International Publishing AG 2017
J.D. Kelley, J.J. Leventhal, *Problems in Classical and Quantum Mechanics*, DOI 10.1007/978-3-319-46664-4

Appendix R
Spherical Harmonics

The spherical harmonics $Y_{\ell m}(\theta, \phi)$ are the eigenfunctions of the angular momentum operator $\hat{L}^2$. They are given by [1]

$$Y_{\ell m}(\theta, \phi) = (-)^m \sqrt{\frac{(2\ell+1)}{4\pi} \cdot \frac{(\ell-m)!}{(\ell+m)!}} P_\ell^m(\cos\theta) e^{im\phi} \qquad (\text{R.1})$$

where the $P_\ell^m(\cos\theta)$ are associated Legendre functions. Spherical harmonics have definite parity which is determined by the value of ℓ. Simply

$$\text{parity of } Y_{\ell m}(\theta, \phi) = (-1)^\ell \qquad (\text{R.2})$$

Below is a list of some spherical harmonics (Table R.1).

Table R.1 The first few spherical harmonics

$Y_{\ell m}(\theta, \phi)$	Spherical harmonic
$Y_{00}(\theta, \phi)$	$\sqrt{1/4\pi}$
$Y_{10}(\theta, \phi)$	$\sqrt{3/4\pi}\cos\theta$
$Y_{1\pm1}(\theta, \phi)$	$\mp\sqrt{3/8\pi}\sin\theta e^{\pm i\phi}$
$Y_{20}(\theta, \phi)$	$\sqrt{5/16\pi}\left(3\cos^2\theta - 1\right)$
$Y_{2\pm1}(\theta, \phi)$	$\mp\sqrt{15/8\pi}\cos\theta\sin\theta e^{\pm i\phi}$
$Y_{2\pm2}(\theta, \phi)$	$\sqrt{15/32\pi}\sin^2\theta e^{\pm 2i\phi}$

© Springer International Publishing AG 2017
J.D. Kelley, J.J. Leventhal, *Problems in Classical and Quantum Mechanics*, DOI 10.1007/978-3-319-46664-4

The Addition Theorem for Spherical Harmonics

The addition theorem for spherical harmonics is a useful formula when there are two specified directions as shown in Fig. P.1. The theorem expresses the Legendre polynomial of the angle γ in terms of the spherical harmonics for each of the directions r_1 and r_2. The formula is

$$P_\ell (\cos \gamma) = \frac{4\pi}{2\ell + 1} \sum_{m=-\ell}^{\ell} Y_{\ell m} (\theta_1, \phi_1) Y_{\ell m}^* (\theta_2, \phi_2) \tag{R.3}$$

Appendix S
Clebsch–Gordan Tables

Table S.1 Clebsch–Gordan coefficients for any value of j_1 and $j_2 = 1/2$

j	$m_s = 1/2$	$m_s = -1/2$
$j_1 + 1/2$	$\sqrt{(j_1 + 1/2 + m_j)/(2j_1 + 1)}$	$\sqrt{(j_1 + 1/2 - m_j)/(2j_1 + 1)}$
$j_1 - 1/2$	$-\sqrt{(j_1 + 1/2 - m_j)/(2j_1 + 1)}$	$\sqrt{(j_1 + 1/2 + m_j)/(2j_1 + 1)}$

Table S.2 Clebsch–Gordan coefficients for two spin-1/2 particles

S	$m_{s2} = 1/2$	$m_{s2} = -1/2$
1	$\sqrt{(1 + M)/2}$	$\sqrt{(1 - M)/2}$
0	$-1/\sqrt{2}$	$1/\sqrt{2}$

Table S.3 Clebsch–Gordan coefficients for $j_1 = \ell = 1$ and $j_2 = 1/2$

$j_1 = 1; j_2 = 1/2$		$j = 3/2$				$j = 1/2$	
m_ℓ	m_s	3/2	1/2	−1/2	−3/2	1/2	−1/2
1	1/2	1					
1	−1/2		$\sqrt{1/3}$			$\sqrt{2/3}$	
0	1/2		$\sqrt{2/3}$			$-\sqrt{1/3}$	
0	−1/2			$\sqrt{2/3}$			$\sqrt{1/3}$
−1	1/2			$\sqrt{1/3}$			$-\sqrt{2/3}$
−1	−1/2				1		

© Springer International Publishing AG 2017
J.D. Kelley, J.J. Leventhal, *Problems in Classical and Quantum Mechanics*, DOI 10.1007/978-3-319-46664-4

Appendix T
The Hydrogen Atom

This appendix summarizes the properties of the quantum mechanical H-atom (one-electron atom, $Z = 1$). The Coulomb potential

$$U(r) = -\left(\frac{1}{4\pi\epsilon_0}\right)\frac{e^2}{r} = -\frac{1}{r} \text{ in a.u.} \tag{T.1}$$

is a central potential so the angular parts of the eigenfunctions are the spherical harmonics $Y_{\ell m}(\theta, \phi)$. The H-atom quantum numbers are therefore the usual angular quantum numbers ℓ and m which represent the total and z-components of the angular momentum. The energy quantum number is n. The energy eigenvalues, $E_n = -1/2n^2$ (a.u.), depend upon the single quantum number n rather than n and ℓ (spherical symmetry precludes m). This "accidental degeneracy" is the result of an "extra" symmetry of the Coulomb potential that eliminates the dependence of the energy on ℓ [2].

When discussing the H-atom the Greek letter α is almost universally used to designate one of the most important constants in physics, the unitless fine structure constant α. This constant is

$$\alpha = \left[\frac{e^2}{(4\pi\epsilon_0)\,\hbar c}\right]$$

$$\simeq \frac{1}{137} \tag{T.2}$$

In Tables T.1, T.2, and T.3 are some important quantities and properties of the one-electron atom with nuclear charge Z.

© Springer International Publishing AG 2017
J.D. Kelley, J.J. Leventhal, *Problems in Classical and Quantum Mechanics*, DOI 10.1007/978-3-319-46664-4

Table T.1 Relationships between the one-electron atom quantum numbers (integers)

Name of quantum no.	Notation	Range
Principal	n	$n \geq 1$
Azimuthal (ang. momentum)	ℓ	$0 \leq \ell \leq (n-1)$
Magnetic	m (or m_ℓ)	$-\ell \leq m_\ell \leq \ell$

Spin quantum numbers not included

Table T.2 Properties of the quantum mechanical one-electron atom

Potential	$U(r) = -\dfrac{Ze^2}{4\pi\epsilon_0}\cdot\dfrac{1}{r} = -Z(\alpha\hbar c)\cdot\dfrac{1}{r}$			
Energy eigenvalues	$E_n = -\left(\dfrac{Z^2 e^2}{4\pi\epsilon_0}\right)\dfrac{1}{2n^2 a_0} = -Z^2\left(m_e c^2\right)\alpha^2\cdot\dfrac{1}{2n^2}$			
Eigenfunctions	n	ℓ	m	$\psi_{n\ell m}(r,\theta,\phi)$
	1	0 (s)	0	$\dfrac{1}{\sqrt{\pi}}\left(\dfrac{Z}{a_0}\right)^{3/2} e^{-Zr/a_0}$
	2	0 (s)	0	$\dfrac{1}{2\sqrt{2\pi}}\left(\dfrac{Z}{a_0}\right)^{3/2}\left(1-\dfrac{Zr}{2a_0}\right)e^{-Zr/2a_0}$
	2	1 (p)	1	$-\dfrac{1}{4\sqrt{\pi}}\left(\dfrac{Z}{a_0}\right)^{3/2}\left(\dfrac{Zr}{2a_0}\right)e^{-Zr/2a_0}\sin\theta\, e^{i\phi}$
	2	1 (p)	0	$\dfrac{1}{2\sqrt{2\pi}}\left(\dfrac{Z}{a_0}\right)^{3/2}\left(\dfrac{Zr}{2a_0}\right)e^{-Zr/2a_0}\cos\theta$
	2	1 (p)	-1	$\dfrac{1}{4\sqrt{\pi}}\left(\dfrac{Z}{a_0}\right)^{3/2}\left(\dfrac{Zr}{2a_0}\right)e^{-Zr/2a_0}\sin\theta\, e^{-i\phi}$

Table T.3 Expectation values of r^s for the one electron atom

$$\langle r^2 \rangle = \left(\frac{a_0}{Z}\right)^2 \left\{ \frac{n^2}{2}\left[5n^2 + 1 - 3\ell(\ell+1)\right]\right\}$$

$$\langle r \rangle = \left(\frac{a_0}{Z}\right)\left\{\frac{1}{2}\left[3n^2 - \ell(\ell+1)\right]\right\}$$

$$\langle r^{-1} \rangle = \frac{1}{(a_0/Z)}\left(\frac{1}{n^2}\right)$$

$$\langle r^{-2} \rangle = \frac{1}{(a_0/Z)^2}\left\{\frac{1}{n^3\left(\ell+\frac{1}{2}\right)}\right\}$$

$$\langle r^{-3} \rangle = \frac{1}{(a_0/Z)^3}\left\{\frac{1}{n^3\ell\left(\ell+\frac{1}{2}\right)(\ell+1)}\right\}$$

$$\frac{(s+1)}{n^2}\langle r^s\rangle - (2s+1)a_0\langle r^{s-1}\rangle + \frac{s}{4}a_0^2\left[(2\ell+1)^2 - s^2\right]\langle r^{s-2}\rangle = 0$$

The last entry is Kramer's relation for r^s

References

1. Arfken GB, Weber HJ (2005) Mathematical methods for physicists, 6th edn. Elsevier, New York
2. Burkhardt CE, Leventhal JJ (2008) Foundations of quantum physics. Springer, New York
3. Boas ML (1983) Mathematical methods in the physical sciences. Wiley, New York
4. Spiegel MR, Lipschutz S, Liu J (1961) Mathematical handbook of formulas and tables. Macmillan, New York

Index

A

acceleration, 18
  angular, 5, 112
  gravitational, 14, 15, 18, 40–42, 50, 52, 54,
    58–62, 64
angular momentum
  classical, 26, 67, 68, 71, 73, 76, 80, 81, 83,
    89
  quantum, 121, 124, 126, 189–196, 198–
    200, 208, 213–215, 227, 233–235,
    237, 239, 242, 270, 271, 289, 317,
    357, 361
    ladder operators, 190, 192–196, 202,
      205, 210
apocenter, 88, 92, 93
approximation methods, 243, 280, 281
  degenerate perturbation theory, 284–286,
    288, 292, 294
  time dependent perturbation theory, 297,
    298, 300, 303, 305, 307, 310
  time independent perturbation theory, 258,
    261–263, 267, 270, 271, 273, 277,
    280, 281
  variational method, 248, 249, 251,
    253–256, 282, 283
  WKB, 243–247
Atwood's machine, 40, 58

B

barrier, 137–140, 142–146
Bertrand's theorem, 72, 89
Bohr, 121
  correspondence principle, 121, 126, 137
  model, 121, 122
  orbit, 122, 125, 127, 317
  radius, 127, 128, 231, 317
brachistochrone, 27

C

calculus of variations, 25, 27, 30
canonically conjugate, 26, 27
center of mass, 112
central forces
  classical, 67, 69–72, 79, 81, 83, 89
  quantum, 223
central potential
  classical, 67–73, 81, 83, 90
  quantum, 223, 226, 227, 233, 234, 341
centrifugal potential
  classical, 67
  quantum, 240
Clebsch-Gordan
  coefficients, 192, 205–208
  table, 206–208, 359
commutator, 174, 184, 193, 205, 348
conic sections, 85, 89, 319
constraint
  equation of, 58, 61–63
  undetermined multiplier, 59–61
cyclic coordinates, 27, 44, 69, 70

D

de Broglie wavelength, 139, 340
deBroglie wavelength, 126

© Springer International Publishing AG 2017
J.D. Kelley, J.J. Leventhal, *Problems in Classical
and Quantum Mechanics*, DOI 10.1007/978-3-319-46664-4

degeneracy
  classical, 94
  quantum mechanical, 149, 223, 225–227,
      237, 242, 289, 361
density
  mass, 13, 15
  of states, 226
  probability, 135, 138
determinant of coefficients, 100, 115, 229, 296
dipole
  electric, 309
  magnetic, 129
dipole moment
  magnetic, 129
Dirac
  δ-function, 142–144, 156, 157, 170, 260,
      270, 281, 288, 289, 333
  notation, 179, 182, 281, 292, 298, 301, 308

E
effective potential
  classical, 67, 81, 82, 84, 87
  quantum, 230, 235, 237, 240
Ehrenfest theorem, 181, 182
elastic collision, 8, 11, 22, 23, 213

G
Γ-functions
  half-integral, 331
  integral, 134, 331
Gauss
  law, 13, 14, 16
  trick, 238

H
Hamiltonian
  classical mechanics, 26, 27, 32, 33, 43–48,
      56, 58
  quantum mechanics, 131, 132, 176, 179,
      181, 182, 203, 204, 209–211, 220,
      224, 225, 227, 248, 256, 258, 276,
      279–281, 283–289, 291–294, 297,
      300, 307
  relativistic, 263–266, 271, 272
harmonic oscillator
  classical, 31
    isotropic, 81–83
  quantum, 132, 157, 170
    isotropic, 292
    ladder operators, 173–175, 179,
      181–183, 185–187, 258

Hermitian operator, 347
hydrogen molecule, 271

I
inclined plane, 18, 52, 60
indistinguishable particles, 217
  boson, 218, 219
  fermion, 218–220

K
Kepler
  laws of planetary motion, 69, 71, 72
  orbit, 92–94, 320
  potential, 83, 89, 320
  problem, 84, 85, 89–92, 319

L
Lagrangian, 3, 19, 25–27, 30–36, 38–45, 47,
      49–56, 59, 60, 62, 65, 68–70, 73, 97,
      104, 105, 109, 113–115
Levi–Cevita symbol, 189

O
orbit, 67, 71–74, 81, 82, 89, 92, 94, 126
  Bohr, 121–123, 125, 126, 155
  circular, 74, 75, 81, 84, 85, 92, 94
  elliptical, 69, 71, 83, 85, 87–90, 92, 94
  equation of, 78–80, 85, 89
  hyperbolic, 88
  parabolic, 88
  spiral, 76–80

P
pendulum, 3, 33, 35, 54, 99, 108
  double, 103
pericenter, 90, 92, 93
polar coordinates, 67, 68, 71, 72, 81, 82, 319,
      320
Principal quantum number, 124
projectile, 47

Q
quantum defect, 239, 240, 242

R
reflection coefficient, 137, 139, 143, 146
Runge-Lenz vector, 89, 92

S

secular equation, 100, 103, 106, 110, 115, 279, 288, 290, 295

spring, 10, 11

superposition
    classical, 16
    quantum, 151, 160, 167

T

transmission coefficient, 137, 139, 143, 144, 146, 244, 247

U

uncertainty principle, 126, 348

Printed in the United States
By Bookmasters